The Soul and the Fabric of the Universe

The Soul and the Fabric of the Universe

R. A. Elschlager

www.theuniverseandbeing.com

Contents

Detailed Contents

List of Figures

Larger Framework of Book

There is a mystery that humans have wondered about since ancient times and maybe ever since they walked the earth. The modern version of the mystery is this. How can we, who have awareness, experiences, feelings, consciousness, sentience, mind, arise out of the brain, which is nothing more than a dead machine?

This book is an investigation of that mystery, using some of the approaches of science. Perhaps the reader thinks, if we use just science, we will not be able to see very far into the mystery. Yet Heraclitus said that if we do not expect the unexpected, we will not find it. Sometimes, even though we are far from fully understanding a thing, still, we can observe and record the logical characteristics of the reality, and from this we can begin to build the required logical framework. This approach has sometimes occurred in the history of science when it has jumped ahead of where it was.

Over time, many words have come to refer to the same mystery. Aristotle spoke of the *soul*. Modern philosophy speaks of *mind* or *person*, and then draws in words such as *awareness, consciousness, sentience, feelings, experiences*. This book often uses the word *sentience*, and also uses the words *I, we, us*, as in referring to we as a being with mind and consciousness and feeling and experiences and awareness. We are a sentience. How can the sentience that is us arise from the brain, which is nothing more than a dead machine?

Many books look for correlates of certain parts of sentience in specific physiological processes of logic in the brain. Some books look for correlates of sentience in specific processes of physics. This book does neither, for the deepest answers lie not in these locals, and that is one of the hardest things to scientifically grasp, that there should be other substantial areas in the fabric of the universe.

The book does at times look at issues that are closer to our conscious experience, for example, at how our conscious experience can be divided up into vision of color or of shapes and the like, or into hearing, feeling, willing, thinking, sensation of the external and internal to the body and the sensation of kinds of thoughts. But the purpose is not to identify these with physiological correlates in the brain, *but to identify these first and foremost* with *logics*, which then could theoretically go on in all kinds of ways in the brain.

If one had to give a brief explanation of an important part of the place the book eventually reaches, it might go as follows. The reason we never scientifically see the answer to the mystery is that we are not material. The reason we can never see the answer is that the physical objects that we can naturally see about us, and indeed hardness itself, are surprisingly not material. The reason that the issue of hardness and physical objects throw us off the path is that such are only hardness and physical objects *for us*, they are not that way independent of us, they are not that way outside our own particular native physical selves. Creating a framework for this, and more, is what the book starts to do, and it does it from a perspective of science.

This is a journey where eventually matter time and space themselves seem to play a somewhat more fluid generic role, where meaning steps forward independent of the human mind and appears to assert a substantiality not visible from our native perspective.

This book does not go into complex theories of what is science. It merely uses some of the approaches of science, then starts from the more concrete parts of physics, combines that with the assertive positing that sometimes occurs in the history of science when science jumps ahead of where it is, when we divine a part of the fabric of the universe and then record its logical properties even though we do not understand why they are there. It is possible that there is nascent material here of a scientific logical character.

A journey to investigate sentience from the perspective of science has powerful advantages over an approach using purely philosophical means, although it also has drawbacks too in that it does not go into certain philosophical issues. Nevertheless, science has a large metaphysical structure, with much basis already solidly asserted. Using the approach of science means that the book can invoke all this structure and basis without much further ado, sometimes with surprising results. It should be noted that the book does involve itself with philosophical issues, just not issues usually given the most coverage in philosophy of mind.

About Author

Why would I write *The Soul and the Fabric of the Universe*? It is a pretty big task, isn't it, to try and answer what is sentience, consciousness, soul, awareness, mind, or however you want to say it? Fairness requires some statement of who is this person who wrote the book and where are they coming from.

I almost always wondered why some things were sentient and others not.

Naturally, when I was little, I didn't use the *words* - "sentience" "mind" "soul". At first, I thought the answer was in movie cartoons, where sheets of paper are flipped through and they produce a moving image. The intensity of the very young is absolute and pure. With excitement, eventually I got hold of a tablet of paper and drew some figures. Trying to flip through even a few pages, it was obvious that this had nothing to do with sentience.

Then I heard about puppets, the kind one operates with strings. Today they tend to be called marionettes. My uncle Norbert was going to bring one over and I waited with such excitement. Finally, the day arrived and he sat down in a big chair we had in the corner of our living room. The puppet was wonderful, how the little figure moved around with the strings. For the rest of my life I was fascinated with puppets. Still, as soon as Norbert took the puppet out the box, and I saw the little thing with the strings, it presented itself as a given that this had nothing to do with sentience, consciousness, soul, or however you want to say it.

This was before kindergarten.

At the end of grade school, through some book club, I became very fascinated by books that claimed to talk about the connection of words and meaning – or some such. In high school, there were lots of involvements. But eventually I became committed to mathematics and physics (I was a disaster in chemistry). I pleased my physics teacher by noticing that our text book equations for electricity and magnetism algebraically implied some simple forms connected with relativity theory, yet later I think I worried him by wondering about, and starting to look directly at, the structure of explanation systems in physics. In mathematics, I even tried hopelessly to understand why there can be no solution to equations of the fifth degree or higher (Galois theory).

Life continued. I went on to college and after that, graduate school. There I moved into science and mathematics even more, but at the start had a strong interest in humanities. As for details, after majoring in mathematics and minoring in physics, at the University of Illinois, Champaign Urbana, I went on to a master's degree and doctoral work, at the University of California at Berkeley in mathematical logic, and after working for a while, a master's degree and doctoral work at Stanford University in artificial intelligence. Areas of specialization were computational complexity and machine theory, at Berkeley, and automatic programming with the investigation of the formalization of natural language for the purpose of specifying computer programs, at Stanford. The natural language work was not intended to relate to sentience, yet privately there were times when I tried to formally peer into our awareness of others and of self, as to what could be in logics taking place in the neural electro-chemical activities in the brain, and it was then that it began to stand out the lack in the classical metaphysics of sentience, language, logic, and the brain, though such specific words as these were not clearly in my mind at the time.

All this was in the late 1960's to early 1980's. The interest in sentience always remained, although many other activities moved into life. Sometimes the issue of sentience would rise out of the burble of activity, when I would think ideas or even make extended notes, but then it would sink back down into all the other activity. Life went on.

Sometimes outside the cares of work and life, over the years, I got involved with thinking through parts of the works of the standard philosophers, as for example, Kant, Hegel, Cohen (Morris R.), Wittgenstein, Spinoza, the forms of Aristotle and Plato, as well as the larger scoped work of Lao Tzu and Chuang Tzu. Later I became involved with that amazing group, the Presocratics, including Thales (and Anaximander and Anaximenes), of course Pythagoras, and especially Heraclitus, and also with the surprising role of Parmenides and Zeno, and the atomists, the main one being the one I read before knowing about the Presocratics, the wonderful Lucretius (along with the brief earlier atomists and Presocratics, Democritus and Leucippus). At other times I dug quite a bit into parts of Gottlob Frege's work, where mathematical meaning resides out in the external world, so different from the area of mathematical logic, where it resides in a formalized clean-cut model. These are a substantial number of the works which I thought through parts of. More recently, while writing the book, I looked a little into the medieval philosophers, such as John Scotus (Eriugena) and Alfarabi (and Avicenna).

Like many, I was interested all my adult life in the intellectual framework of science. The bibliography of *The Soul and the Fabric of the Universe* lists books that are long on the details of the great history of the human development of science but briefer with parts presenting the "metaphysics" just illustrated in the historical details. These books tend to be from the earlier part of the twentieth century, and I feel lucky that I got hold of them, probably while at Berkeley years ago, roaming through all the used book stores.

Here I am today older. Yet the questions from way back remain. One cannot give full answers to such questions, but one can start to present answers as to what kind of framework follows from a certain mindset based in the approaches science uses, combined sometimes with something like the logicization of reality.

And so the book was a journey of exploration to list some of the characteristics of this aspect of the universe we call sentience, and which we still find so difficult to grasp. The journey is also an exploration that delves into understanding, for the mystery of sentience leads out into the nature of the universe itself.

Sometimes one wants to know what an author has already published.

Biographical Notes

"The representation and matching of pictorial structures," M. Fishler and R. Elschlager, IEEE Trans. on Computer, Vol. C-22, 67-92, 1973. This research was done by Fishler and myself at the Lockheed Research Labs in Palo Alto, CA.

Martin Fishler, who was my manager, came up with the special representation of pictorial structure. I eventually came up with an algorithm to match such structures to a photograph. It was one of those algorithms that people would at first say, that can't be done, but on looking through it carefully, they saw that the algorithm actually worked.

In spite of the algorithm making possible what had been computationally impossible, even then some heuristic approximations had to be introduced.

It is not that I did not think and see deeply into logical implications of the representation of the pictorial structure. I had to in order to eventually come up with the algorithm. But Martin is the one who had all the ideas about the representation.

The Handbook of Artificial Intelligence, vol 2, Avron Barr and Edward A. Feigenbaum, general editors, 1982, William Kaufmann, Inc.

Somewhere around 1980, I was responsible for putting together the final form of chapter 10, on Automatic Programming, of *The Handbook of Artificial Intelligence*. Eventually we decided to have major research groups from around the country send papers on their research, where different researchers sometimes had different ways of thinking about and conceptualizing their work. I struggled and got the whole thing in a single conceptual framework. Then I rewrote everything to fit into that framework. (This is not to say that there was not significant contribution and work in other ways and parts by others.) (This was when I was a graduate student in the Computer Science Department at Stanford.)

Summary of Book

Should a book start with a summary? Doing so could give away a little bit of the ideas, could it not? Yet five hundred pages seemed too much without some kind of map, even if just a partial one.

This book is a journey to understand what is sentience in terms of the fabric of the universe.

Early in Part 1, the book discusses the idea of perspective, an idea that arises over and over on the journey as a literary device, but interestingly, also as a technical conceptual device, which maybe should not be surprising considering that from the perspective of science, the mind is much connected with the brain, and perspectives are in the brain. Shortly the book introduces us to the *zoomoscope*, a physical device that gives us a perspective on reality different from our native one, a device that alters our visual perspective in ways that force us to be aware of how much we take for granted as to what is out in reality yet which is actually computed in our brain but outside of consciousness. Whether there are issues here for the science of ethology, this does introduce metaphysical and ontological (what *really* exists) concerns about our perception of reality.

Part 2 of the book looks at the logics going on in animals, especially going on in their brains, because one of the principles of science is to look out into what we call "external reality," and to look at it for that which we are investigating. So it is natural to look at these logics, because they bear a special connection to sentience.

The book looks in considerable depth at one particular logic that goes on in the brain of one particular type of creature, the electric fish called *Eigenmannia*. The logic is called the *JAR* logic. It detects whether there are interfering electrical waves from a nearby electrical fish. This part of the journey is in line with another principle of science – we want to get both breadth and depth of coverage of external reality around the phenomenon we are investigating. Included in this principle is that this "external reality" exceeds what we are able to construct purely mentally on our own. That is why we look out at this external reality.

Toward the end of Part 2, the book presents conclusions about the characteristics of logics going on in brains, not just in fish but in any creature.

And this leads to one of the distinguishing characteristics of the book. The book first cuts out a larger problem than what is sentience. What is the nature of the going on of meaning – or of logic – in the dynamics of masses of matter, whether that mass of matter is thought of as atoms or as larger pieces of matter? Sentience is clearly profoundly intertwined with this larger problem, because after all, from the perspective of science, sentience is a certain kind of logic "going on" in the dynamics of the matter in the brain. Not that the book does not also deal with sentience itself, but some of the metaphysics and ontology of the larger slice must be grasped in terms of what reality presents us, before we can more solidly take the journey of what is sentience.

Part 3 opens with the chapter on the "I", wherein is enumerated possible categories of what we can be aware of – what we can feel, experience, be conscious of. Just as with the preceding material in the book, there is an emphasis to continually think about these as, or as corresponding to, logics or meaning going on in the dynamics of matter – neurons – in the brain. This opening chapter of Part 3 ends with a look at some of the remarkably sophisticated logics that go on in our brain, logics that we are not conscious of, and that we take completely for granted. Yet these sophisticated logics feed their results into consciousness (or as we might alternately say, into awareness, feeling, experience, or sentience). The going on of all this takes place in that most intimate space between our very ears.

The next chapter is one of the more important in the book. It brings forth the metaphysics of logics. No, it does not do this completely, but it is a large step in bringing forth important parts of our feel of what "logics are." The chapter opens centuries in the future, when we are admitted to a series of operating rooms for brain surgery. We are shown the advanced – at least it appears that way to us from the past – technology that is used to look into the brains undergoing surgery in the different rooms. In fact the technology is so advanced, that it shows whole moving pictures on a display screen, pictures that show what is going on the brain at that very instant, that show the logics going on in the brain, and most importantly, that show the logics from different perspectives. By the time we finish our tour of the brain surgery center, we have a new grasp of the idea of logics, which, strangely, on one hand are completely tied up with the dynamics of material systems, and yet on the other hand, can almost present themselves as fully separated from matter.

A certain aspect of the fabric of the universe comes to play a central role across the book, an aspect the book decides to call *logics*. We may briefly give some idea of it here.

In our regular way of speaking, we might point to some area, and say, a logic is going on there. In other words, there is a certain logic going on in the things going on there. For instance, when a certain logic is going on in pieces of matter in a certain place, we say, "there is a car." A car is a certain logic going on in matter. Or for instance, in huge clouds of molecules or atoms, when a certain kind of logic goes on in those clouds, we say and think, there is a tree. A tree is a certain kind of logic going on in matter. A stone, a chair, a cloud, a light bulb, the physical human body, an eye, a shoe, a cat, a volcano – all these are logics which when they go on in a collection matter, we say, "there is a stone," or "chair" or "cloud" and so on. These are all logics that go on, and they are all logics that go on in vast clouds of atoms. Or when a certain logic goes on in a set of electrons, neutrons, and protons, we say and think, "there is an oxygen atom."

When a certain kind of logic goes on in the molecules of air, we say, "there is a sound wave." The same with the color red – which in the terms used by physicists, a certain logic is going on in the dynamics of photons.

Some logics go on in environments for such long periods of time that they affect – even eventually determine – the creatures in that environment. The book names such logics, *evolution logics*, because of the cause and effect role they play in the framework of thought that humans refer to as *the theory of evolution*. By the way, just because the going on of such logics is huge in terms of space and time, does not mean they are less real. Their hugeness is only by comparison with the relatively tiny time and space that our own body occupies. Size in itself does not bear on degree of existence. This is from the perspective of science. And it is obviously from the perspective of reality too.

One particularly important logic is a *sentience logic*. That is a logic which when it goes on in matter, gives rise to a sentience. For instance, there is a certain logic going on in our brain right now, that gives rise to us – gives rise to our "I", consciousness, awareness, being, mind. This is a fairly standard assumption of science – that our sentience arises from some logic going on in the brain. (Even if the going on of this logic should in fact extend in some sense beyond the brain, there are powerful scientific arguments to the effect that the logics in the brain are a kind of total gateway. And by the way, as stated earlier, not all logics going on in the brain are part of the sentience logic, not by any means.)

As stated, Part 3 opens with a chapter on the "I" which is followed by the important chapter "Devices to See Logics." There are many ideas in the sections of the book, some developing in one location of the book, some across many chapters. Indeed, one idea developed here and there across the first three quarters of the book is the concept of a *wave* as it is in science, because the book at times makes use of the idea. Waves occur on the surface of water, and they can also occur under the surface as pressure waves, and they can occur in the air as pressure waves (we usually anthropomorphically call these waves *sound waves*), and waves can occur as waves of electrical voltage pressure in water, and as motion waves in the ear drum and in the little bones behind the ear drum, or pressure waves in the fluid in the cochlea further behind the ear drum, and so on. All waves are common shared logic going on in the material world. At one point the book follows the different waves, from the source of a sound, then through the air, then eventually into the brain where a variety of logics result, logics that are very different from the logic of a wave, and this following out is done in the opening chapter of Part 3, on the "I".

Thus, the "I" chapter is followed by the important chapter "Devices to See Logics." The latter chapter contains the following important conclusion, here stated in abbreviated terms. Perhaps a logic can be displayed in several ways. If it is displayed in one of those ways, then that is *a* valid view of what the logic is. For instance, the logic of sentience could be displayed as a logic which had other logics entering into and out of it – corresponding to whether various things entered or left our awareness (alternately expressed as, entered or left our experiencing, consciousness, and the like). This sentience logic is going on in the brain, and this is one way it could be displayed, and so it is *one* of the legitimate views of this logic, a logic that is literally going on in our brain.

The chapter on machines explores scenarios of how humans far in the future will finally scientifically solve the mystery of sentience. The chapter ventures estimates of how long it will take, and why machines – and this certainly includes computerizations – will be required. What power and sophistication will be required of such machines? What about whether a logical system is capable of understanding itself, and what is understanding, and so on. The chapter closes with examples of logics of amazing depth taking place right now in the space between our two ears, outside of consciousness but feeding into consciousness. Someday we and our machines will not only see all this in detail in the brain, but will eventually copy these logics into our human-made neural net machines as part of building the sophistication needed to assist us in ultimately understanding – with the assistance of our machines - sentience.

So far the book has been looking at logics going on in brains, as is quite appropriate for our journey. But over the next several chapters, we notice how all is logic, from physical objects to many abstract objects, and that in fact many abstract objects have full ontological existence in the material world no less than physical objects. To illustrate these ideas better, one chapter is even about creatures laid out differently in space and matter than we are.

In the journey through these chapters, issues of logic, space, time, and existence are starting to appear. For instance, consider three notes of a harmonious chord, played on the piano, with a very short silence between them. Does that harmony exist in external reality, or only in our brain? The question seems limited but illustrates larger issues.

The chapter on evolution has parts deep and others not, but all of some interest. It opens with an exceedingly brief statement trying to lessen the metaphysical conflict between the perspectives of science and theology. Further on, the chapter introduces the concept of evolution logics, which in their relation to the logics in the brain – or alternately, as they are logics in the brain – are involved with time transformation. Maybe there are mathematical research problems here.

One theme that appears in Part 1 and is further developed in the chapter on Newton in Part 3 is the magic of science. The idea of magic is used here simultaneously casually and seriously – what might be called the magic of reality, and how science, in its own way, eventually absorbs it undigested. The chapter on Newton is on the basis of the current massive amount of quality historical research.

Part 3 closes with a rough look at the seeming difficulty of what is the form of the logic that corresponds to sentience and that goes on in the brain, though possibly it goes in a region larger than the brain. From one of the standard perspectives of science, this logic encompasses the deepest metaphysical meaning of sentience. However, when such a logic is viewed from the mathematical and philosophical areas that study formal logic, there seem to be barriers as to what its form could possibly be. It should be kept in mind that any logic that could be expressed by any program in any programming language falls too within being a formal logic.

The last part of the book, Part 4, becomes somewhat more technical and moves into the whole of sentience, logic, space, and time. For a sentience, what are the time aspects of physical death? The opening chapter looks at an interesting theorem, possibly surprising.

The next two chapters look at questions brought up by teletransportation, from several different views, including its relation to evolution logics (which may be the larger logic that we are) as well as its relation to mathematically slicing up time.

What if one lets go, for one chapter, the constraints of trying to deductively base sentience on approaches out of science? This is the "Brain Mirror Theory" chapter.

Later, chapter 37 investigates whether what we perceive is real. The chapter focuses on the example of what it is like for humans to see the color red.

Chapter 38 is a longish multifaceted chapter on logic, trying to wrap up all the considerations when one moves logics into a mathematical material perspective.

Finally chapter 39 briefly touches on the relation of the contents of the book to other material in philosophy. At one place it wonders if techniques from recursion theory can be applied to the issue of whether function cannot explain sentience.

After that, the final two chapters close the book's five-hundred page journey to understand what is sentience in terms of the fabric of the universe.

No summary can capture everything. Many items are not mentioned above. For instance, logics exist outside of the mind; they are not creations of the mind. Why? Because among other things, the book is from the perspective of science. Basic fully accepted principles of science are that most logics were going on long before the human brain – mind – was around, so these logics cannot have been created by the human mind. Another item is endnote 26, which looks at the fallacy of the widely held belief that we know that we know. There is even a several-sentence observation about the value of different reproductive strategies (see index, under "asexual"). Hopefully, other items will be thought-provoking too. As for this summary as a whole, it should be kept in mind that some mentioned items are not developed till later in the book, after more solid buildup.

The book opens with the main character roaming about a rocky desert and repeatedly thinking about how we ourselves, our body, and our brain, are nothing but bouncing around rocks - the atoms, molecules, and interacting pieces of matter in the brain are the bouncing around rocks, and consequently so are all the thoughts and feelings going in that brain. At least this is so from the scientific perspective. Soon the person meets a god who makes a special journey possible, the journey to investigate the mystery of sentience.

Issues of Language

A few technical issues regarding the use of language should be mentioned.

The book employs some casualness of language, but hopefully in a reasonable manner. For instance, the book might roughly state that if a human were large enough so that we could see its individual atoms as points, then a human would be perhaps 36,000 miles high, and that would be over four times the size of the earth. Yet if this were phrased with complete precision, it would be stated that the size of the human in this new perspective would be over four times the size of the earth in our usual perspective. Furthermore, the word "human", would be replaced with a more precise phrasing, such as perhaps "the human body" or "the physical form associated with a human sentience", and "size of the earth" would be replaced by "diameter of the earth". In a few places the book does use such extra precision of language. But unless it seems required, typically such precision is not used.

Or another example. The book speaks of atoms as common sense objects, which they are because they have the characteristics of such objects: they have shape, size, and they move around like we think objects would move around, whereas, for example, the internals of atoms are subject to the laws of quantum mechanics and do not move around at all like common sense objects. If the book wanted to be completely precise in phrasing this, a number of issues would have to be looked at, including a long distracting discussion on statistics, even getting into philosophical discussion of what does it mean when a mathematical probability gets closer and closer to the mathematical zero.

Reasonableness in the casual usage of language is one issue in the book's language. Another issue is how various

xxxvi   Issues of Language

sciences differently treat, possibly even differently conceive, concepts that may be described in the same way.

We can easily fall into the belief that the way of using the same "idea" in different sciences is the same. However on very close sight, there is not always such a simple identity, and in fact, a bit of assertiveness is needed to pull everything into a single view. For instance, in neurobiology, in the studies of the *Eigenmannia*, the electric fish, one reads of the idea that "this particular neuron computes the area of a particular ellipse," where the particular ellipse is one that can be mathematically defined in terms of certain electrical wave properties in the watery environment of the fish. In Heiligenberg's book, this idea is backed up by different kinds of graphs, including histograms, showing probabilities determined from experiments of a neuron firing, across many cases of selected neurons in some selected fish, in a certain structurally identifiable areas of the brain. These graphs appear along with other repeated drawings of ellipses in graphs with certain logical axes. A philosopher, or a formal logician, or a mathematician, or physicist, or many other readers, might assume that "what is intended by the neurobiologist" is as precise as mathematical equations on a piece of paper, regarding the ellipse. After carefully reading parts of Heiligenberg's book and also combining those readings with that of other neurobiology books, I somewhat assertively decided on my own that this assumption is simply not the case. The neurobiologist does not have that kind of logical absolute precision of meaning in mind. There is not the absolute put-together of data and classification. Nor is this wrong. Nor is this false. Certainly not in this case. But it would take a fair amount of philosophical analysis to clarify these issues. It might even require a fair amount of thought about human capability of observation in general.

So on one hand there was this nature of "meaning" as used by neurobiologists. On the hand, was the problem of the "true" ellipse of logic going on in the electrical waves at points in the water about the fish. From my very general awareness in figuring out the way two waves of different frequencies can combine to form "beat" waves, I could see that it was indeed probable that in terms of mathematics and physics, ellipses were going on in the logic at points in the water, but I am not at all an expert in these matters.

I would not be surprised that in order to confirm these things, it would take a fair amount of discussion with neurobiologists who were specialist in this particular area, and discussion maybe even with several kinds of engineering and mathematics specialists to get a solid idea of these ellipses. And this would be assuming that the fish are point sources of the electrical waves, whereas the electrical waves emitted by the fish are not from a point but from much of their whole long tail that presumably is emitting the waves. (Throw into this that there are cases where the summation from the relevant neurons "clearly" gives the wrong answer, and we see the difference here from the logically absolute precision of mathematics and formal logic.)

Hence, I put the coverage in the book into one framework encompassing everything from the neurobiological data to the engineering and mathematics of the ellipses in the logic of the electrical waves in the water to the idea of a higher level of meaning going on that they all referred to. Toward the end of the discussion, I pointed out that it would be nice to have, to the degree that someone could figure out what it was that we wanted to mathematically physically formulate, a theorem that would encompass all these aspects, a theorem that non-mathematicians could use fairly easily in these kinds of matters.

Version History and Pagination

A rougher version of this book was copyrighted August, 2003. I think I sent copies to three parties. Later I reworked the entire book, with new sections, chapters, and ideas. That version was copyrighted March, 2004.

After that, the content probably stayed substantially the same, but I made repeated changes over time as to how the content was expressed. At one point an Adobe PDF file was temporarily made available. From then to the present, I made a number of changes of word, phrase, or sometimes even sentence or paragraph. But I maintained the same pagination of the material, as a way of enforcing a kind of identity with the PDF copy. Finally that identity of pagination was broken when I moved chapter 32 back to its natural place (for a while it had been located at the start of the book). Some further slight shift in pagination occurred at the end of the book when I re-proofed the Endnotes, and there were some additions to the bibliography and index. Later I made some smaller changes on a few pages here and there that might have shifted the pagination a little.

For this 2005 version of the book, chapter 32, "All Souls are Waiting Right Now," was completely rewritten. The underlying meaning of the theorem has been kept the same, but it is hopefully explained in a clearer way.

The material before chapter 1 has been completely rewritten. The content itself is all new, almost.

The only section titles that has been changed from chapter 1 onward is toward the end of Part 2, where "Structure of the Logics in the Fish's Brain" has been changed to "Structure of the Logics Going on in Brains." Elsewhere in the book four other headings have been slightly rephrased for formatting purposes.

Part 1
Background

1 The Desert

It was a simple morning. The intense sun in a cloudless blue sky cast its heat over the forever expanding open spaces.

The car kept going up and up over an undulating swell of land, then down and down, and now up again, toward the last ridge. The barren land, with rocks and occasional sage brush and Joshua tree cactuses, blurred by the side of my eyes, all of it looking as if it were a given of the universe.

On that final ridge, in the great immenseness, my little car rounded past the few pine trees next to a mountainous boulder, and it was always there on the road that the universe itself seemed to open up with the naked desert unrolling up ahead, as if I looked down on a piece of the planet so dear and yet hard to the eye. Above, the blue dome of the sky stretched outward to the mountainous walls of the valley, possibly ten miles across and forty miles to the south. Below, the yellowish floor of the desert, broken by the rust red cliffs just to the north, stretched off to the mountains, rising up like great gnarled alligator skins of sandy reds and snaggy grays.

This was my land, maybe even my home. I wondered.

Today I would breathe the sun and traipse through the dusty trails over more rocks than one could ever count or imagine, trails that went all over, up through increasingly rough terrain, or down to little lost cliffs overlooking a stream burbling through the rocks in the vegetation below.

My car slowly floated down the string of a road, like a little dot of paper drifting to the desert floor.

My father passed away. That was five years ago. As the cactuses and shrubs darted past my side windows, my thoughts went back. I again saw myself lying on the living room floor, phone to ear, for hours listening to him. Mainly we talked about all his illnesses. But we talked about life too, and even about the nature of the universe, and about the neighbors around his little Arizona home. And I thought about him talking about the goat that got so mad, it tried to attack the donkey. But mainly we talked about his illnesses. And now and then, briefly, about the universe. Then I saw myself lying on the floor asleep. I did not know how long. When I woke, his voice was still droning on.

All that was some time ago.

Today was a good morning. The earth is different out here from back in the West - or the East. Here the colors of dirt are that ubiquitous light sand-color and rust . The rocks are in a forever variety of sizes from gravel to pebbles to stones to boulders to mountains. All is mixed with gritty yellow and red desert dirt

Hills like great dry ocean swells rolled in the shadows toward the horizon, and beyond that, in the good-morning hello, a bright yellow of our sun, stretched in the blue without even a wisp of a cloud

Farther on the trail, cactuses rose with weak brown and gray, and some were greener who had managed to get hold of a little more water during the last rain. There was still plenty of pebbly dirt one could use to walk between most of the growth.

God, there was something refreshing about all this, the distance from humans and their world, from the city which seemed like a never ending civilization that encompassed the being of the universe. When you got out here, you discovered that all that stuff there was, well, nothing. It had no meaning in the greater scheme of things.

Stumbling along the trail across all those stones, I wondered again is that all we are? What I mean is that when you look close enough at our physical bodies, it is all atoms and molecules. In effect it is only rocks bouncing around, pulling each other this way and that, but all bouncing around. And that is it. That is us. That is even our brains, all our thoughts, everything.

As I hiked up and down the winding little trail, I kept looking at the yellow reddish soil and different sized rocks. I continued thinking about how our whole body - from the scientific perspective - is just bouncing around rocks – the molecules we are made of. The scientific perspective places a lot of emphasis on the brain. Our being is in the brain. Yet the brain too is just bouncing-around molecules, also like bouncing-around, dead rocks. Well, a lot of electrons flowing among the molecules. But in essence, just bouncing around rocks. And that is what I saw beneath the glistening sun. Rocks. Stones. Lots and lots of them. We should be dead. Nothing more than bouncing around rocks. Why aren't we dead?

To the right, I looked up at the cactuses marching off in great random waves toward the mountains. Clearly we're not dead, because, for one thing, we can see, and something that can see cannot be dead.

Don't you think unbelievable that something dead is alive? I do. But there it is. We are the rocks, yet we're not the rocks.

Ah, but what we really are is something about the extremely abstract dynamic form that is going on in the bouncing around of these things. However this is a very abstract indeed. It is almost a paradox.

I shook my head and bounded on.

And so I walked and thought about rocks and our being, amidst these thousands of pebbles and stones, this direction and that, all glittering under the blue heavens and yellow sun. Close and far, North and West, sky and earth, all of them, and it was then the first idea marched out of the desert to me.

Zoomoscope

What is reality? You think you know? Well do not be so sure.

We humans see in three-dimensions. Our vision shows us that some things are closer to us and some farther. We don't have to think about it. This automatically comes to us as we look around. That is just a beautiful, inherent part of who we are; it is a kind of perception we are born with. It is part of our *native perspective*, our birthright as part of being born a human animal.

But now consider the following device. There are a couple of mirrors fastened to some kind of stick that goes to the right and left of your face. Your eyes look in. A little mirror for each eye immediately reflects the sight of each eye in opposite directions, the mirror on the right, directing the right eye's sight, at a 90 degree angle to the right, and the mirror for the left eye doing the same to the left. After the sight of each eye goes in opposite directions for a while, there are two more mirrors that direct the sight forward again.

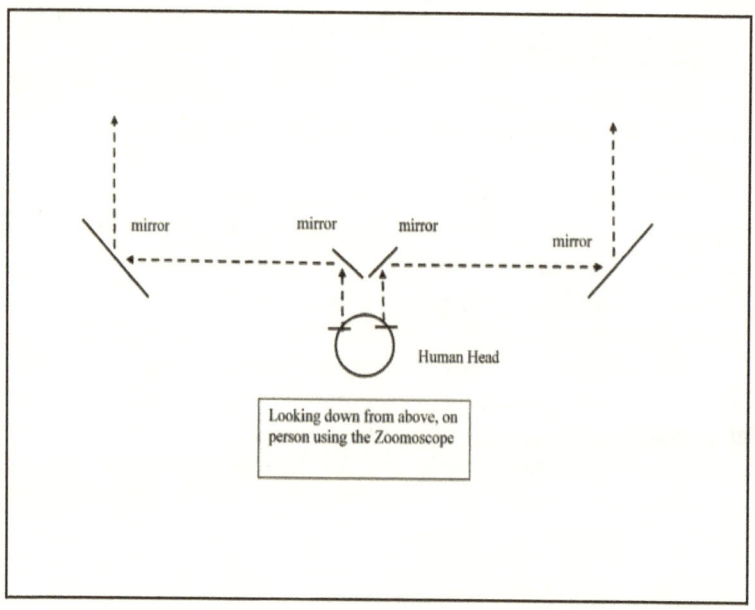

Figure 1 The Zoomoscope

I sketched the device, which I came to call the *zoomoscope*.

When we look through this device, we get a fun view.

Keep in mind that there are no lenses in the zoomoscope, just mirrors. So it is not like a microscope or telescope. It's just like looking into a mirror, except that your vision goes forward. If the mirrors are of good quality, it's like they are not there and you are looking at the real thing. In fact, it would be looking at the real thing regardless because when you look in a mirror that is just what you see, the real thing. I pictured lifting it to my eyes.

To describe what I experienced, and what you would experience too if you looked through the zoomoscope, we have to touch on some numbers. Science always goes back to numbers, and this book is from the scientific perspective. Let us say humans have 2 ½ inches between the centers of their eyes (some have a little more, some have a little less; you might measure your own; do not hurt yourself.)

I made the two outer mirrors ten times farther apart than my own eyes – so that is 10 times 2 ½, or 25 inches apart.

To understand what we would see through such a device, one has to understand a little of the logic going on in the brain circuitry behind your eyes, going on right now, and all the time.

Since our two eyes are a few inches a part, they see a slightly different view of things, and the circuitry in your brain uses that difference to compute the information about three-dimensionality. That is how our vision comes to show us the three dimensional.

With the mirrors 10 times farther apart than our eyes, the device will cause our perception to be ten times closer. Closer than what? The things in the zoomoscope perspective appear ten times closer than they do in our native perspective. When we look with the zoomoscope, that circuitry inside our skull, in the brain, is computing the distances as ten times closer than it is computing them when we look without the zoomoscope. That is all in that circuitry.

And there is another aspect of the logic going on somewhere inside your head even right now.

In science there is something called "angular size" – it is the angle "subtended" by the object you are looking at. With the zoomoscope, that angle doesn't change. So some other part of your brain circuitry notes that a thing you are looking at is ten times closer yet has the same angular size, and therefore, the way the circuitry works, it computes it as being ten times smaller, which indeed is just how things look when you look through the zoomoscope. Through this zoomoscope, we perceive things as being 10 times closer *and* 10 times smaller than we perceive compared to our native perspective. These circuitries are working all the time that we have our eyes open.

Now I looked about at the Joshua tree cactuses, and some of the smaller Prickly Pears, first without the device, then with. How fascinating.

In fact, then I perceived the whole valley with the zoomoscope, with its cactuses, and shrubs, and hillocks close and far, and the path winding onward, and the close and far cliffs, everything was ten times closer. It was *exactly* as if the whole valley had been reduced to a cute little model ten times smaller. Yet I was looking at the real thing. Certainly nothing like through lenses. I was startled to see that the Joshua cactus tree which was about 4 car lengths away and another twice as far, were now just 1/2 a car length away, one behind the other, and all like in a cute little diorama, but absolutely real and crystal clear. In fact, when a breeze came up and wiggled leaves on a shrub, the absolutely crystal clear leaves in the little diorama also wiggled. A desert bird angularly darted from one shrub to another, and there was the little thing, clear and bright, in the diorama – tenth size of course. But the diorama was reality. For the next hours I alternately looked at the desert through the zoomoscope, then only through my native eyes.

Without the zoomoscope, I looked at two desert bushes. Because of the lay of the land, and because they were both pretty far away, I couldn't tell if one was farther away than the other. I put the zoomoscope up to my eyes. Instantly I saw one bush at about 10 car lengths and the other, about twice the distance, in the diorama, at about 20 car lengths. (Outside the zoomoscope, they were about 100 and 200 car lengths away and completely impossible to tell if one was farther away than the other).

I thought I would try looking at some of the things right around me, all within about 3 car lengths, some little hills, rock mounds, little cactuses, a small willow tree. With the zoomoscope, it was within just a third of a car length away, all quite cute. I had to grin. So I played quite a while. I tried walking about while looking through the zoomoscope, and except for almost falling several times, I got the feel of it, and the word "feel" was right, for after a while I found himself developing a kind of altered awareness. At one point, I reached out and wiggled a branch on a little tree. With the zoomoscope, I had to look cross-eyed at my hand held at such a short arm's length.

Thus did this "little" reality appear before me – little only by comparison with our native perspective. But apart from one being our native perspective, and the other not, they are both – well – perspectives. They are both a picking-up on reality. They are both real. One is our native one. So we are used to that. The other is a different perspective and that we are not used to.

(Maybe a scientist would explain it this way. In the zoomoscope, the angular size of everything is the same. But the *parallax* changes, giving the same parallax as if the thing were ten times closer. Computations in parts of your brain that are outside your awareness therefore compute the object as ten times closer. Most importantly it should be added that the result of this computation is passed on into your awareness, and that is why you see things as ten times closer.)

After a while I looked at the cliffs (through the zoomoscope). It was hard to see a difference. They were still far away. So I decided to move the outer mirrors much farther apart; I now adjusted them a 100 times wider than my eyes, so now everything was a 100 times closer than from our native perspective (and a 100 times correspondingly smaller). But it confirmed something new but which I was suspecting. Everything was now 100 times hazier in the little diorama. It's as if someone blew smoke into the little model. Everything, except what was almost on top of me, was misty.

I thought how even out in the desert, there is a haze – if it hasn't rained for a while. The dust gets permanently in the air, and that gives everything far away a hazy look. Of course, what was far away in one's native perspective was close in the diorama perspective.

So the zoomoscope not only brought everything so many times closer, relative to one's native perspective, and made it all so many times smaller, but whatever haze was in the atmosphere, it also made that many times hazier. And that too was due to what was going on in the circuitry behind our eyes but outside our consciousness!

For a couple of days I played on and off with the instrument. My mind wondered. What if the end mirrors were hundreds of feet apart. Why then, even the distant cliffs and mountains would seem close up and, more than that, you could see the distance between those far away cliffs and those that were farther away yet. After thinking about the gray cliffs being close up, I thought about the mirrors being really far apart. Why not put them at one end of the Earth's orbit, and a half-year later, at the other side of the Earth's orbit, so now they would be 186 million miles apart! Preposterous fantasy. Wait minute. No it was not. In effect that is what astronomers used to do to measure the distance to the nearby stars. In effect, they looked at the stars from one mirror, and then half a year later, when the Earth was way across on the opposite side of the orbit, 186 million miles, they looked again, and they could see how the stars moved, just like when you look through a little zoomoscope, or through no zoomoscope at all, and you open and close alternate eyes. You will see things closer and farther jumping around with respect to each other. But then I thought how even the closest star, Alpha Centauri, was already a 146 thousand times farther than the 186 million miles between the two imagined mirrors. So the astronomers had to measure the jumping around positions pretty carefully. But they can do that sort of thing.

Returning my thoughts back down to earth, I also thought about how haze is due to dust in the atmosphere, and water vapor. If you're near a city, then you also have industrial and auto pollution. But out in the desert, that was not a factor in haze.

Then it dawned on me why mountains in the desert will sometimes look many times closer than they actually are. It's an optical illusion. If there is not much dust in the atmosphere, the desert air is so dry, that there is no haze from water vapor either. The air will be extraordinarily clear. And our brain must use haze as one of the cues for distance. And if something has no haze whatsoever, then we tend to get a perception of it being closer. And that's why mountains which we know from past experience when we had to drive up to them, and how long it took, will appear so close. In the desert.

Perspectives

Thus, a zoomoscope gives us one perspective while our natural visual system that we are born with gives us another. And a microscope will give still another, and so will a telescope, as well will a telescope tuned to light that our natural eye - our native perspective - will not pick up. All these various perspectives show truth, but they show different aspects of truth, they give us different ways of seeing reality.

Perspectives can be due to as small a thing as the angle with which you look at something. If you look down at a dining room table you get a different perspective than if you get on your hands and knees and look up at it from underneath. Both perspectives are true, both perspectives are of the external real world, but in one you see the glistening top of the table spread out smooth, while in the other, you see rough-grained wood, with many breaks due to cross members and edges and supports for the table legs.

Perspective can be due to distance from the object and angle. This changes what light that comes into your eyes. Pushing your hair back in the wind, you can stand on a cool mountain top looking down on a city with the multitude of dots that are buildings far below. That's one perspective. From another perspective, you can be looking at the city from down inside it, perhaps from a sidewalk along one of its boulevards, and in this perspective, what were little dots are whole single and multi-story stucco and brick buildings and store fronts, and what you saw as below you, now you see as all around you.

Perspective can vary in important psychological ways across just a few feet. You can look at a friend across the room and marvel at their fine skin complexion. But if you have good eyes, and you move close to them, very, very close to them, and they let you, now the light coming into your eyes shows not a smooth, flawless complexion, but a surprising number of ranges of mounds and depressions between the mounds, almost like smoothed out mountains and valleys from a great distance. And in some places on the face, though you say nothing, the mountains and valleys are not that smoothed out. Though we tend to view such changes of perspective as inhering purely in social considerations, the perspective is fully scientific and real. The perspective that you get of your friend from across the room and the perspective that you get close up are both equally, scientifically valid, perspectives.

Perspectives may not be due only to angle and distance. They can be due to the emotional state of some part of your brain. That too determines what enters into your awareness or consciousness. In deep moodiness, hopelessness, and dejection about life itself, if you stood on the windy, cool mountain top, looking down at the city, it would look frightening, and would exude the hopelessness of all humanity and universe itself. The dots that were the buildings down there would have the hard steely blue color of the harsh blue sky above. That's how the city would look. That's the city you would see.

On the other hand, we could go to a different mood and suppose that you had a swimming year, that people liked you, that your job went great, you got a salary increase, maybe you found a love. Now you stand on the mountaintop looking down at the city. The whole view is vibrant, happy, and bathed in the friendliness of the beautiful yellow sun above this windy mountain top. In one case you perceive the hard harsh steely city of buildings; in the other you perceive the soft warm yellowy city and friendly buildings.

There is a predisposition to see these "psychological" perspectives as not scientific, not literal. But that is not true. There are definite precise signals going through our hundred thousand million neurons. These signals *are* this perspective. We think of such a perspective as figurative, but it is scientific and literal. Each of the above "psychological" perspectives has its own kind of picture in the mind dominating what you perceive of the city – that picture is also precise brain signals, though it may take centuries before our science is able to "see" the signals and to understand their meaning.

Perspectives originate not only in the visual. We inherit other sensory channels by virtue of being born humans. If you are sitting next to a desk top computer, you hear the incessant fan of the machine, not loud, but always there. It is a definite perception that one has of the machine. This perception will change if you move to the other side of the room. There you perceive a quieter "noise" from the computer, a noise that is not "right up close but a little more mellow and distant" (though it is your brain computing aspects of this aural information such as no longer having the quality of being up close, a quality that then goes on to enter your consciousness, all based in the signals in the brain).

If you close your eyes, that will accentuate this perception through the ears. You will get different perspectives if you get up and walk away from your computer, or if you walk back and then sit down at your computer, keeping your eyes closed (and being careful).

For a blind person, the aural perception is even more accentuated than that of a sighted person with their eyes closed for a little while. For visual absence going back to birth will strengthened the other channels even more.

From the mountaintop, possibly with your eyes closed, you hear the din of the city, a din perceived as both distant and at the same time spread out in all directions. Or is it perhaps a roar? And if the city is neither too big nor too far, within the noise, you will hear punctuated sounds, perhaps of a car horn, or a factory whistle, or sometimes one cannot be sure what. After being on the mountain top, you can change the aural perception by going down into the city, and getting yourself back on that sidewalk next to the busy boulevard. Now the roar is loud, and there are many, many pieces to the roar, cars and trucks and buses and vans all the time going by, each producing for a little while it's own kind of roar, building up, then fading, each one, but so many. And voices, talking, occasionally yelling, many close, some far away in a punctuated surprise that just as quickly disappears. The aural perception on the mountain top and the aural perception that you have done here on the boulevard, both perceptions are true, both are perceptions of reality.

There is perspective we get from touch. For people who have neither hearing nor sight, the tactile perspective is central. Helen Keller's whole world was the tactile perspective. From birth she neither heard nor saw. The tactile *was* her world, well, of course, with emotions and communication overlaid on top of that.

We have feeling from all over the surface of our body. And from much of the inside too. We can feel hot and cold almost anywhere on our skin, though on some places more accurately than others. We can feel pressure from light to the heavy, to the pain of a cut or injury. We can feel even the lightest of touch from a slight breeze or a tiny bug crawling along the skin, and if there is a breeze or if the bug is in an area that has hairs, we can feel the hairs being disturbed too. If we have a stomach ache, or a head ache, or a foot ache, we know that too. Or an eye itch, or nose itch, or lip itch, or ear itch, we recognize that too as a unique kind of tactile perception.

Perspective may originate not only in individual senses, or from emotional parts of the brain. Such can also come from vast arrays of knowledge through which we "look" at something with our "mind". For instance, one might think or perceive something from the perspective of science, alternately, from the perspective of religion. For many people there is a perspective which is neither of science nor of religion but just our day to day moving about and pushing through life and existence.

These perspectives are due to whole systems of ideas, feelings, and emotions, in our brain, with the electrical and physical structure that realize the logics spread over areas of thought, and like any perspective, they determine how certain things enter consciousness.

The journey of this book is to look at consciousness from the perspective of the ideas of science.

The pebbly path started to curve to the right. For a while I used a walking stick I had picked up under large shrubs like small pine trees. The stick was on top of a bed of rocks that went underneath most of the vegetation. A piece of paper blew by, which was a bit annoying. Paper should not be blowing around in the wilderness of the desert.

For a while, the land fell forward toward a rocky wash and more shrubs, and eventually the trail moved on into the wash, now completely dry, as most things are most of the time in the desert. After a while I was back on higher ground, winding through all the rocks and more of the pine-tree like shrubs.

I started to think again how our being is just like dead rocks. That's all the atoms and molecules were. What would one find out from a close perspective of that? What we are, what is our awareness, is already in those rocks just lying around. Just form. And motion. Especially the logic going on in that motion.

As I sauntered along the trail I started to think back on the details of my father long ago.

In some ways, he was lucky, though he had no shortage of ailments. Worrisomely, their treatment came to consume much of his day. Yet he never became bed-ridden for long. Oh, toward the end, he was in hospitals for a week or so.

I thought of those interminable multi-hour phone conversations, where he mainly talked and talked and talked about all his illnesses, and I mainly listened and listened and listened. But sometimes in the middle of that we would touch on issues of what does it all means, a lifetime of effort followed by more effort at the end with health issues, and is there something after.

He was not part of any particular religion. It was then that I started to tell him some of my own ideas. Not much. Just a little. That science does not really know anything about sentience. It is a great mystery. There is much to the universe that is *scientifically* far beyond our current understanding. I don't know if I used the word "scientifically", but the idea was there that we know very little and it is wrong to conclude with certainty that we end with the end of the physical body. I'm not sure I used the phrase "with certainty", in fact, I probably did not, since our conversations were extremely casual.

For a long time in my life, even before that, I thought now and then about such things. I always had an indefinite feeling of obligation that those ideas should be dug into.

But I never did.

The god

The air was hot and dry, something I had got used to long ago in the desert; cliffs and sand, with vegetation lying thinly across all.

And what was that? Another piece of paper! Someone must have let go several sheets of paper. They seemed to be down a row of low willow-like shrubs. I looked harder at one of the sheets at the end of the row of little trees. That was not paper. It was a piece of cloth, slowly billowing in the wind. This was odd. Since there is easy walking space between most vegetation in the desert, I walked toward it. As I got closer, I could see it was pure white. That was interesting considering how dusty the desert is. But possible – I guess. Now I moved toward it with more curiosity. It had pleats, and they were swaying easily in the wind. Now I saw that the cloth was not small, all of it was pleated, and billowing. Of course, this was a surprise. It was seemingly draped over a body. Was this a person? Here?

In the warm light wind, I stepped around the last Mesquite tree before a little opening, and I saw the whole figure. It was sitting on a rock. Even after I appeared, the person did not turn but kept a fixated stare off into the distance, for at this point you could see the red and sandstone cliffs in that direction.

I stopped, almost prepared to turn fast. This was all unusual. There are no people out here, or at least very rarely. Who could I call?

To the person's side was a vaporous willow shrub, which waved in the gentle breeze. The pleated cloth seemed the purest possible. Maybe the person was depressed. The pleats in the cloth rode like silk on the billows of the warm desert air.

"Don't worry. I'm not a strange thing or a terrorist," the figure said, but continued to gaze forward not at me.

"Oh, I was just hiking." What else could I say? "Is everything alright?"

But there was no response. It just sat there. Some of the long strands of its hair were also moving in the wind, along with the cloth. I found myself looking intensely at the cloth. Surely one could analyze this if one only looked hard enough. That graceful cloth caught in the wind so quickly.

They never answered me if they were alright. Finally they spoke again.

"Time."

I had no idea what they were talking about, nor did that surprise me. Obviously it was not my watch I needed to look at.

"Time answers all questions."

"Questions?" Now I could only look at that face. It was a fair amount of beauty in it. At ease. Calm. As if not a care in the world resided there. At least not in my world. I finally thought I would just leave this rather strange situation. Who knew what this was about?

"What?" I said.

"Time. You need time." The head moved, naturally, but not much. The eyes continued to stare ahead.

"What do you mean?" I guessed I might as well exchange some words with this ... this being. A vague knot of apprehension was in my stomach. Knots it would be nice not to have.

"Time is needed. For you. To answer questions."

"What ..."

"You need to answer things. I can't answer things. Except. If you were to answer them anyway. In some time. Not now. I can encourage you. Then."

I nodded, more out of manners, than comprehension. The situation was not in order. But my manners could not work because of the twisted look on my face.

I prepared to get out of there. Still, the creature just sat there. Well, I was going. That's a bit of cheeky nerve, I thought. I need to continue with my usual. For a quarter of a second an image of talking to my father on the phone fluttered through some part of my brain.

"Well, good bye," I said.

The creature just sat there.

"Well," I nodded again, prepared to repeat what I said. That cloth was incredibly fine, more beautiful than I had ever seen. And that face and hair. How special. "Well," and I added quietly, "good bye."

The creature turned now, about halfway toward me.

I looked at its face.

The dry air of the desert blew across my own face. The billowy white cloth fluttered. Ten seconds passed, then twenty. Silence. Strange, I thought. Then I became aware that it must know it was strange too. That alarmed me. We both knew this was strange and yet we kept looking in the same directions. At the end of the minute, I incomprehensibly kept looking.

There was something beautiful. The - what should I call it - the entity, the being, was just sitting there, flowing robes, behind mesquite bushes, the twiggy stuff so typical of desert. Nevertheless the situation was alarming.

I started to back up to leave, but instead stumbled partly forwards into the mesquite bushes, in spite of my fear of thorns and bugs and scratches. I caught myself in the nick of time, partly got upright, but then fell completely over, backwards into the Mesquite bushes, stopping about two feet above the ground, whereupon then they gave completely away, and I crashed to the ground, through the remainder of the cutting, scratchy twigs.

I started to wriggle to get up. Now the god looked in my direction, with a different expression, which I definitely noticed as I fought the tree.

After pushing several branches in nasty directions, I had successfully righted myself and stood outside the smashed branches behind me, all the while keeping an eye on whatever this creature was in front of me.

To my surprise, the entity, or whatever it was, said, "what you experienced when you fell backwards, that was ..." Here the entity just stopped talking, as if it were accessing some information source. After another strange lull, it uttered the word, "hurt."

"What you experienced when you fell backwards, that was *hurt*."

"Oh, yes, that was indeed hurt." I felt embarrassed.

The entity was quiet again but increasing energy seemed to be building in it. Suddenly it blurted out.

"Human 'hurt' and 'pain' – that is amazing. What an experience!"

Oh no, I thought, now looking intensely at this 'person'. "What do you mean, that is amazing," I asked warily.

The god laughed a little.

"Yes. Fascinating."

What is going on here, I asked myself.

"That is an amazing experience. These things are always new for me. But it is fascinating, what you humans feel in that situation."

This is all getting too weird. Terrorist or not. Something isn't right here.

"I said I was not a terrorist."

"Have a nice day. Good bye."

"Don't leave." And here, this person or whatever seemed to be accessing some information source again. Then it said, "You need my help. You are struggling for a long time, and I can help you. I can offer guidance and assistance, of a sort."

"About what?"

And thus began a discussion between myself and this thing – for I am still not sure what to call it. Since I can't always call it a thing, or an entity, and since I can't in any sense of good conscience call it a human, because some of its nature is way beyond, often I just think of it as a god. Somehow a god was there. It was real enough. No question about that.

After I got up from the twiggy mesquite brambles, the god loosened up quite a bit. And so did I, after a time. We talked about all kinds of things and about feelings. But how can one put feelings into words. The next day I came back to the same area, and we talked even more. But then the god got single-minded.

"Now we need to get down to work," the god said.

"Oh sure. What do you mean?"

"The chit chat is ended."

"What do you mean?" I grew less chit-chatty.

"You know quite well why you are here."

"I do?"

The god nodded. I was caught.

"I am here ...," I hesitated.

The god kept looking at me, sympathetically but without mercy.

"... to learn ..."

The god nodded for me to keep going.

"... to study ..."

Another nod.

"... to find out ... to finally find out ... about ... the ... human soul."

"Yes."

That is why I was here. That is why I had been hiking for so long in the desert.

"I can help you," the god continued. "I know much of the answers, and of the future. But I cannot tell you anything that you do not know already, or ..."

"Then what good, er, I mean, what help can you be on my quest or journey. I don't mean to be insulting. Just truthful. How will your presence help? Oh, wait. I think I know. You will provide emotional support, and ..."

"I cannot tell you anything that you do not know already, or that you would not eventually find out on your own. But, I *can* tell you some things that it might take you a while to figure out. I can tell them to you only if you would find them out eventually on your own anyway."

This was interesting indeed, I thought.

Most of this situation was so strange that I did not react as I would have with another person when the god's smallest finger, the pinky, fell off. The god immediately cleared its throat and between my looking at its face and back down at the hand, the hand was back to five fingers. Trying to unobtrusively glance at the ground, I could see no finger anywhere. The god just continued.

"Also, I can give you hints to speed up what you would eventually figure out on your own. I can do things as long as it does not give you information that you would never have found out eventually anyway. In short, I am here to speed things up within the boundaries of what I can do."

"Hmm. I have thought so long about these things. Especially after my father passed away. But in general, too."

"I know."

"What do you mean you know? Were you watching me in the past?"

"No. I didn't know anything about you or your planet till recently. But I do know about you. Not everything. That's why the experience of your hurt when you fell in the twigs was interesting."

"You could feel what I felt."

"That time, yes."

"It wasn't pleasant."

"No it wasn't. You human have your own particular way of feeling it. That was interesting."

"We humans?"

"Oh, pain and hurt, and for that matter joy and happiness. Exist inherent in the whole universe. But things have unique aspects. Like for you humans."

"Pain is never nice."

"No. Nowhere, never. Never is, suffering."

For a second the god looked thoughtful, with the beautiful cloth about it wafting in the desert. Then the god seemed to see me again.

"Let's get going."

"You mean ..."

"Yes, tomorrow, with the journey."

The next day I hiked around some more, thinking about the usual issues. Then there was the god again. We talked for a while.

"It is necessary to get moving, to do something," said the god.

"Well, of course, that is what I have been doing all along."

"No. You must do some*thing*."

"I am," I asserted.

"What must you do?"

"I am doing everything I can."

"Not the right things though. At least, at this point."

I felt a little surprised. Could be, I thought. Well, what did the god think I was supposed to do.

"Well, what do you think I should do?"

"Well, what do *you* think you should do?"

I thought, if this weren't a god ... But presumably a god knows something of what it is talking about. Yet after a little while, I spoke.

"I am going to write this as a book."

"Fine."

"I guess, at the beginning of a book, or at the beginning of a journey, one needs to define the basic concepts."

"Exactly."

And so, on the next part of the journey, I did.

2 *The Soul*

What is the soul? As part of opening this journey, we must give some definition of it. I hiked further into the desert, getting closer to beginning mountain slopes. One could talk forever about the soul, or sentience, or consciousness, or awareness, or mind, or whatever you wish to call it, and still one will not fully understand it. Philosophers and others have been writing about such issues since ancient times, and no doubt, even in prehistoric times, thoughtful people must have wondered, as they stared carefully down at some rock, how is it that the rock has not sentience or awareness while things like me do. Much has been written over thousands of years. Our journey cannot present all of that. We must present only a limited statement for what defines the beginning of the path for this book.

In this book, the words "soul", "awareness", "consciousness", "mind", "sentience", and "mind" are used as equivalents. It is true that they have slightly different emphasis of meaning. But we want to get at the essence of all these words; we look at the problem from a perspective large enough that it includes all of them. The word "soul" in ancient times was used in the sense of life, in the sense of awareness or consciousness, with the meaning at times blended in with theology, and often with no distinction made as to the different possible meanings or varying ways to emphasize those meanings. In modern times, even from the perspective of religion, soul is something that has a kind of awareness or sentience. Perhaps a deep investigation of awareness from within the scientific perspective might offer ways for theological concepts to come closer to science, and vice versa.

The Perspectives on the Soul

We have two perspectives on the soul or consciousness. One perspective is to see consciousness as just what we are. We are a consciousness, we are conscious, we have awareness. The other perspective is to see – or try to see –

consciousness from the perspective of science. I say "try" because it is hard to see the part about how something like the brain, basically one vast, dead machine, can be conscious, how it can have soul. But for now, let us point out those things that we can say about the soul, sentience, consciousness, from the perspective that we have by virtue of our being a soul?

Experiences are a part of consciousness. We can see the color red. This is an experience. Our consciousness experiences red. It is impossible to tell another person with words what it is like to experience red, though everyone who is not color blind knows. We can experience cool, and we can experience ice cold. Or we can experience the warmth on coming inside from a cold day, or the reverse, entering an air-conditioned house leaving the hot sun outside. And pressure. We can experience pressure from almost every part of the outer surface of our body – the skin. And we can experience pain, pain on the surface of the skin, pain from a mild cut or a deep cut, and pain from inside our physical body, such as from the stomach, appendix, back, and from headaches mild to migraine. These are all part of consciousness or sentience.

We can experience the beautiful soft smell of roses, or of steamy roast beef, or of the pungency of animals on a farm on a spring morning. If a space alien came to Earth, we could not explain to it what such experiences are like. Yet all of us humans have them, or can have them. It doesn't seem that a vast, dead machine, such as the brain, could have such beautiful experiences. These experiences are part of being a consciousness, of having awareness.

There are other things that we speak of feeling or experiencing. We can experience happiness. Or on a different day we can be sad. And the experiencing or feeling of happiness can have different aspects, from a bouncy light happiness, to a deep happiness, or joy, where one feels that it will go on forever, with the world opening up for one, a feeling of lightness and excitement. (From a scientific perspective, all these must correspond to different kinds of signaling patterns taking place in various parts of the hundred thousand million neurons in the brain, each with their hundreds to thousands of connections to other neurons.)

Likewise we humans experience grief, and it too can be of various kinds. It can be due from mild to severe life loss.

Or we may experience the warmth of kindness, either given or received. Or the fear of pain or of loss or of treachery. Or a feeling of confusion. Or a feeling of clarity. A feeling of peace, or a feeling of disaster. A feeling of love, or of hate. Of anger, or exasperation, or desperation, or of lust, or frantic greed, or a healthy desire to obtain or achieve. We have all these feelings and experiences and more. (All these are certain forms going on in the massive electrical signaling in the hundred thousand million neurons, when looked at from the scientific perspective.)

There are more parts of being a sentience of our sort. There is the experience of "causing," or "willing," something to happen. You reach for a glass of water. You will your hand and arm to go to the glass of water. Within your consciousness there is the *experience* of willing that to happen. When "you cause" your body to do something, such as reaching for the glass of water, you experience in your consciousness what it is like to will yourself to reach for the glass of water, and we will call it the experience of willing, the experience of willing your body to do something. You can will your hand to bring the glass of water to your mouth, and so on. You will your muscles to move in the right way to get up and go for a walk, or to talk, or to close your eyes, or open them, or to type at a keyboard.

(In the perspective of science, we see that some recent brain research on humans suggests that *possibly* the brain starts taking an action before you think that you "willed" it. Whatever the reality, we do have the *experience* of willing things to happen with our body.)

We feel or are aware of many kinds of experiences. We are almost constantly experiencing some aspects of memory of the immediate past, the intermediate past, and in some ways, the far past. It is hard if not impossible to put in words what the experience is like, but such experiences go on for most of us most of the time.

We can experience having a thought. And finally we can experience love, hope, and beauty.

All these kinds of experiences, from simple to complex, are a kind of miracle. After all, a rock doesn't have any experiences, for it is not aware, it is not a consciousness. But somehow, we, who, in terms of molecules and atoms, are just like rocks bouncing around, do have experiences – we are aware – in the massive electrical signaling flying around in the brain.

Philosophers are much involved with carefulness in using language, and with much benefit. For when we talk about more difficult things, and consciousness is certainly one of those, our language itself can trip us up. One good way to phrase all of the above experiences is described in Lowe, 2000. If we want to talk about the experience of feeling warmth on the skin, we might speak of "What it is like to feel warmth". Phrasing it in this manner presumably avoids implying that consciousness is a thing.

(The "what it is like" language apparently goes back mainly to Thomas Nagel's article on "What it is Like to be a Bat." Some overview of this, as well as further references, can be found in Lowe's *An Introduciton to the Philosophy of Mind*, 2000, page 52, and in Chalmer's *The Conscious Mind*, 1996, page 4.)

Sometimes we will use language such as, "What it is like for a *human* to feel warmth." After all, how do we know that what a cat feels when there is warmth on its skin is the same thing we feel when there is warmth on our skin? And so we might speak of "what it is like for a human to feel warmth" versus "what it is like for a cat to feel warmth" or "what it is like for a dog to feel warmth". Thus we could rephrase the first of the experiences mentioned earlier as, "what it is like for a human to see the color red". It is true that by a stretch of the imagination this might be misinterpreted as merely a listing of some set of physical conditions. But by what it is like for a human to see the color red, we mean some inner quality of what it is like. In fact, since only humans are reading this, we all know what it is like to see red, we know this inner quality (unless the cones in the retina at the back of your eyes are color-blind). We know what the experience of seeing red is even though for the life of us we cannot explain in words or actions, except to utter the word "red" and point to something that is red. Yet we are unable to explain in words what is that inner quality that we experience when we see red, we cannot explain in words "what it is like for a human to see the color red."

We would similarly speak of "what it is like for a human to feel something that is cool", "what it is like for a human to feel something that is ice cold", "what it is like for a human to experience the pain of a mild cut.", "what it is like for a human to experience willing that a hand and arm do such and such".

In this book, we use various ways to say the same thing. Thus, we have "what it is like to feel the pain of a cut," or "what it is like for a human to feel the pain of a cut," or "the actual experience of the pain of a cut," or "the sensation that a human will have accompanying a mild cut," or "the pain of a mild cut enters consciousness," or "the pain of a mild cut enters awareness," and so on. In this book, all these may be used to describe the same event.

The understanding sought for in this book goes beyond the specific way of phrasing or emphasizing such concepts as well as beyond variations on the concepts.

There is another feeling that we can have. It is vague and hard to describe but no less real. We have a unique, powerful feeling, or sensation, or awareness, or experience, of I, of being an I, of being a ourselves. Thus when we look out a window at rain (or at sun or when we do anything) we have an absolute sensation of I, me, myself, an I, a me, as looking out on the world. Like other experiences, it is impossible to convey the sensation in words, other than to use the word "I", to others who are able to use the word "I".

We can talk about what it is like for a human when they say things of the sort, "I just have a powerful feeling of an I, of me. It's hard to explain. I have an intense feeling of me, or of a me. In fact it is the most real thing, obviously, in the world." But we can phrase this as, what it is like for a human when they say these sorts of things, what is the inner quality they are experiencing. Thus we can give "I-ness" a concrete basis by speaking of what it is like for humans when they are using the word "I".

Indeed, one could apply the same technique to any self-reporting. If a person says X, then no matter how strange or vague or seemingly difficult to define are the words used in X, one can still talk about what it is like for a person when they say X – what is a person experiencing when they say something like X. In this way, words and ideas that we find very hard to scientifically pin down, even though we *feel* we have good grasp of, can still be pushed back to a concrete basis by talking about what it is like for a person when they are saying such words. Thus we give a definition to X.

Looking at any of these from the scientific perspective, they all correspond to some form going on in the electric signaling flying through the hundred thousand million neurons in the brain.

A sentience is such things.

From the perspective of science, the whole gamut of our experiences, from the shallowest to the deepest, is certain form going on in the oceans of electric signaling in the brain inside our skull. The journey in this book is to understand sentience from the perspective of science, thus bringing together the two perspectives, one of what it is like to be a soul or sentience, the other of what is sentience in the "external reality" that science studies, a reality that science takes as a starting point.

For many of us who believe in the reality of the perspective of science there this a mismatch the nature of which we cannot fathom. Yet scientific reason dictates that everything comes out of our brain. Everything we speak comes *solely* out of this ocean of electrical signals; and this is an important fact. All our thoughts, no matter how deep, could be spoken about in a conversational way. So looking at the physical world, with this brain in it, even the deepest thoughts are simply – when spoken about – an outflow from the ocean of signals, and hence these deepest thoughts, whether or not spoken about, are in the ocean of signals.

There is a massive physical, biological, cellular, and molecular substrate of support for these electrical signals. But everything about our sentience or soul derives from the signals and how they go around in the brain. This in turn derives from how signals are combined and channeled about at individual brain cells.

Along with the author, many people have a hard time imagining that their I, their inner most being, their sentience, there consciousness, their soul, their "what it is like", their very self, is nothing more than oceans of electrical signals, notwithstanding how vast those oceans are. Remember, this is the I which in the reader at this moment has the sensation of seeing the letters on this page and seeing the paper on which they are printed and the color of the paper. This is the I that right now senses the room we are in and has the experience of being in the room.

Or if we are outside right now reading this book, this is the I that experiences the blue sky and clouds. This is the I that feels the beauty of the world about, or horror. For myself, I cannot *see* that the brain itself, or a bunch of electrical signals, is the I or the experience itself. Yet from the perspective of science, it is.

Sometimes some science people feel there is no mismatch between there being a true I versus there being nothing more than oceans of electrical signals in the brain. They feel that the brain, a machine, just does its thing, and we are it. And the word "feel" is appropriate here because these are feelings not backed up by any solid understanding, for at the present time we do not have a solid understanding of sentience.

However, even for these science people, sometimes they look at the brain and wonder how can we be nothing more than a machine?

At present, scientifically, we know almost an infinitely small amount about the *details* of the logic going on in the oceans of those signals, though we have some very general information from technologies such as PET scans or magnetic resonance. But these technologies are like standing on the moon and looking at the oceans on earth.

You are hundreds of thousands of miles from any detail. It is difficult to get our bearings on these issues. Yet some day, whether it is a hundred years or a thousand years away, science will see these oceans close up, completely close up, and that will – if you accept certain scientific beliefs – allow us to fully understand the mystery of sentience. Then we shall understand exactly how soul, consciousness, I, being, experience, can be "in the brain."

Kittens and Mirrors

It was early morning. I was hiking. The sun had not come up over the eastern cliffs.

I saw the god again. Right off the bat it made an odd statement.

"Not all of sentience is 'in' the brain, though the brain is the overwhelming gateway."

"What?" I cried. I know I could not help looking confused and frustrated. The god seemed to be accessing for a while, then said, "Reality complex."

"Good morning," I said uncertainly. The god gave a slight nod. I continued.

"There are additional things I wanted to bring up. Two of them."

"Go ahead. By the way, the material on the soul yesterday was good."

"Thanks."

"But this is only the beginning of the journey. The issues are complex. You mentioned that our – your human – experiences themselves are fully in the signals that are sweeping in your brains right now."

"Yes."

"The reality is complex."

The god fell quiet. I shrugged a little.

"You wouldn't tell me this if I would not find it out eventually on my own."

"True."

After a little silence, I brought up the two points that I had thought about overnight.

"The first item is just a little test. It adds to what I mean by something being sentient."

The god nodded.

"Suppose that you are a person who feels bad if another person experiences some kind of pain. If something was sentient, then you would feel bad if it was in pain." Then I added parenthetically, "of course, only a sentient being could experience pain. What I am saying is, one test of whether something is sentient is if you feel bad when it is in pain – this is assuming you are the kind of person that feels bad when something is in pain. For instance, if a dog or cat or human is in pain, you feel bad about that. If a rock crashes into another rock, or a car into another car, you do not feel bad that the rock or car was in pain. Indeed, there is nothing there to experience pain."

"And the second issue that you wanted to bring up?"

"As far as I am concerned, sometimes scientists in certain areas get into a funny state regarding what is awareness or sentience. There is this idea that an animal being able to recognize itself in a mirror is a good indicator of consciousness or sentience.

"By sentience I don't mean anything to do with recognizing oneself in a mirror (although I have no grudges against such activity). In truth, one of my happiest times, perhaps when I was eight or nine, was when a few of us kids were out in the front yard.

"We took our two cats and put them on the small front porch and built a throne for them with sunny dandelions pulled from the bright green grass.

"I loved those cats. There was so much light-heartedness around them. That was idyllic play.

"They sat up on this little rickety gray porch – we lived out in the country – and they propped themselves as two little sphinxes overlooking the waves of golden dandelions in the grass. We placed dandelion on their heads, like little crowns."

"It's amazing they didn't run away."

"No. They just stayed there surveying their domain. And we found white daisies, the real small ones that grow wild in an overgrown lawn, and placed those on them too.

Eventually we put flower crowns not only on them but all around them.

"There was a happiness for me in those days that, to be honest, doesn't exist anymore. That happiness would mean more to me than being able to recognize myself in a mirror, or being able to figure out scientific or philosophical problems. All the sun long of that day of gracious fun was lovely. It was more real than any of this abstract knowledge that I now have.

"So when I define sentience, the definition must include such immediacy and happiness. If I had that kind of sentience, that would be more than enough, would be better than now."

"Alright."

"So for me, that would be a more important definer of sentience, or at least happy sentience. And what would really be bad is defining sentience in such a way that it excluded that day of little heaven, or defining sentience in such a way that it relegated that kind of high spirits. Because that is more important than all these other things. And please note, I am not saying that science and philosophy are not important. They are. Extremely. It's just that in the grand scheme of things a little bit of heaven is more valuable."

"You want to be careful," the god responded, "not to end up in a situation where you have neither."

"True.

"That fun back then is plenty good enough for me for what is sentience. I'm sure back then, when we were those little kids, that we could recognize ourselves in a mirror. But who cares. If there is something like being that transcends, in some sense, the existence of our physical body, I am not interested in it such that it can recognize itself in a mirror for eternity. The capability of experiencing joyfulness of being is all that I'm concerned about in sentience. In fact, as I am now much older and am so involved with many concerns, and with the challenges of science, still, I really don't care about all these things as much as I care about that sweetness of that early experience.

"Hmm. What am I saying? I mean, oh look, one of the reasons for concern about what is sentience is to see more clearly about the possibility of life after death, or some such thing. And frankly I don't really care if all these concerns and all of the reasoning and thinking I do in my daily life is part of such a future – if there is such. And even about the awe of science, I feel the same. It is wonderful. But in terms of the really important things, I don't care if it is part of any future sentience of mine. That joy and fun on that day regaling the cats with flowers and dandelions, overlooking the fierce green grass, that feeling is plenty enough, in fact it is better than all the others, as far as any of what I want to experience.

"So therefore these things are requisites in my idea of sentience. And I certainly don't care about if I can recognize myself in a mirror."

"Not unimpressive," the god said thoughtfully. "Though *if* you can have both the beauty you were talking about *and* the introspection into science and the universe, that can be even more wondrous."

Walking along the rocks away from my meeting with the god, I thought back on those days of early childhood, with the royal kittens. That was one of the moments that I was far happier, carefree, unburdened, and uncrushed than during most of the rest of my life. I would think of that as beautiful and good. I didn't have all this reasoning, analyzing stuff in my mind so much. If I can analyze that earlier state, that's good enough for me. If ideas about sentience don't incorporate that state, then they don't deal with the most important things in sentience's being.

3 Science: Introduction

I was talking again with the god.

"I think this time I know on my own what should be next on the journey," I said. "That must be science. Science must be explained. The book is from the perspective of science. Science is an incredibly important concept. I need to explain what I mean by science."

"What *you* mean?"

"Yes, what *I* mean. Because science means different things to different people. Even for people totally devoted to science, they see different things."

"Science is a large topic," the god said with a slight chuckle.

"True. But just as with the soul, I am not going to cover all of if but only a limited amount for what is relevant to the journey."

And thus begins this segment on science.

Reality is magical and science is the study of reality. But science has another image, a totalitarian force insistent on suppressing all other conceptions of reality. Perhaps when one considers the imperfect way in which science comes about, this image may have validity here and there. For myself, I think well of the goals of true science, and the journey of this book is from that perspective. Yet science is not the only perspective on the universe, and later we will touch on how science fits into other human ways of understanding.

We will also look at two examples of the magical in reality with science reaching out to them. First though, we look at the special origins of the way of science. Then we look at a chart of the physical universe as seen from the perspective of science today.

4 Science: Non-sentient Origins

"Let's turn to the beginnings of science."

"Why?" the god asked, "Time is relevant and here I am. Help motivate and speed up your journey."

"Because from the scientific perspective, this question of what is the soul is relatively new. For thousands of years people tried to figure out some of these issues about the soul, from non-scientific perspectives. I should say, what we *call* non-scientific."

"What do you mean by non-scientific perspectives?"

"Nothing negative. But this book is from the scientific perspective, and so that's where the journey is moving."

The god nodded.

"I want to look at the beginning of science. It's appropriate that we should understand the beginning. We're starting an enterprise of a kind that science generally has not done."

"Some parts," the god interrupted, "of the scientific community have started to look at characteristics of the brain that would seem to be correlated with sentience, but"

"But," I broke in, "we also want to make sure we back off to a larger perspective, the perspective of humanity in its journey through the universe and history, and look at how what we call science – how humanity first started on that road. It's natural to do that for our journey. What were the first actual appearances of what today we call science. Of course back then the word science wasn't used. There was only a somewhat new kind or type of thought, and maybe at first people didn't even recognize it as new."

The god had recommended certain books on ancient history, for instance Guthrie's *A History of Greek Philosophy*, but other books too. So I plowed into them, turning the pages of the books, imagining traveling back through the mists of time, far, far, through the clouds, and the blue sky, and nights and days, passing backward even past 1800 when there were no electric lights anywhere.

Then I was stunned for I was no longer in front of the book. I was flying, through the clouds, without a plane, through the mists and days and nights, and there, down there was 1800. I looked up again and for an instant I thought I saw the god far in the distance behind several clouds, but it was gone before I was sure. And down on earth I see, no washers, driers, automobiles, trucks, not even motors, nor television nor radio nor telephone, and I see when it took two months for news to travel across the Atlantic Ocean. Around 1800. I kept going back, further back, in time. It wasn't in my imagination, it was all real.

I saw Newton's published discoveries in 1687 (Newton, Cohen, Whitman, The Principia, page 11) that turned the physical universe into something like an intellectual machine of clockwork precision by introducing the whole precise science of motion. And still I kept flying further back through the clouds – flying through time and space.

A series of rough jolts hit. First, I saw the printing press invented by Gutenberg. Then I was jolted way back earlier in time to the discovery of the printing press in Korea. Then I was jolted way forward in time again to the discovery of gun powder in Europe and its fast use for military matters. Then I was lurched again to the past in Korea where gunpowder was discovered long before but for a long time never used for much other than fireworks. I was lurched once more forward, now to the great three-some, Galileo, Kepler, and Copernicus. In the early 1600's Galileo was the first human to turn a telescope to the heavens. Also in the early 1600's, Kepler figured out that the planets went in ellipses around the sun, not circles. Then quickly I flew back another hundred years, and there in the early 1500's Copernicus was declaring that the sun was the center with all the planets moving (in circles) around the sun. What a trio people.

And I was whisked further back, through the clouds, now over the middle east, where for three centuries from

750 AD to 1050 AD, the Golden Age of the Islamic world greatly flourished in culture and science and mathematics: the Arab world as the place of knowledge and intellectual culture of the earth, with its wonderful discoveries of algebra and its great works in astronomy – though astronomy was always kept absolutely a part of, and subjugated to, astrology.

Flying through the clouds, I saw the appearance in the early 600's of the great religion of Islam

Then below me, in the mid 400's was a great commotion from far beneath the clouds, with the sack of Rome by the Visigoths – what destruction! – though I could see it only in an unclear way.

Then roughly about 159 AD I saw the great Greek and Alexandrian astronomer Ptolemy who created a precise – precise for that time – description of how the planets moved across the heavens, a system with the Earth at the center – reasonable for that time – and involving so many circles upon circles going around other circles, all called *epicycles*. Then it was around year 0 and there was a birth of a great historic religion of Christianity. At about the same time I saw the pyramids of Egypt, even then over a thousand years old, and I saw the collapse of the Egyptian hieroglyphics into history, replaced by a new writing.

Then I launched into an extremely vague period when I had little idea what period of history I was in, and I was over the continent of India and saw the discovery of the concept of zero, along with the numeral for zero.

Then, without even traveling, I was suddenly over ancient Rome and there was Julius Caesar, around 40 BC, and in 27 BC the awful burning of the great library at Alexandria with possibly 400,000 scrolls of knowledge from the whole range of the ancient world, current, and past and further past – all gone – oh woe!

Then I was another century earlier. 150 BC. There was the good and wonderful Roman philosopher Lucretius, who carried on the work of even earlier Greeks, Leucippus and then Democritus (about 400 BC), and who had proposed that the physical world was composed of enormous numbers of incredibly small particles – atoms –yet the

Romans could see nothing other than poetry in Lucretius' works.

Further back I was taken through the clouds. The very air and sun appeared slightly different. 300 BC. Rome was an area of not much importance.

320 BC. Greece. There was Theophrastus – the student of Aristotle – who took over the Lyceum in ancient Athens and who participated in the great compilation and writing down of all knowledge, and Theophrastus himself responsible for 18 volumes on even more ancient philosophers and "physicoi". Two or three of the volumes survived for a time and it is to those that we owe almost everything we know of the even early philosophers before Plato, Socrates, and Aristotle. (For information on Theophrastus, see Wheelwright's *The Presocratics*, especially the note on page 277 and 285. A more technical presentation of how we know about the Presocratics can be found on pages 3 to 7, of Kirk and Raven's book, *The Presocratic Philsophers*.)

And then I started going even further back, moving over the Mediterranean shores and mountains of Ancient Greek and I saw below people in tunics picking olives, and I also saw ancient kinds of ships, even earlier than Triremes, and I saw the great Aristotle, about 350 BC, and before that Plato, about 390 BC, and further back yet, Socrates, about 435 BC.

Then out of the corner of my eyes I saw the great books of the Tao and the Chuang Tsu in the ancient Chinese regions, and the great philosophers Pythagoras and Heraclitus in the areas around Greece; all these in the 500 BC region of time.

Then I flew further, but I was slowing down, only a few decades further back, a few more, slower slower. And finally the flying stopped. I was in Miletus, an unusually prosperous Greek settlement on the coast of Asia minor, across the Aegean-Mediterranean sea form Helos (what we now call Greece). Philosophy and science have always required some prosperity. I saw one man, Anaximenes, about 535 BC, and then roughly two decades before that, an

Anaximander, 555 BC, and about another three decades even before that I saw the first one, a lone man, Thales, 585 BC. That was our man.

The three were making statements about how everything – and we do mean everything - was composed out of just one thing. One of them was saying that everything in the universe was made of air. Another was saying of water. One was even saying everything was made of the "unbounded indefinite" – an advanced thought for that time.

But it was the first one that I started to see better, Thales. Since this was so far before our time it was hard to see anything too clearly.

He seemed to be quite accomplished at, I couldn't tell, maybe having built bridges, or other things, and he also had a reputation for being a thinker. Ah, now I could see a little better. He seemed to be talking about how things were made out of ... again, I couldn't tell. But I could clearly see from the various things around him, over years, though it was all condensed for me all at once, that around him people had always thought of everything as having mental states and they had always made everything full of gods – well here I am using modern words and concepts. "Mental states," "the physical world," "the universe" – we humans did not have clear ideas like this. Not 2500 years ago. All was a continuous undifferentiated world view. We didn't have the foggiest idea that the earth was a planet. What today we call science and non-science, or the physical or non-physical world, they did not have concepts such as these.

But here, Thales, for whatever reason, was talking about things being made out of something. And when he was talking like this, he did not talk about gods or mental states. This type of talk was different from everyone else.

Here were the beginnings of science, on earth, a spec of a planet in the universe. He removed the anthropomorphic qualities from things, removed the gods, all the deities – at least when he was talking *in this way*.

Ah, but now I saw him a little over there – and this time he was talking about other things and he talked more like everyone else, he retreated from that far out way of talking, and he said all things are full of gods.

Hmm, he was still part of his times. I guess we have no choice. Not even the greatest of geniuses are fully separated from their world.

But again, when I saw him talking about things being made out of ... ah, he was the one who said they were made out of water. Maybe he said this because water seemed so widespread – moisture, clouds, the wetness inside our body and inside animals and some plants, and in the lakes and rivers, and seas, and rain.

It really didn't make any difference what he said things were made out of. His contribution to human thought was the idea that, all of what is about us, is *composed* out of a something more basic; and in this he did not rely on deities. These were his break-through ideas. (Given the primitive thought of that time, it is impossible to precisely translate into our ideas, what he stated.)

Here was the beginning of science.

I looked at him for a long time.

I strayed around a little, getting an idea of the three, Thales, and later, Anaximander, and later yet, Anaximenes, all in Miletus. But I knew I had to get going on my journey. What a pity.

Suddenly I quickly moved up to the clouds again, so fast an elevator ride I left my stomach on the ground. Now I was flying in the reverse direction, forward in time. Aristotle, then the ancient Romans, the religions of the world, the rise and fall of cultures and systems of thought, the appearance of all the technology that surrounds us.

The clouds spread. Instead of my home where the books where, was the desert. As rapidly as I had ascended from ancient Greece, I plunged back down to my world.

The god was there.

"You liked that, didn't you," the god gazed into the distance.

"Yes I did. Thank you."

I joined the god, looking at the mountains. An hour of silence later the sun was close to the horizon.

"What I don't understand is why did you send me back there, when the message was no gods."

"To see the truth."

"No gods? That's the truth?" I looked at the god.

"No, to see the mental place where science started."

"The point where it started, and then spread out from."

"Yes."

(For reference works on ancient Greek philosophers, see endnote 3.)

5 Science: Chart of the Physical Universe

The insides of an atom – the particles it is made out of – the electrons, neutrons, and protons that make it up – these are not common sense objects. Atoms are. That is why, in this book, atoms start off our chart of the physical universe.

Roughly speaking, almost all the atoms about us – including those that make up our physical body – are made of about 1 to 40 of the even smaller electrons, protons, and neutrons. Atoms really, fully exist down there in that very small world. From the right perspective, you could actually see them as objects moving around. This fact gets lost in science books because there are so many complex issues. Too bad such books are so much work to read. Yet there is no choice. That is just how the universe is. It is like our everyday life – the more details you get into, the less simple it is. Fortunately for purposes of our book, there are not large amounts of detail.

THIS	IS MADE OUT OF THESE	COMMENT	COMMENT
An atom is made out of	often about 2 to 90 electrons, protons, neutrons.		
A molecule is made of	often 1 to 5 atoms (*but* see next row)	water is a molecule made up of one oxygen atom and two hydrogen atoms	
But! molecules in living creatures - like bacteria, viruses, insects, fish, plants, cats, dogs, humans, are made of	usually a few hundred to thousands of atoms, sometimes even millions, in each molecule		Called *organic* compounds. The area of science that studies these molecules is called *organic chemistry*.

We will not be using this information in this book, but it is good to know in general way, in order to have some connectivity and understanding to what you will hear in the news and see in science text books. The above molecules in living creatures are usually divided into various categories, such as, carbohydrates, lipids, proteins (lots of these - they're chains of amino acids), nucleic acids (DNA stuff - in us humans, chains of 3.2 billion 'nucleotides')			
Organelles (these are the organs of a cell - just like we have organs - a heart - lungs - liver - eyes - skin - and so on, a cell has organs - called organelles)	Many millions of molecules make up one organelle	Examples of organelles: nucleus (at most one in the cell), Golgi bodies (lots in the cell), vesicle (lots), mitochondria (lots), endoplasmic reticulum (lots)	Organelles have technical names because non-scientists rarely talk about them

The Cell. This is the basic unit of life. The physical body of any form of life is made out of cells.	A cell is made out of a few thousand organelles	A human is made out of about 65 trillion cells. On average, a cell in a human is made out of about 100 trillion atoms.	
Tissue (an organ is composed of different kinds of tissue in the organ)	is made out of cells. A specific kind of tissue of a specific organ will have millions to billions of cells.		
An organ - examples are heart, lungs, pancreas, liver, bone, blood, skin, brain, eyes. An organ is divided up into the different kinds of tissue that make up the organ	An organ will be made up of several kinds of tissue		
A creature is made up of organs. Examples of creatures (on Earth): a human (us), a dog, a cat, a bird, a tree, a fly, a bacteria, an amoeba.	Very roughly maybe a hundred organs make up us human creatures (What is an organ and what are different types of tissue in an organ is not, and probably cannot, always be precisely defined).		
Planets, planetoids, asteroids, comets	Made of numbers of molecules beyond counting. Our planet has creatures on it.		

Solar systems	Perhaps a bunch of planets going around a sun (also called star). Actually, a bunch of planetoids, asteroids, comets, perhaps planets etc going around a sun.	There is quite a bit of structure to these solar systems. For instance in ours, beyond Pluto, is the Kuiper belt and then the Oort cloud, extending possibly half way or so to the next star, which is pretty far away.	
Galaxy	A galaxy has maybe on average about 100 billion solar systems.		
The region of the universe within range of telescopes from Earth	Estimated to have about 100 billion galaxies. Has about 10 thousand, thousand, million solar systems for *each* man, woman, and child on the planet. (This is not to imply anything about issues of "ownership")		

Figure 2 The Spatio-Temporal-Physical Hierarchy of the Universe

Moving to the next row in the chart, we have that molecules are composed out of atoms. There are a lot more kinds of molecules than kinds of atoms. Many molecules are made of just one to five atoms. For instance, the salt we sprinkle on our food is molecules made out of two atoms each, a sodium atom and a chlorine atom. Water is made out of three atoms, two hydrogen atoms and one oxygen atom. The hydrogen atoms are small and lightweight. Oxygen is hefty compared to them.

Of course, it was humans who came up with the names "hydrogen" and "oxygen".

It is not only that we humans have come to call these two kinds of atoms "hydrogen" and "oxygen". It is obvious that these words are arbitrary. Even so, we easily drift into feeling that the words themselves are more real than the types of matter they are describing. Little mistakes follow. Consider the word *atom*. Some science books will describe how the word comes from ancient Greece and that it meant "indivisible" – an atom can't be broken into smaller pieces, and hence is the smallest thing there is. But then the same book will go on to say that the Greeks were wrong because modern science has shown that the atom *can* be broken into smaller pieces. This reasoning is invalid.

The ancient Greek philosopher Leucippus (435 B.C.), followed by Democritus (410 B.C.), introduced the theory that all of matter was made out of very small indivisible things, which they called *atoms*. About four centuries later, the Roman Lucretius (in about the same time frame as of Julius Caesar, about 70 B.C.) took up the theory and developed it much further. Then about 1900 years after that, John Dalton proposed the modern theory of atoms, about 1800 A.D. (see for instance, Wightman *The Growth of Scientific Ideas*). This modern use of the word *atom* was no doubt in honor of the ancients Leucippus, Democritus, and Lucretius. If there is any scientific mistake, it might be that of scientists who chose the word "atom" a few centuries ago. However the one definite mistake is when books say that the ancients were wrong to say that the atom is indivisible. The ancients developed a theory based on an assumption of smallest pieces of matter. They made no mistake in calling such hypothesized pieces of matter, "atoms".

Many of the molecules about us are made of about 1 to 5 atoms. But as for the molecules in humans, that is a whole other story. In fact, it is a whole other story for all living creatures, including plants, bacteria, viruses, animals. These molecules in living things are often incredibly complex, having thousands of atoms in one molecule, and sometimes even thousands of millions (the DNA molecule for instance in humans).

The building block of all plants and creatures is the *cell*. All cactuses and willow trees are made out of cells. So are all trees and plants and cats and dogs and humans and crickets and insects and amoeba and bacteria. All of them are composed of cells.

The amoeba and bacteria are just one cell all by itself. They are small. The only way we gigantic humans can see them from our native perspective is with the help of a microscope.

Humans are composed of about 65 thousand billion cells (see Appendix: Numbers). Our brain contains about a hundred thousand million neurons (nerve cells), and these have stringy cable-like branches going between them, and signals travel over these cables. From the scientific perspective, those signals would seem to be our consciousness. *From the perspective of science, the gateway to understanding sentience (that is, awareness, or soul) is the electrical signals in the brain and nervous system; in particular, the gateway is the logics going on (in those signals).* There are tens, hundreds, often thousands of branches going out from just one neuron. And the same is true for the number of branches going into a neuron. So, indeed, our brain and nervous system is quite complex.

(The number of neurons is not known exactly but is estimated at from one tenth to ten times the number stated above. See Appendix: Numbers.)

Though the cables going into and coming out of a neuron have the shape of tree branches, they are frequently much longer in shape.

The cells of which all living creatures are made of have become specialized for the different parts of the body they are in, so that while neurons (nerve cells), stomach cells, or retina cells (cells in the back of the eye) are basically similar, they all have differences as part of their specialized roles.

Just as fleas, cats, and humans have organs, for instance, stomach, brain, eyes, mouth, intestine, skin – these are all organs – so too do individual cells. The organs of a cell are called *organelles*. Some of these organ*elles* in a cell are called mitochondria, Golgi bodies, endoplasmic reticulum, and vesicles. There are lots of each of these in each cell. More advanced cells have an organelle called the *nucleus*. All the plants and animals that you can see are made only out of advanced cells.

Amoeba and bacteria are made of only one cell. Long ago all life was only one-celled creatures, but then they started getting together to form larger creatures. Many one-celled creatures, and that includes all bacteria, are not an advanced cell. They have no nucleus.

Moving down in the chart, we have solar systems, consisting of a sun (a star) with various quantities of matter revolving around it: dust, pebbles, rocks, asteroids, comets, planetoids, perhaps planets, as well as regions further out from the sun, which regions are called the Kuiper belt and the Oort Cloud. Planets like our Earth are loaded with creatures and plants. As of the year 2000, there were 6 billion humans on the surface. Interestingly, now that we humans know more about the various environments of bacteria, we see that if you took all the bacteria together, they would outweigh all humans, mammals and birds, taken together, possibly by a factor of 10 to 1 (see Appendix: Numbers). By *weight*, there is ten times as much bacteria as all these other animals put together. That's quite an accomplishment for such small creatures.

Toward the bottom of the chart we have galaxies, each of which is composed of maybe on average a 100 billion solar systems. The current estimate of the number of galaxies within the range of our telescopes is about 100 billion (see Appendix: Numbers). There would be about 10 thousand, thousand million solar systems for each man, woman, and child on earth (see Appendix).

Such is the size of the universe.

From the biggest, the galaxies, let us move back to the smallest, and ask, how does the smallest join together to make the larger.

One wonders how all these little pieces of matter — atoms, molecules, groupings of molecules and so on — stay together. Why don't they just fall apart? Just the weight from gravity by itself should make everything fall apart. What keeps them all together so that they make up the tremendous structures of trillions of atoms and molecules in a chair, or a rug, or a car, or the human physical body? One of the main answers is electricity, as in static electricity.

Sometimes you may notice how a little thread becomes electrified and will jump to your shirt, or maybe just the reverse, it will try and jump away from your shirt. Or sometimes that will happen with little pieces of paper, or Styrofoam. The reason is that they have become electrified, which means nothing other than they got some extra electrons or lost some electrons – electrons are a kind of loose canon in the atom – in some situations they easily leave the atom and go to another atom – or – just roam around free. The little thread that will jump up to your clothing is so light weight that gravity means nothing to it compared to those electrostatic forces. And it is even more so with atoms and molecules. That electrostatic force is so strong compared to gravity. That is one of the main things that holds atoms together to form molecules, and molecules together in various way to eventually form all the hard objects you see about you, including your physical body. As for the details of *all* the ways that atoms can be attracted or repelled, or that sides of a molecule can be attracted or repelled from other sides of other molecules, you better believe they are complicated. But still, electrostatic forces are one of the main ways, the same thing that you see with jumping threads or pieces of Styrofoam or paper.

Let us briefly attend to a point made at the start of this section.

We began the chart of the physical universe with atoms because they are common sense objects, but smaller things are not. We have certain common sense ideas about what a physical object is, and atoms fit in with those ideas. At any point in time, they are basically in one single place (unlike electrons, neutrons, protons, and other particles of that sort and smaller – at any single point in time they are *not* at one single place – that's what makes them non-common sensical). Atoms move around like little pieces of matter, *exceedingly* little, but little pieces of matter all the same. They become attracted to or bounce away from other atoms. In short, they really and truly are little objects that move around.

Once you start to look at what atoms "are composed of" – electrons, protons, neutrons – you move into an area of probabilistic rather than definite locations in space. The area of science that includes a study of this is called *quantum mechanics*. But our investigation of what is sentience can be done without invoking the knowledge of this science. (The god informs me that later in our journey there will be a few brief points, mainly of comparison with issues in quantum mechanics, but that quantum mechanics will not be used as a source of sentience.)

The chart illustrates the "compositionality" of the physical universe. This was not always known to the degree that it is today. Things are made out of smaller things. Those in turn are made out of smaller things yet, and those out of even smaller, and so on for quite a distance. This grand compositionality, along with all its details, is one of the secular miracles of the universe; humans took several millennia to arrive at our current ideas about this mosaic.

Space and Time Increase

Space and Number (Multiplicity)

Let us do some thinking comparing different levels of the chart. As we move downward in the chart, objects occupy more of space, there are fewer of them (less multiplicity), and they move slower. This section presents some examples.

Can we get a feel for how big our red blood cells are? They are not that tiny. You could lay out 65 red blood cells end to end and you would get about the thickness of a finger nail (other than the thumb nail which is a little thicker – that would be about 80 red blood cells end to end).

2000 of these little physical things – these red blood cells – could lie across about ½ inch, which is about the width of either a finger or thumb nail, so if you look down on your finger nail right now, and picture it as if you were looking downward from an airplane to far below, you could have 2000 of these little things next to each other.

That is if you could get them lined up and get them to stay still. It is not easy to get anything that small to stay still - everything in that world is vibrating and bouncing against each other all the time.

Cells and red blood cells are not the smallest things in the chart. Even smaller are atoms.

Recall the size of cells compared to us – 65 trillion cells in our body – or for instance 65 blood cells will lie across the width of a finger nail. Lots of cells in the body. Interestingly, there is about the same number of the little atoms in a cell. Roughly. 65 trillion cells in our body. 100 trillion atoms in a cell. Roughly. On average.

The size of a cell compared to the human body is roughly the same as the size of space allocated to an atom compared to a cell.

Take all the people on earth (as of 2000 there were about 6 billion). Consider all thee teeming people, all those masses of people even in one city, like Moscow, or Beijing, or Tokyo, or New York, or London, or Paris, or Berlin, or Warsaw. Now take all the people in all those cities, plus all the rest of the people on earth. For each single soul in that teeming mass, replace the person by ten thousand new people. Then you would have the number of cells teeming around in your body right now, or about the number of atoms teeming in a single cell in the human body. We cannot perceive any of these things because we are way too monstrous a behemoth to see them, these cells and atoms, from our native perspective. But they are all there.

Time Shifts between Levels

Not only do size and quantity change between levels of the chart. Each succeeding row has objects that move more slowly. They take more time to cover relevant distances in their world.

To walk around the room you are in right now takes about 10 to 40 seconds. Other typical places that humans walk around might take seconds to hours to walk. The same is true for most creatures as long as they are not too small.

Let us move on in the chart to the next larger level. Here time takes much longer - compared to our natural instincts of time. The planets go around the sun anywhere from a few months to hundreds of years, our own planet taking one year. So we might say the average for planets to go around the sun is roughly 50 years.

At the next larger level in the chart, there are 100 billion solar systems in our galaxy, all swirling around like a mammoth pinwheel. The time it takes our solar system to go around the galaxy is about 300 million years. (By the way, according to the current scientific estimates of when life started on Earth, our solar system has gone 12 times around the galaxy since then.)

creatures: an hour
planets: a few years
solar system: a few hundred million years

We have looked at time in that region of the chart of us humans, then time in the region of planets, then of the solar system, and finally the galaxy. Now let us look at those regions of the hierarchy where we are large.

Cells in our body, for instance in our blood, fly along, covering many times there size in a time, that is for us, extremely quick. In larger blood vessels, red blood cells move in *one second* maybe 60 thousand diameters of a red blood cell. In other words, for a red blood cell to cover say 10 times its size takes only about 1/6 of a thousandth of a second. See how fast things are moving in this part of the hierarchy – fast compared to *our* native sense of time – just as the planets and solar system were moving slow in our native sense.

When molecules move in cells, they likely move extremely fast. However, let us look at something where we have more uniform information: how fast the molecules are moving in the air around you right now.

Well, of course the molecules in the air are going at all kinds of different speeds, and bumping into each other, bouncing in another direction, and so on. On average, though, their speed is about 2000 miles per hour. Yes, that is how fast they are moving about you right now. On average, they travel a distance of about 200 to 300 times their size before colliding with another molecule, which means they travel about 1/10,000 of a millimeter between collisions, so that they collide *billions* of times in just one second. If you consider that in one thimble of the air around you there are about 25 billion, billion molecules, you realize that what appears so peaceful and quiet in our native perspective, is a wild melee from a close-up perspective.

By the way, the molecules in the air are only two or so atoms a piece, and so the molecules are close to being atoms.

We have been looking at individual air molecules, and we want to consider the related phenomenon of "sound waves." Actually, the "sound" part of the term is a misnomer. They are in fact waves of high and low pressure in the air, spreading out from the source, and eventually impacting the ear drum. High pressure means denser, and that means where air molecules are more crowded together. Low pressure means where fewer molecules are crowded together. It is just the way it works, but these waves of more crowded and less crowded fan out from the source. Although the molecules are flying around at an average speed of about 2000 miles per hour, these pressure waves move through the air at a uniform speed of about 760 miles per hour – this is called the speed of sound though as I have said the word "sound" is a misnomer because these are actually waves of crowded and not crowded.

Here are very approximate times of common events in different sized objects:

air molecules: one ten-millionth of a second
red blood cells: one 6000-th of a second
creatures: an hour
planets: a few years
solar system: a few hundred million years

Perspective – Airplane – Behemoth – Brain

As a person high up in a plane looking down far below at a behemoth, so is our native perspective of ourself.

We have gone up and down the chart in terms of size and time scale. How might our own physical body fit into this? What if you were flying in a passenger jet at a height of 30,000 feet, a height many travel at? You look out the window and there far below on the ground is an accurate model of a hand, probably two miles long. The fingers are stretched out.

We will also suppose that either you are in an expensive seat with a good window to look out, or that you are traveling years ago when the windows in the planes were better. You have a good view of the hand on the ground.

You notice the finger nails, and the skin wrinkles, and some of the hairs. If it were not for the fact that it is six miles below, the hand has the proportions and texture of your own hand if you held it somewhat at arm's length.

At 30,000 feet of height (6 miles of height), our native perspective does not pick up any three-dimensionality; the hand might as well be painted on the ground for all we can tell. Other people on the plane, along with yourself, study the hand. It is so unusual to see such a thing from an airplane window.

Now you get hold of a zoomoscope, with end mirrors so far apart that it makes the hand look three-dimensionally exactly like the way you see your own hand when you hold it at arms length. In fact, you are astounded. And so are the other passengers who look through the zoomoscope. The gigantic model hand below now is seen exactly the same as your own hand held out at arms length, even the same three-dimensionality, and, it is even perceived as if it were only three feet away! Yet there are no lenses in the zoomoscope.

There are many perspectives into reality, into that which science calls the external physical world. And they are all equally real.

Now let us suppose that we are on a jet liner, with the same other people, but the whole jet, with all the people, has been shrunk 10,000 times smaller. To us on the shrunk plane, three feet now looks like 30,000 feet. Actually the whole jet liner, which was 200 feet long, is now only a ¼ inch long (200 feet long jet liner /10,000 = 1/50 ft = ¼ inch). Suppose this tiny, shrunk jet is flying very near the face of a regular-sized person who is holding their hand downward at arms length, about 3 feet from their face. To us shrunk people on the shrunk plane, the hand looks about 2 miles long and looks to be 30,000 feet below, and we see no three-dimensionality to it. Everyone on the plane is surprised to see such a hand 30,000 feet below.

Now you and the other shrunk passengers look through a zoomoscope that makes the hand look only 3 feet away (which it is, but without the zoomoscope, it appears as 30,000 feet to your shrunk self), and you can see the three-dimensionality of the hand. In fact, now it is perceived in everyway as a regular hand held out at arms length.

All these perspectives are equally real. Don't be mislead thinking that your native perspective is more real than others.

Let us think again on the chart, and on the cells and atoms that make up our physical body, and let us move in closer and closer, smaller and smaller, until finally we perceive the cells of our body as more or less 1 cubic inch each. Oh yes, there are cells of all different shapes and plenty of different sizes too. But now we have a perspective where, on *average*, the cells take up about a cubic inch. You could take out a ruler now and see what that is like.

Perhaps you are looking inside the lungs, or maybe the surface of the hand, or even inside the brain.

Most of the cells are all lined up, all structured, this direction and that direction. You see them going in all directions. The human lies down on the ground. You see that its body is 3 miles long.

If on average the cells in our body were about a cubic inch, we would be about 3 miles tall.

Not only that, the person is over ½ a mile wide and more than ¼ of a mile thick, and all made up of all these little 1-cubic-inch cells.

You make your way over to one of the thinner finger nails. You see it is five feet thick. If you took red blood cells end to end across the thickness of the finger nail, there are about 65 of them, with each being about 1 inch. The finger nail is about 170 feet or 11 car lengths wide. All this is with most cells being about 1 inch.

Now let us roam over to the center stage of our whole journey. The brain. We want to look much closer at it because it is so important. Anyway, the neuron cells that make up the brain are not at all like other cells in shape. If we get really close to them, they look quite a bit like trees. And being amidst all of them together, we are like in a jungle. They're not exactly like trees. There is a small center part, called the cell body, and there are generally two trees that come out in two different directions from that cell body. Usually electrical signals are coming in over one tree to the small cell body, and if the cell body should "fire", then a signal goes out over the other tree, down to the end of every branch of the other tree. This is the typical behavior, but there are always exceptions.

Close up, they indeed look like trees, though they go in all kinds of directions. There is another difference from the trees we see in the ground. For a given diameter of branch, the branches of the neuron go an incredible distance. So the trees can sprout out a huge amount and go a long way even though their branches are not too thick. There is one other difference with trees in the ground. There are no leaves, or if one considers something called the *boutons* to be the leaves, there is only one maybe every five feet of branch. That's what most neuron cells look like, two trees pretty much coming out of opposite sides of a small cell body.

Now our perspective moves in much closer. From this closeness, many neuron cells have one tree about 150 feet long coming out one side (about 10 car lengths long), and another tree, also about 150 feet long coming out the other side. Some trees are going up and down, some side ways. There is a lot of other supporting vegetation too. Everything is all intermixed, jammed and twisted together, with no space between anything. You have never seen a jungle like this. Some neuron branches are much shorter than a 150 feet, yet a few stretch on for miles!, winding through the dense morass of vegetation.

With incredible effort we hike to the center of the brain. "This is amazing!" you whisper.

Here is a jungle of squished vegetation all pushed up against itself. You figure out that it extends for 35 miles forward, 35 miles backward, and the same distance to each side. At the center you stand on several branches scrunched together. Unlike any jungle on earth, this one extends 35 miles up, and another 35 down, which by the way would be 17,500 stories up and 17,500 stories downward. All of it dense branches and supporting structure, squished all together, on and on, of every size and direction. From this perspective, the brain is pretty much like a 70-mile diameter sphere! Of jungle.

Indeed, if you look under a microscope, this is just what it looks like, except that one technique used to stain the cells fortunately only stains some of them, and those it stains completely to the tip of every branch. So what we see are only some of the trees scattered out in the jungle. And that is complicated and beautiful enough. If all the trees were stained, the microscope view would be impossible. On the other hand, if we look through an electron microscope, we see what a totally jam-packed mess the whole place is. That is our brain. A 70-mile sphere – from a good perspective. As we will see later on, there are, nevertheless, kinds of order running through that mess.

Electrical signals are going on all over the place, all the time, in this three-dimensional jungle that puts to shame the greatest jungles on earth. Let us suppose that our perspective is further changed so that we can see these electrical signals – spikes of electrical energy – traveling over the branches. By the way, to perceive this, our time perspective also would also have to be changed, for otherwise the electrical signals as well as any cell changes would happen like lightening and would be a hopeless blur. Perhaps if the time perspective were changed so that what used to be perceived as a second is now perceived as several hours, that would work.

So if you could somehow see these electrical signals, then when you looked out at the jungle around, you would see sprawling regions beautifully filled with electrical spike signals traveling everywhere along so many of the branches, and you would see signals coming into neuron bodies, along many branches, and either causing the neuron body to fire, or causing it not to fire, and if it did fire, a bunch of electrical spikes would dart out on the outgoing tree which eventually splits up into ever more branches, with the electrical spikes traveling down across all to the end of every branch.

This is your brain. Somehow from the perspective of science, your consciousness, your soul, is going on in all this. But how can something which is just bouncing around electrical signals be conscious?

By the way, from this perspective the human is a 1000 mile high behemoth.

If we move in and see the neurons as trees, say two trees from each neuron, each tree roughly about 10 car lengths long, then we can really see what the brain is like. It is an incredibly dense jungle, a three-dimensional jungle, with the vegetation pushed together; it has the same volume as a 70 mile diameter sphere, which it somewhat is. The human would be a 1000 mile tall behemoth.

Let us move closer yet, to the point where atoms, on average, have a piece of space that is about ¼ inch by ¼ inch by ¼ inch to move around in. We will call this a ¼ *inch cube*. Remember, on average there are about 100 thousand billion atoms in a cell in the human body.

In *this* perspective, cells would be about two full city blocks long, almost a quarter of a mile in length. Actually, most of each ¼ inch cube is empty because there is mostly empty space between atoms. They would look like vibrating bouncing around points in this empty space. So as you did your quarter mile walk across the cell, you would be walking through these masses of teeny vibrating points, with all these points being on average ¼ inch apart from each other. Well, they wouldn't be still. Whole bunches would be moving together one way or another, all over the place.

Such would be the cells and atoms from this perspective. It is a fully real perspective, as real as our native one. All these atoms are little physical objects moving and bouncing around. They are totally real: they are there, right now, throughout your whole body.

And what size would the human behemoth be in this perspective? About 36,000 miles tall. Considering that in the standard perspective the Earth is only an 8,000 mile diameter rock, the human would be about four times taller than the Earth. That's if atoms were about every ¼ of an inch. So if you ever wonder why we are complex, keep that in mind. (As stated earlier, it is not till you go inside the atom that we leave the world of common sense objects – but the atom itself is indeed a full-fledged common sense piece of matter.)

You look around in this new perspective where atoms are ¼ inch apart and humans are 36,000 miles tall. Though the perspective is real, you have never seen it before. You study it – atoms moving all over the place, gamboling about, in incredible quantities, going on seemingly forever into the distance, right, left, forward, back, up, down, springing all over the place against each other, going this way and that, going in loose constellations, and different constellations going this way and that, and those constellations joined in loose other larger constellations, and those too going every which way, and constellations built from constellations built from constellations and so on.

Everything is moving around, influenced everywhere by what is around it.

Our native time perspective would have to increasingly dilate at different levels if we were not to see everything as an impossible blur. At the level of looking at the atoms we would need one time dilation. From that perspective the atoms would be moving around so that you could see them, but the constellations of atoms would be moving somewhat slow, and the constellation made out of constellations would be moving so slow it seems it would take forever for it to move in any significant way.

To see the movement of these constellations our time perspective would have to be shifted so that the atoms were perceived as moving awfully fast. Now we would see the movement of constellations more clearly. And if our perspective moved up to larger constellations of those constellations and so on, they would now move so slow they would seem to be standing still. We would need to change the time perspective even more. Also we would need to change the spatial perspective more so that we would see the constellations, otherwise they would be too big and we would be too close to them. At each succeeding level everything smaller would have to appear to go faster and faster so that we could see significant movement at the current level, and also we would have to perceive things to be smaller, otherwise the current level would be too big to easily see.

All these perspectives are fully, totally of reality.

Our body, our nervous system, and our brain are made of nothing but things bouncing and moving around, with a lot of flow of electrons (the electricity, the means of the electric spike signals). Maybe we could imbue the perspective with color, some atoms are sea green, some radish red, others corn yellow or dark leaf green, so we could see things more clearly. All is bouncing around. Something is going on here. But it is all a mess of a vastness before which you can only feel humble. Behold the human.

If we move in to where the atoms are on average only a ¼ inch apart, then a cell would be about a quarter mile long, and we would be a 36,000 mile tall behemoth.

6 Science: Fizeau and Light

The universe is magical.

In 1849, for the first time the human race measured the speed of light, on earth. Now this speed is not some distant abstraction nor some number stated in a dreary text book. It's right here, it's what light is zooming around us, right now. Right now as you read this page, or look up at the wall, or out the window, this is the speed the light is flying at you. And the pieces of light are flying, well ... like bats out of hell.

It was a Monsieur Fizeau, in Paris, who measured the speed, in 1849. (See endnote 1 for references on Fizeau.) He took a cog wheel with 120 gear teeth around the edge. Fizeau shown a bright source of light through one of the gear openings, to a mirror on a hill 5 miles distant, and through a small telescope looked through the same gear opening. Now, as the wheel starts to revolve, the pieces of light flying out from between the gear teeth, fly to the mirror several miles away, and fly back to the telescope But, as the wheel twirls faster and faster, the next gear starts to get in the way before the returning piece of light can get back, and it starts to block the light. Fizeau sped the wheel up, and still could see the light, and sped it more, and still he saw the light, and faster and faster, and finally, wildly spinning around at 12.6 revolutions each second!, the light disappeared. Those flying pieces of light couldn't make it back in time before the next gear got in their way.

The speed of light had been measured!

Going through the logic and mathematics, Fizeau determined that the pieces of light were going like little bats out of hell – he did not use that phrase – going at an unbelievable 186,000 miles per *second*. How could something be flying at that speed? Never mind. There it was. It was reality. The wheel with the 120 gears was spinning around at 12.6 revolutions per second, or,

120*12.6 = 18,600 gears were flying across the telescope eyepiece each second, and the light had to travel 10 miles round trip and make it back through the same gear opening, so that's 10 miles * 18,600 gears per second = 186,000 miles in a second. That's how fast the light was going. I don't know about you, but to me that is amazing. The light flying around us, from this book as you read it, from the walls around, from outside through the window, it's all flying like little streaking particles at 186,000 miles per second. No matter how amazing or incredulous or un-explainable an observation is, science accepts it (of course, it must be a valid observation).

So that was Fizeau's accomplishment. Magical.

Just as in your everyday life when you give a fast explanation to a friend, you can't possibly state everything exactly, so too in the above presentation. Fizeau's full name was Armand-Hippolyte-Louis Fizeau, and he was born in 1819, Paris, France, and passed away from jaw cancer in 1896, Venteuil, France. He was thirty when he measured the speed of light. Fizeau began his school studies in the footsteps of his prosperous, physician father, but after a period of illness his interest had shifted to the physical sciences, and later, after his father died, he received enough inheritance to pursue his scientific interests without concern of income.

Naturally, Fizeau used meters, not miles. In fact, in addition to stating some distances in meters, he calculates the speed in terms of leagues, and came up with a value about 5% too large, or about 315,000 meters per second (the speed of light is 299,792 kilometers per second, or 186,281.7 miles per second, and today even these numbers are known with still more exactness). In 1849, it was not easy to make such an apparatus work. There were several lenses at various places, and a semi-reflecting mirror so that Fizeau looked through a telescope that looked out the same gear opening as the light was going out of. The large gear-toothed wheel was connected to clockworks driven by weights. The hardest problem was estimating where the light was brightest and where it disappeared. The light didn't suddenly disappear as the wheel went faster and faster. Instead it slowly got dimmer as more and more of the light that left the opening crashed into the ensuing gear.

And where were M. Fizeau and his fast spinning wheel of gears? They were on the top of Montmartre – on the top of which hill is the landmark cathedral, the Sacre Cour (Sacred Heart). The reflecting mirror was 8.633 kilometers away in the belvedere of his father's house at Suresnes.

Though Fizeau was the first to measure the speed of light on earth, the Danish astronomer Olaus Roemer, almost two centuries earlier, in 1676, used information about the moons of Jupiter and the size of the Earth's orbit to come up with a speed of about 150,000 miles a second, a good measurement for the time, even though it is 20% too low. Fizeau's measurement, being done completely on earth, was more *concrete*, and hence gave people a much greater sense of confidence that they had really got down the actual speed of light. Abstract mathematical computations concerning hundreds of millions of miles across outer space don't stack up to a measurement done totally on earth over a mere 10 or so miles.

In fact, about a hundred years before Olaus Roemer, in maybe about 1600, Galileo tried to measure the speed of light, on earth. He had an assistant stand on a hilltop about a mile away, and when Galileo uncovered a lantern, the assistant was to immediately uncover his lantern, and Galileo could see how long it took for the light to travel the two miles to and back from the other hill. What a joke! Since the light would have taken barely one one-hundred-thousandth of a second to cover that distance, poor Galileo obviously couldn't tell anything. From the perspective of our knowledge today, this seems humorous. But to the people of the time, they did the best they could. His idea for measuring the speed of light was fully valid, but the available technology was hopelessly inadequate.

How will the future look at us?

As unbelievably fast as is the speed of light, it is a part of the magic of reality. It is there, right now. It is real. Being unexplainable and incredulous cannot be used as an argument against what is validly observed. This will also apply to delving into the soul.

7 Science: Newton, Gravity, the Laws of Motion

(For references on Newton, see the beginning and end of the chapter on "Deep Physics, Newton, Some Religion," in Part 3.)

The universe is magical. Here is another example.

In 1687, Isaac Newton published a set of laws that were of both a conceptual and mathematical nature, that explained the motions of *all* the heavens: the sun, the moon, Mercury, Venus, Mars, Jupiter, and Saturn (in 1687 they didn't know about the other planets); the laws explained how all of these planets course through our sky, and predicted their motions to a degree of accuracy and precision way, way, way beyond that of any previous theory. Not only did the laws explain the motions of all the heavens, but, they explained the motions of the tides on Earth, and indeed explained the motions of most of the things on Earth that "were not" sentient, though it would take humans a long time to work out the mathematics of these motions as derived from Newton's laws. The laws gave a conceptual, mathematical, predictive framework for motions of tides, clouds, gases, fluids, solids, from the large down to the molecules down to the atoms – the motions of all these.

Now there is this thing in Newton's theory which he named "gravity." *Every piece of matter anywhere in the universe is pulled toward every other piece of matter anywhere else in the universe.* You don't feel the pull between you and, for instance a tree, because the tree is too small. Nor do you feel the pull between you and the moon or sun, because they are too far away. The moon is distant, and the sun is about 350 times farther yet.

But our planet the Earth is another story. It is huge, and you are right up against it, so here you really feel the pull gravity. You feel it at this instant, and you feel it every second of your whole life. You feel it on the bottom of your feet when you walk, and as you get old you feel it in every bone of your body when you move about.

This miracle of gravity is everywhere about us, all the time. Our extreme familiarity with this miracle has "de-miraclized" it, but it is a miracle all the same.

With Newton's theory, thinkers became aware of how strange, how unexplainable was this thing called gravity. What "magical" thing was going on between all the pieces of matter in the universe? What, were they sending little telegrams to each other, "pull this way"? Another magical aspect was that whatever was transacting between all these pieces of matter throughout the universe was going through *empty* space (since all the planets and stars are pulling each other together across the vacuum of outer space between them). Newton was honest about how amazing and unexplainable this was. Further, in spite of pressure to make his theory and ideas sound better, to the end of his life Newton resisted framing a hypothesis as to what gravity could be. This is his famous pronouncement, "hypothesis non fingo" – "of this I frame no hypotheses."

Gravity is still a much studied and questioned thing today. Let us just say, "Behold gravity." It's a part of reality, it's a secular miracle, it was totally unexplainable for a long time after Newton, and even today some feel that the explanations from science as to what gravity is are not full explanations. Such is our universe.

"Feeling" Gravity in Our Everyday Lives

Let us look in more detail at the reality of gravity in our everyday life. Since we are right up against this huge Earth, we feel the pull of gravity plenty, constantly for our whole life. You feel it in your seat when you sit, in your feet when you walk, and in your bones as you age. Step on the scale. The scale may show that you and the Earth are pulling together at 180 pounds.

When you lean back on a sofa, you feel the results of gravity on your back as well as your seat, as gravity pulls you toward the Earth, and the sofa pushes back. And if you let your arms go, just rest them, your hands and the bottom half of your arms feel the effect of gravity as they rest on your lap, or on an arm rest, or pillows. In outer space you simply would not feel any of this continuous pressure of gravity. You wouldn't even stay in a sofa unless you were fastened to it. There would be no need to sit on anything soft because no part of your body would be stressed by gravity anyway. All that the softest of pillows and feather down and beds and furniture do is to distribute the pressure of gravity across the surface of your skin so that no point is uncomfortable from being pressured too much. But in space there is no pressure at all, so nothing could be softer.

In more than 99.999% of the space in the universe, one would not feel gravity. But we are in the other part that is right up against a planet.

Again, let us suppose that you step on a scale that says 180 pounds, which is to say that you and the Earth are pulling together at 180 pounds. If you stand on a box and jumped off, the Earth and you are pulled together with 180 pounds of force, and *both of you* go flying toward each other. "What!" you say, "the Earth doesn't move when I jump off the scale." Oh, but it does, but not by much. Since our home planet is just a bit larger than you, you will be the one doing just a little more of the moving. To be precise, the Earth has about 120,000,000,000,000,000,000,000 times more mass than you, and so you move that many times more than what the Earth moves. Nevertheless the Earth will move to you, even though by an incredibly small amount. (The details are complicated: if the Earth were a solid globe incapable of stretching or bending, this would be true. However, the Earth is made of fluids and dirt and so on; I leave it to the physicists and mathematicians to work out the details of in what way the various parts of the planet move toward a person who jumps off a box. Nevertheless, overall, the Earth does move toward the person, a little.)

Gravity Magically Goes Through Everything

In Newton's time the people who thought a whole lot about things were aware of how strange, how unexplainable, was this thing Newton called gravity. What "magical" thing was going on between all the pieces of matter in the universe? Not only was it going through unbelievably great distances of empty outer space, but it was also going through everything that was in between the two pieces of matter. So if there was dirt, air, water, or outer space between the two things pulling each other together, the pull went through all of that, and it seemed to go through all of it basically instantly, and without the slightest weakening for having gone through all the stuff in between. Now that is odd.

This was previously undreamed of. For instance, you would think that thousands of miles of crushed rock lying between two objects would hinder gravity in at least some way. No.

There is an attraction between you and all parts of the Earth, but the net result is that you are pulled toward the center of the Earth with a "force" of 180 pounds. That 180 pounds is due to a mutual attraction of all parts of your body being attracted to all parts of the 8000 mile diameter planet beneath you. Now some of those parts of our globe are down below thousands of miles of densely, massively crushed rocks, the equivalent of thousands of Mt. Everests. Yet the attraction of gravity takes place exactly with the same force as if the thousands of Mt. Everests were not in between. Whatever gravity is, it goes right through all those thousands of miles of rock with not an iota of difficulty. Yet there it is. And that is a secular miracle. All of that constantly, all of that going on right now.

Whether today's physicists are able to look at gravity as particles or as distortions in 4 dimensional space, it still is magic. The Earth contains 10 billion cubes of matter, each 2 miles by 2 miles by 2miles. Right now, every molecule in your body is pulling together with every one of those 10 billion cubes, the pull going through all the cubes in between. Every piece of matter in the whole universe throughout the billons of galaxies some with hundreds of billions of stars is attracted to every other piece of matter, the attraction not interfered with by *anything* in between. All of that constantly, all of that going on right now.

Gravity Affects Our Perceptual Space

Gravity affects our perceptual environment in more ways than you might think. It gives us a "false" conceptualization of reality. We are born on Earth, spend our whole lives on Earth, and die on Earth. Gravity is so much of our existence that it doesn't occur to us how unusual such an environment is. Indeed, gravity is a never varying part of the existence of the whole human race, and it wasn't until Newton in the 1600's clearly saw that there were situations where no gravity felt was felt. Since gravity is a pervasive non-varying, never interrupted part of individual human lives, and of the human race, and of pre-human ancestors, and of the whole experience of all life ever on Earth going back to the one-celled creatures billions of years ago and before, it follows that the issues of gravity must be hard-wired into the brains of all creatures on our planet (if we are talking about creatures with brains), giving the creature the impression that gravity is a metaphysical existent of all parts of the whole universe (if we are talking about brains advanced enough to have impressions).

Yet, in the last few decades, for the first time in the history of life on Earth some creatures have in effect left the confines of Earth's gravity. Astronauts. For periods that they are in space, they do not feel Earth's gravity. Much of our fascination of watching them float around on our television screen is that the situation contradicts the hard wiring in our brain, which has come from the eons of our evolving on the surface of a planet. There would simply be no concept of up and down, if we had not evolved on the surface of a planet.

Change in Human Concepts Describing the Universe

Some say that the discoveries of Isaac Newton changed the world from the pre-science to the science era. The idea of gravity is Newton's contribution to history. Indeed the very idea of force, as moved into a definite scientific perspective, is his too.

Modern physicists take the conceptualization of something called force as so fundamental that they don't realize what an intellectual breakthrough it was at the time. Just prior to Newton the metaphysical predisposition was that there should be only physical things and only motions of physical things, and that there should be nothing else. For example, even two millennia earlier, the ancient atomists Leucippus and Democritus and Lucretius assumed that objects could only affect each other only if they were touching.

However, from his own mind and apparently from whiffs of partial ideas floating around in his intellectual environment, Newton early on came to accept that, conceptually speaking, something like this abstraction of force, was of the *first* value in explaining things and their motions. In other words, in terms of a system used to think about and to explain and to predict and to analyze the motions of physical things in the world, this abstraction, this concept of force, was near the center. It was what all explanation went back to. Whenever one wanted to analyze the motions of objects, one needed to first bring into one's thoughts this concept of force. That was a new way of thinking.

And one shouldn't judge too harshly the prevailing ideas up till Newton's time. Things and their motions, those are hard, real, in the here and now. The idea of an abstraction like "force" was not something concrete.

It is true that this shift in Newton's thinking – this abstraction called "force" – occurred to him early in his investigations. It is true that he went far beyond this, when he developed specific formulas of gravity, specific equations of motion and force, and a whole tool box of intellectual techniques for using this idea of force. But without the initial shift, he would never have made it to the next base, and he never would have opened up the new system of the universe.

To be sure, one must absolutely have a firm footing in the concrete, but here and there in human intellectual history, it is necessary to branch forward with a new idea that at first seems "abstract," that at first seems not so real.

Thus goes the work of the human mind in advancing its understanding. Thus advances the human race in its eternal quest of grasping the meaning of the universe. Thus too goes our own journey toward the soul, from the perspective of science. Of light, gravity, and the soul, the last is the most wondrous, and real.

Scientists Not the Final Arbiter of Fact and Language

There is a slight danger when scientists set themselves up as the ultimate arbiter of the universe. The facts that scientists discover are not owned by scientists but rather inhere in the nature of the universe of which and in which we are all beings. Scientists are like explorers in a heretofore unknown geography. They do not create that geography, nor do they own it, for it belongs to the universe. Hence we have as much right to speak of it in our own choice of words.

The legitimate disquiet scientists feel with terms, for instance, such as "magic", may have to do with such words being used to deny genuine truths that scientists have introduced into human knowledge. Further, explorers of a new region of the universe have the right to give names to the new things they perceive. This is fitting and honorable. Nonetheless, let us bear in mind that the things in themselves are independent of any names humans provide.

8 *What is Science*

Generally, what this book uses from science's conception of the world is well-accepted by almost everyone who agrees with the way of science. There is such a thing as a "physical" reality external to our thoughts, and this reality is composed of atoms and quite a bit of etc., and it is structured in certain ways via concepts of time and space, and there is a rich set of mathematical and logical ways for intellectually grasping this reality, and there is "the experimental method." Though not explicitly acknowledged as such, the experimental method is the assertion of the *a priori* precedence of this external reality over our imagination. This is why you often see scientists looking through their microscopes, or telescopes, or looking at nature out in the field or in the laboratory. The scientist is first and foremost looking hard at that external reality, because that reality is the presenter of all. When doing science, that reality is the primordial confirmer. This book is the journey that starts from the point of science.

If we can prove or derive something on the basis of this external reality, this external world that seems so strong to us, why then so much the better than if we proved it from some purely mental explorations.

If I get some insight into whether we exist after death, I would be delighted if the insight were rooted in the physical real world and in knowledge about that world. In short I want the insight to be as real as the reality of engineering and chemistry and physics and science.

Science versus Philosophy

Science and philosophy have each other as friends. They are not perfect friends.

Philosophy will jump to the largest of questions about the fabric of the universe. Science on the other hand has built up over centuries an extensive, careful system, and only branches out from there in a strict manner. As for that system that constitutes science, any rational person will concede that it is not only valid but is also one of the greatest accomplishments of human kind, an accomplishment for the human species to be more than proud of.

A colossal difference between science and philosophy is that science accepts external reality, no matter how unexplainable it seems. (Critics of science will see this matter differently, complaining that science does not allow certain parts of reality into its area of consideration, and maybe sometimes these people are right. The back and forth polemics about the intellectual activity that constitutes science can be endless.) To science, reality is reality. There it is. Whether or not it can understand reality, science can't argue with it. Philosophy, on the other hand, wants to go into the subject matter and explain it deeply; it has reluctance to have large unexplained areas. This difference gives science strength over philosophy, for it can get at some truths faster, and then go on from there, even though it may not understand the basis, just as for a long time it did not understand the basis of light or gravity, and perhaps still does not.

Science versus Religion

Science and philosophy are friends, of a sort. But there are other players about too, a little more cordial with philosophy than science, but interacting, at times, with science.

The universe is a big place. Science starts at one point and branches out from there. Religion starts at another place, and branches out from its point. They both explore the same universe. There has been some attempt in some scientific circles to speak of two realities, one of science, the other of religion. But there cannot be *two* realities. There is only one, there is only one universe, even though there are different perspectives or views into that one reality, and even though it may be a long time before we can explain how various perspectives of the one reality can appear so different.

We might well note the famous Newton, who had no conflict between religion and science nor between religion and knowledge. Nor did others. (See endnote 2.)

Science versus New Age

"New age" has a variety of meanings. Generally science is a much more precise type of thinking, more tightly based in the external reality of science, than is new age.

Science versus Mysticism

The word "mysticism" can have a wide variety of meanings. Some parts of mysticism are not at all of a precise kind of thinking. Other parts are simply an acknowledgement of how intellectually profound is the universe, which universe includes the external reality of science.

9 *Our Universe*

These are the things that make people stop and wonder just what kind of universe we live in. Is it possible, nay, is it conceivable that billions of photons are flying into our eyes, traveling in incredibly straight lines over great distances? And even more challenging to believe is the speed of their flight. That speed is truly momentous. But it *is* real.

Then there is gravity. Going on right now, all around our body. Every single piece of material in our body is being pulled to every other piece of material in the whole universe. Even when millions of cubic miles of dirt and rock are between, the pull is still happening. And the pull between us and the Earth is something that we feel in every part of our body from the moment of birth till the minute of death (unless we are an astronaut). Unbelievable. Yet there reality is.

These examples would be unreal − except that they are real. Conceivable or not. Understandable or not. They *are* the universe. Human, behold the universe.

Part 2   Animals and Brains

10 Discussion with the God

"Well that's good. You defined your basic concepts. That section on perception was good too."

"Maybe."

"It was important."

"It was? I guess it is. But I don't see the whole picture yet."

"It is. You humans are misled by your native perceptions. You tend to think they define reality. But they are only *one* view of reality. There are many other views, and they are fully and as totally real, *scientifically*, as your native perception. And yes, this idea will be important."

Interesting, I thought.

"Now what?" the god asked.

"What do you mean, now what?"

"What are you going to do next?"

"I hadn't given that any thought. Frankly, I thought I would, well, just enjoy for a while having got the last part done."

"What will be the next thing you need to do? On this journey?"

To tell the truth, this was a question I did not need. I thought I had worked hard – well, somewhat hard at any rate. I needed a break.

"What is the next thing?" I asked, "The next thing is I am going to leave for a little while. I'll be back. Of course. I just need a bit of a break," whereupon I turned and walked out of the little desert grotto. Frankly, I was happy to get away from the god, at least for a while.

The next several days I hiked, in the sun, and peace. It seemed I was always getting farther and farther every day, down and around this little trail into a wash with pine-tree like shrubs, and back up again, and on.

After several days, it did become clear what had to be the next part of the journey. Anyway, I had my break.

"I know what the next part is," I said to the god.

The god nodded, and remained silent.

"What does a scientist do? We must look very hard at reality. We must look quite carefully at this external physical world. That is the way of science. This external reality transcends our ability to think up ideas. It transcends us in almost everyway. So we must look determinedly at it. There are many answers there."

The god nodded.

"Therefore, this journey being from the perspective of science, we must look – into the physical world – at where and how sentience takes place. And that is in brains – of animals – we are an animal – we are a human animal. We must look out into that physical reality, we must look hard and carefully at what is the nature of the logic going on in animals and their brains."

I continued, "So we will look at a series of creatures where the logic going on in them is increasingly advanced. That's what a good scientist would do."

We talked a little about what this part of the journey might be like.

"That is good," said the god. "In addition you have some pieces still to pick up from standard existing science."

"I do?"

"Waves, zero crossings, and neurons."

"I do? OK."

"You will have a separate section on each of these. Except the neuron material, which you will mesh with the coverage of one of the creatures, even though it applies to all animals."

This god sure knew a lot, I thought.

"I can see into the future. To a degree. I can see that this is how you will handle this part of the journey eventually. And that you would handle it this way even if I said nothing. I am just speeding things up a little by telling you."

"I know." I wasn't sure how I felt about this. "So waves, huh?"

"Waves. Waves occur all over in nature. And you will need to cover them. As in waves on the surface of water. As in sound waves. As in light waves. As in pressure waves. As in waves of electromagnetic radiation. As in pressure waves of electrical voltage in water."

"Pressure waves of electrical voltage in water?" Where in the world would that come from?

"And since waves on the surface of water are the mother of all ways of thinking about waves, you will use that as the main example of waves."

I nodded.

"And neurons. You will need to cover neurons. Those are the cells that make up our brain. And they also make up the nervous system outside the brain too. Neurons are the cells over which all the electrical signals travel around in your head and also travel to and from all the senses and muscles in your body. Those are the neurons. And they're in humans and all creatures – except not the simplest of animals like one celled microscopic creatures. And you will want to put almost all this neuron material in with the chapter on one particular animal."

"I'm sure I'll find a good arrangement."

"You really need to cover all this. But don't worry. It will come naturally at the right places – given that you do the work to find out the information. As for the neurons, it's important to get into some of the concrete. It really makes a world of difference. You can talk forever in abstractions like 'signals' going around in the nervous system. But you don't get the right feel for the situation unless you get some concrete detail."

"Oh, I definitely agree with that," I said with strength and without hesitation. "But tell me. What on earth are zero crossings?"

"Owls, and bats, and other creatures too ..."

"Owls and bats? What has this got to do with owls and bats?" I asked.

"Owls and bats - they make incredible judgments, based on the sound waves that reach their ears, about the location of a tasty mouse or insect. Then they fly right to where the tasty thing is, whether a little closer or farther, or to the right or left, and catch it."

"Nature is a little on the cruel side."

"They don't always catch it. Anyway, there is a remarkable thing going on in their nervous system that is the basis for figuring out where the mouse or insect is. For each ear there are neurons that are hooking into how fast the sound waves are coming in. Maybe some waves are coming in at 200 times a second – just think of water waves next to a wall in a lake, and the waves are going up and down 200 times every second – well sound waves are very fast – 200 times a second. Now as the wave goes up and down on the wall, at one point it is at its high point, the crest, and at another point it is at its low point, a trough. Half-way in between is called the *zero point*, and every time the wave is at the half-way point between the high and the low, a neuron is detecting that and emitting a signal. Those are the zero crossings. In this case it would be 200 zero crossings a second. That's quite a bit."

"But not really," I interrupted.

The god looked at me.

"It's not really," I continued. "It's only from our native perspective that it *appears* fast – 200 times a second. From another perspective, it would seem lethargic indeed, where 200 times every second might appear like once every five minutes. And that is the best perspective in this case because you would see how there is plenty of leisurely time for an electrical signal to move here or there in the brain, and for this or that to happen, before the next five minutes. Our native perspective misleads us in this situation because it makes it seem like the waves are hopelessly fast and that the brain could never do all that it needs to with that kind of speed."

The god continued. "The neurons in the nervous system of owls and electric fish compare these zero-crossing signals from the two ears. For instance the owl uses the comparisons to detect the location of prey. You are going to explain zero crossings."

"I am? Alright. I am! Zero crossings! Owls. Electric fish?"

"As for the report you will write, the reader could skim some of the technical material, if they wanted to. Especially on the advanced example. Also some side philosophical issues that you will bring in, on detection devices and volcanoes."

Volcanoes, electric fish, zero crossings. And so I went back to the hikes. Different creatures' brains. A throw-in of material on waves and zero crossings, a further throw-in on neurons, and that was it.

But as I hiked onward, into a particularly rocky area of great beauty under the piercing sun in the sky that seemed to go on forever, and as I started to enter the foothills of the mountains, I realized that all my discussion with the god had said nothing about the main issue. What would really be found on this part of the journey?

Hours passed. I found I remembered less and less of what the god had said. I knew we had talked about this part of the journey, but the details slipped more and more from memory.

Animals

The way of science is to look out into the world at what you want to study. That world, that external reality, has more than we ever can conceive of on our own.

So on the journey of this book, we now look out at the external world, carefully and hard. Sentience occurs in animals and so it is at animals that we look, from the simplest to the more complex. And we will look not just at sentience, but at a larger cut, so that sentience is only one part of that cut. We must grasp this larger cut as it pertains to the nature and the fabric of the universe, in order to then move onto understanding sentience.

The difference between knowing something in the concrete versus knowing something as airy abstractions is great. (This may be a contributing factor to why Galileo got into trouble with the Church in 1633, whereas Copernicus and Kepler, whose work was to the same end, which is that the sun is the center of the planets' movement, and whose work was more abstract than Galileo's, experienced some friction of a religious nature, but it was nothing compared to Galileo's experience. Galileo was the first to turn a newly discovered device, called the telescope, to the heavens. Suddenly all kinds of abstract ideas became concrete – you we could now see them with your own eyes. For more information on Copernicus, Kepler, Galileo, and on Galileo and religion, see endnote 12.)

This desire to look at the concrete is another reason we look at actual animals and the logic going on in them, and this will culminate in looking in the brain of the electric fish at some of the most sophisticated logic that we can currently understand in detail.

11 Waves

Group of Hikers, Water Waves, Strength and Frequency

When I returned to the desert the next day, I was hiking in the same area again. I had to push through a little avenue of twiggy mesquite. Turning the corner, I heard people.

It was two hikers, and after introducing, it seemed to me they were slightly uncertain about their location. In fact, from their general demeanor I wasn't sure they even knew what they were doing on this path. It seemed to me that at some rational level they knew they shouldn't be here, but it confused them that at an emotional level it didn't bother them.

They decided to go in the direction I was heading. In fact they weren't sure in what direction they had been going. When I mentioned, in an indirect way, something about being and the soul, I was surprised that they responded. I ended up telling them what I had written about the waves. In fact, we sat down in a little area of trees, and I read it to them.

The quintessential origin of understanding for waves is ... well ... waves on the surface of water. That is where people got their first training in a thing which later was found to be out there in reality in many different forms. The same idea occurred over and over, but in different ways. All the different kinds of waves are logically like water waves.

Drop a rock into a quiet pool. The waves spread out in ever increasing circles, on and on. If you stoop down and get close to a little muddy puddle after a rain, drop a small pebble, then as it plops into the water, you can see little circles of waves fanning outward.

You can also drop a tiny pebble into a pot of water, but the waves fly out so fast that for our native perspective it's hard to see. Additionally, the waves reach the side of the pot and bounce off, and run back to collide with other waves bounding off the rest of the sides of the pot. Furthermore the waves are decreasing in strength fast. All of these, the speed, the collision of waves from all directions, the fast fall off of the strength of the groups of waves, all these make it hard for us to see (with the way we are laid out in time).

Nevertheless, if you wiggle your finger gently up and down on the surface of water in something larger than a pot, you will see circles continuously spreading outwards.

On the sandy beech of a little lake you can see little waves rolling in on the shore, maybe only half of a foot high. But on the shore of the Pacific or Atlantic Ocean, on a grey stormy day, monstrous waves the size of small buildings, roar and crash onto the sand, washing far up the sea shore, perhaps as you have to run away from them. These noisy behemoths could be as high as six-story buildings – sixty feet. The height of the waves is how we measure their strength.

Actually, the height of a single wave is measured from where the water would be if everything were calm, and that's half way up the wave, because the other half is half way below where the water would usually be. Thus the roaring white-capped behemoth would be not sixty but thirty feet of strength.

People who do a lot of looking at the reality out there get to make up names for what they see. That's reasonable. Are those the real names? Of course not. They are just the words we humans attach to things we see out there. But we need agreed upon words or talking sounds to talk about the going-on's in our environment. Otherwise when we all talked, it would be a tower of babble.

Somehow, over time, the people who were concentrating their attention on waves, decided to call this measure of strength, this half-way point, this thirty feet, *amplitude*.

There are various forms of the word *amplitude*. Your stereo has an *amplifier* section. It is increasing the strength of the sound waves coming out of the stereo. Yes, sound is also a kind of wave. We'll touch on that in a little while.

So there is the 60 foot mountain of a wave − 30 feet above and 30 feet below the regular water level. Such a beautiful wave coming in over the steely ocean under stormy clouds. Or, there's beauty even in sixteenth of an inch wave as you look close into a muddy puddle after a rain, and let go a little pebble, watch it drop into the water, making little circlets of waves only a sixteenth of an inch high.

I put down the papers I was reading and looked at the two hikers.

"Why is nature always measuring things?" one of them − Jack − asked me.

"Nature or people? Well, that's just where it's at. It's that relation between matter we call distance. If you start looking at huge ocean waves and little lake waves, and tiny waves from a pebble, you might start thinking of them in terms of their height too − thirty feet, half a foot, a sixteenth of an inch."

"What about in a swimming pool where waves are all over the place and the water is real choppy?"

"Well, exactly! Waves are all over the place, there are waves going in every which direction, and they are all over the place, and they're colliding all over the place. So the whole thing looks like one choppy mess."

"By the way, when two waves meet, from whatever direction, do they get even higher and deeper?"

"Sure, if you see two sort of big waves come together, the water will get even higher where the crests meet."

In addition to the height, or strength, or *amplitude*, of waves, there is another thing that people who work a lot with waves are always looking at. For water waves, that is how frequently the water is going up and down as the waves pass a specific place. For instance, let's suppose you are on a cliff looking down at the ocean, out at sea, not right by the shore, and as you look down to the swirly, churning depths outward and below; let us suppose that you concentrate on one place. The water is rising up and falling, rising up and falling. Suppose it does this three times in a minute − high point, low point, high point, low point, high point, low point, in one minute. Then – going back to the words humans make up and then use to describe what they see out in reality – the words are – these waves have a *frequency* of three cycles a minute. *Frequency* means "how frequently," and *cycle* is a way of referring to the whole cycle of one up and down. There are three cycles of waves going by every minute in our big wave example.

"And what about the little waves in a mud puddle when you drop a little pebble in? The ones that are sixteenth of an inch high – I mean *amplitude* of sixteenth's of inch."

"Oh gosh, those waves go up and down so fast. I don't think you could count them in our native perspective. If you video-taped them and played the video at slow speed, then you might be able to tell the frequency."

"We could estimate."

"Yah."

"But best if one had a video tape that you could play at slow speed."

"Right."

"Well, I would guess the waves go by about five times a second – five cycles a second. I don't know. You should try estimating for yourself."

"And it would be a lot harder than you think."

"Nature always is."

"It depends on the circumstances."

"And, too, if we could step into a different native perspective, where we were very fast, we would see the waves go by slowly and easily. Then we could see things easily. But from our native perspective, 5 times a second is pretty fast to be able to see much."

We've got enough about water waves. Let's touch real briefly on some other waves, just to see how similar very different parts of nature are, just to see how similar are the logics that go on in very different parts of nature.

Waves. Sound, Air Pressure, Ear, Thingness, Speed

Yes, sound is waves. If we could step outside our blind native perspective, and see much quicker, but also be able to see pressure in the air about us, we would see something like waves radiating out from your vocal chords when you talked. Or from the stereo. Or the sound of the wind blowing outside. You would be able to see waves of pressure moving through the air. Just like water waves fanning out from a dropped rock, sound waves fan out from something making noise. Water waves are high and then low, high and low, and so on. Sound waves, are higher pressure, then lower pressure, higher pressure, lower, and so on.

Sound waves are waves of air pressure fanning out from your vocal cords, or from a cricket rubbing its legs, or from a whirring motor, or the from the rustle of a tree in the wind, or a note struck on the piano. All of these produce small, very fast moving waves of pressure radiating out through the air. For instance, your vocal cord itself, or the rubbing legs of a cricket, or a piano wire when struck, in one way or another, all these shake or vibrate – if you had the ability to see them fast enough and close enough, you could see that. And as they vibrate back and forth, every time they move in one direction they push air molecules ahead of them, which causes the air molecules to temporarily get scrunched up, which is high pressure. Of course the molecules spread out as fast as they can, but by then the vocal cord or cricket leg or piano wire is moving in the other direction and pulling air after it, which causes a low pressure after it. Any way, this whole thing makes vibrations or waves of pressure spread through the air. What happens to the air overall? Waves of pressure go outward from the vocal cord, or the cricket leg, or the piano string.

"What's pressure?"

"Oh, that just means the air is crammed together more. High pressure is when all those molecules in the air are more squished together, and low pressure means they have a little more space between them. Like people in a room. If you had a lot of people crammed in a room, in a sense they would be pressured together. And if you suddenly removed all the walls, then people would fan out away from the pressure. It's similar if you burst a balloon, all the air molecules inside, under high pressure, suddenly have the walls removed, and they expand out. The "pop!" sound you hear is the pressure wave set in motion by the initial release of all these poor crammed up air molecules. It's like dropping a rock in a pool of water.

"Well, tell the truth, popping a balloon is more like slamming a rock – fairly decent sized – into the water. The initial *big* wave that goes out – that's the 'pop!' sound that hits your ear. The air molecules jump out fast, but they don't get too far before they crash into the molecules in front of them. And then those in front of them get shoved forward and crash into those further in front of them, and so on. It's just like the big rock smashing into the water. A wave of pressure fans out, fast, from the popping. When that pressure wave hits your ear, your ear sends electrical signals into the big hum of electrical activity that is always going on inside your brain, whereupon you, your consciousness experiences a loud 'bang!'

"On the other hand, if there is a continuous sound at some point, perhaps an organ key, or a voice that keeps going, or a radio, all these are producing repeated waves of air pressure coming from that point. This corresponds to rapidly tapping your finger on the surface of the pond or rain puddle. But where the puddle waves are only on the surface of the water moving out like circles, the sound coming form the tuning fork or a voice or whatever, is not on a surface but instead is spreading out in all directions, as spheres of pressures spreading out in the air."

"I have a question for you. *Why* are we blind to sound waves?"

"Well, we can perceive them. We *hear* them. In that sense, we are not blind to them."

"Alright. But we cannot see them with our eyes. They are not in our native visual perspective. Why?"

"In trying to see these sound waves, our blindness consists of two sorts. First of all, we can't see from our native vision higher versus lower pressured areas of air. That is the first way in which we are blind in our eyes. But even if we could see pressured areas of air, these waves move too fast for us to see them in our native time perspective. We simply are not laid out in time in the right way to be able to catch this perspective. But this other perspective is real, as real as our native time perspective, and in that perspective, we would see these pressure waves moving easily and slowly out from the sound source, spreading in every growing spheres, just like the water waves fanning out. First the spheres would start, for instance, at the television speakers, then slowly and gently spread out to the first chair, and eventually they would reach the back wall of the room."

"Why aren't the waves bouncing off the walls, why don't we hear a mess of echoes."

"We're roaming from the topic.

"But briefly - " and I motioned to him, "first of all, I don't know, and second, you would have to take some kind of physics or engineering course to find out what are the specifics of when waves bounce back (echo)."

"The strength of water waves is measured by their height, maybe an 1/8 of an inch high for a little pebble dropped in a puddle, maybe 30 feet high for a giant storm wave in the ocean. Those would be the amplitudes, 1/8 of an inch versus 30 feet. But what is air pressure measured in?"

"Air pressure is measured in something called decibels. Just like the amount of height is measured in inches, or feet, or stories high, well, the amount of pressure is measured in decibels. Just like the water waves, you look at the high and the low (and divide by two) and that is the strength of the waves. The human ear can pick up as little as 2 decibels – dB – tenths of a bel. Seventy decibels we hear as very loud. Over 110 decibels of these air pressure waves is probably damaging your sensory mechanisms, which are neurons that are picking up the vibrations and translating them to spike trains sent into the big hum. The amount of damage primarily depends on decibels, frequency, and duration of the loud sound."

"Heavy.

"What about frequency? How fast are the pressure waves going high then low pressure at a point, as the pressure waves go by?"

"The standard pitch of A on a tuning fork, or piano, or organ, is 440 cycles *every second*. As you can see, or could see if you had the right vision, those waves are vibrating back and forth incredibly fast – from our native perspective.

"What about the ear? How does the ear fit into all this?"

"The pressure waves from air are funneled a little bit by the ear into a membrane about ¼ inch circular (the ear drum). Those air pressure waves going back and forth between high and low pressure cause the membrane to go back and forth with the pressure, and that in turns causes some of the smallest bones in the body, on the other side of the membrane, to vibrate and transfer the waves into a liquid, which then starts to vibrate too according to the waves. The liquid is in a little pea shaped area (called the *cochlea*), about a quarter inch big."

"If you go into your ear, and then a little in back of the membrane"

"Yes. One for each ear. In that liquid are about 15,500 little hairs, tuned to different frequencies of waves. When a hair wiggles, a neuron at its base fires an electrical signal off into the big hum – and you, your I-ness 'hear' 'noise' or 'hear' 'a sound'. Some thinkers would speak of this as sound being purely a subjective sensation, it's correlate being these waves of air pressure."

"Why do you say *some* thinkers?"

"Well, the devil is in the details. What do you mean by 'sensation', or by 'subjective'? These words are exceedingly problematic to give a careful scientific definition to. Nevertheless the sentence as a whole conveys a valid idea."

"Just don't get too confused over what it means" The person who said this looked a little confused why they blurted this out.

"That's right. We're getting into two perspective here. One is from our I-ness. The other is seeing these as pressure waves."

"And the confusion is?" This time the person showed no confusion in asking the question.

"The confusion is in thinking one perspective is true and the other is false. For instance, if you are in a certain thoughtful mood, thinking about the universe and what not, you might think, how interesting, what we perceive as "sound" is actually waves of air pressure. You're thinking of the *sound you perceive* as not so real, but the air pressure waves as more real."

"Right!"

"Of course."

"Not so! They are both real. They are both real in as deepest sense of real as possible. There are pressure waves hitting your ear drum. There is sound to your I-ness, to you. You do hear. That is just as real, in everyway, ontologically, as the pressure waves. And the pressure waves. They too are just as real, in everyway, ontologically, as your perception or experiencing of sound."

There was a little silence in the group. Finally someone barked up.

"This topic is something to be left, at this point."

"Right," I agreed."

This is not a bad point to take an initial look at "thingness." What is a thing? Let us look at a particular sound wave, say the single chirp of a bird. That will be the "thing" we look at.

Is that a thing? Yet we all have plenty intuitive feel for what that chirp is. One of the muddles thinkers get into – and that means all of us at times – is, what does it mean to be a thing. For instance, Heidegger spends a whole book on *What is a Thing?*, and only gets into ever deeper issues.

Let's sort the issue for the chirp, *using the ideas and perspective of science*. That's not to say that there might not be other really true (ontologically speaking) perspectives. In the twenty-first century, we humans have this immense system of ideas from science. That is the perspective of this book/journey. So let us now look at the single chirp of the bird, using this perspective of science.

The vocal chord or it's equivalent in the bird starts vibrating. It goes this way, pushing some air molecules, then it goes that way, pushing air molecules on the other side while at the same time allowing those air molecules on the other side to start coming back. This going back and forth of the bird's equivalent of the vocal chord causes pressure waves in the air to fan out from the bird. (For instance, if this single chirp of the bird were an 1/8 th of a second long, and the frequency of the chirp was for instance the same as the middle C note on a piano, then this one single chirp of the bird would consist of (264 cycles per second for middle C, for an 1/8 th of a second would be 33 single waves) would consist of a band or train of 33 up and down's of pressure waves moving out from the bird.

A philosopher is concerned about what is a thing and about not falsely reifying on the basis of simply using a noun when the noun doesn't really refer to anything like a thing. All one can say in response to this concern is that the chirp is, or consists of, the train of 33 pressure waves fanning out from the bird.

As to the process whereby the train of 33 pressures waves goes from the air into multitudes of spike signals sent off into the big hum of electricity in our heads, encoded with much else, we eventually have a perception, a perception that we label with the word "chirp". And that is the chirp thing. In reality there is no clear demarcation of it being a this or that. It is many things. Some of them being the interpretation that there is a bird making a certain "noise." And maybe more.

I have a hunch that if this physical world is amazing, then the further world of logic in the brain is going to turn out way more amazing.

"Hey, just out of curiosity. Say the waves on the rain puddle are moving, oh I don't know, about two feet a second out form the rock or from your tapping finger. How fast are the air pressure waves moving out from a TV, or from your voice?"

"I'm not sure I want to say."

"Why not?"

"It's a little too weird."

"What do you mean?"

"The way we're laid out in time, it's a little freaky. Which of course is only due to how *our being* exists, how it is laid out in time. If we were laid out differently, it wouldn't be so odd, and speed would be as real as how it appears to us the way we are."

"How fast?" he insisted.

"Oh, about 760 miles an hour." I could see his eyes get round. "Those spheres of pressure waves are moving outward from say a radio at 760 miles an hour. 1100 feet a second. Fast, but that's the way it is."

"Ah hah! That's why if you slowly count after a flash of lightening till you hear the kaboom, then mulitply that number by 1100 feet, that's how far away the lightening was."

"Yes."

From our native perspective we may recoil when we picture how "unrealistically" fast such things as these are happening. But that is a misperception due to our native perception. If we had a different native perception, and we could talk about one wave of pressure going slowly by, taking its time, as the up side of the wave took its meandering time to go by, the crest of the wave dallied for what would seem to us like 30 seconds or so, and then the falling side of the wave or swell would slowly go by for the next few minutes, and then we would be in the valley for a few minutes and so on. And if this perspective, we looked at our previous perspective, we might chuckle, and if we were naive we would have an inner feeling of how the other perception wasn't actually real.

Why go into all of this? Because the goal is to get into reality. Reality in various forms – physical reality, I-ness reality, true reality, what really is, what really exists. This journey is about the soul, or I-ness, and it is from the perspective of science, which is physical reality. All of the issues are a part of reality.

To get psychologically into reality, you have to look at details, otherwise you aren't connecting with reality. Without details it's not concrete. Reality is concrete. It starts from the concrete for us. These details are a necessary part of prepping the mind for the journey. It's a required rung in the ladder we have to climb. This is the way of science, to start at the concrete, and go on.

Electrical Waves and Electric Fish

One place where we will use these ideas is to look at waves of *electrical* potential spreading out in the water. A certain kind of fish – appropriately called "electric fish" – uses such waves as a kind of sight. One of the most advanced logics that we shall look at going on in a brain is something called the "JAR" logic that goes on in the brains of these fish.

When we say 'waves' we don't mean the waves that move along the surface of water, as in the waves being blown across a lake on a warm summer evening, nor as the threatening gray waves of the gigantic swells that sweep over the surface of the ocean in a terrible storm, nor the waves on the surface of a cup of coffee when you tap the side of the cup.

These waves are like sound waves. They are pressure waves of crowdedness and uncrowdedness, but of electrons in the water; they are pressure waves – waves of voltage – fanning through the water. The fish generates these voltage waves, maybe a few hundred per second, which spread out into the water, and then the fish notices how the waves fall back on its body, and from that deduces information about what is around it! It is a kind of vision – electrolocation.

This JAR we will look at toward the end of the animals, as one of the most advanced examples of currently understood logic going on in a brain.

12 *The* Euglena

Hello *Euglena*, little dude. And little it is, about half the diameter of a human hair! Unlike most animals we see around and which contain billions or even trillions of cells, the *Euglena* consists of one cell. That's right. One cell is the whole creature, and it swims around in liquids, though it is found in other places too, such as in the soil. Oh, by the way, it is always twirling around as it swims forward.

The logic that we look at in this creature is exceedingly simple. *Euglena* has a whip-like tail. By lashing out its tail repeatedly, it more or less swims forward.

Individual *Euglena* cannot be seen in our native perspective (the perspective we are born with, our unaided eyes). A microscope is needed to overcome our blindness in this regard.

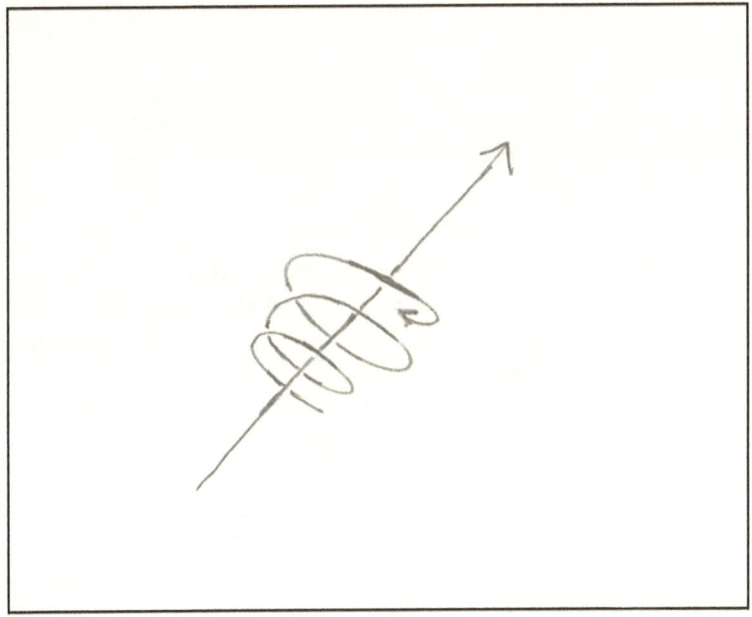

Figure 3 *Euglena* Twirling as Swims Forward

Even without a microscope, sometimes you can see their effects. They are green and if some pond water has a lot of them, the water will look hazy green. Further, if you put a light to one side of the pond, the *Euglena* will swim toward it, and eventually the green will be toward that side of the pond.

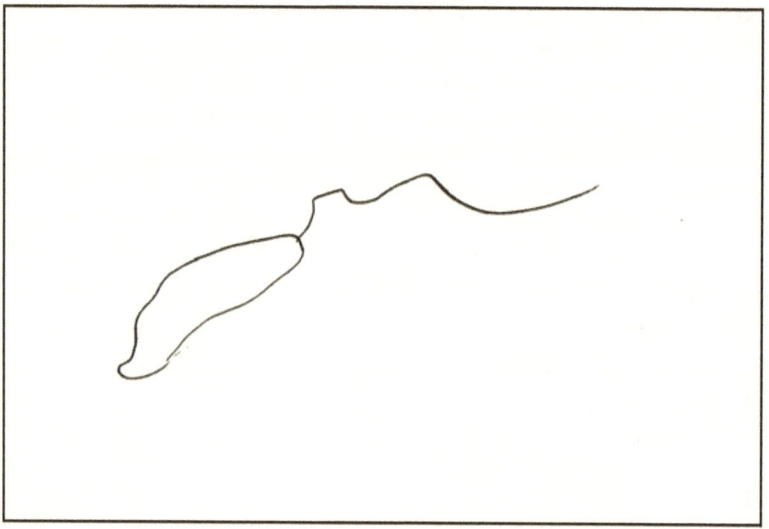

Figure 4 *Euglena*

There must be some evolutionary reason for this – possibly because their food is more likely found near light than away from it. Such is some of the logic going on in this little, itsy-bitsy creature.

Interestingly the whip-like device is not really a tail, but is in the "front" of the *Euglena* whereby it uses to "pull" itself forward.

How does this tiny dude swim toward the light? What logic is going on inside that makes it do this?

This little *Euglena* is like a machine that an engineer might design. There is a light-sensitive red spot inside the *Euglena*, in the area of the "tail." There is a dark area arranged over a certain portion of the red spot, such that when the *Euglena* is lined up in certain ways toward the light, the dark coating shadows the red spot.

The amount of light falling on the red-spot affects the motion of the "tail," and this, combined with the *Euglena* always twirling around, eventually results in the tiny creature swimming toward light.

That is all there is to its moving toward light. Very mechanical. Nothing more than the animal continually rotating about its main axis, and the amount of light falling on the red spot near the base of the tail affecting the motions of the tail. Nothing is going on inside the creature about figuring anything more in this logic. Just a machine, like an engineer might design some cheap, simple little toy contraption, cleverly utilizing a light photo-diode, a little screen off to one side of the diode, and an animal that somewhat wiggles in all directions. And that does it, for this piece of logic in the *Euglena*.

Did you at first imagine something deeper?

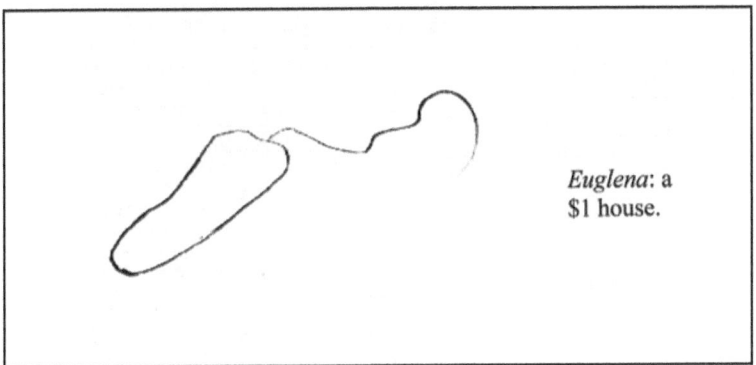

Euglena: a $1 house.

Figure 5 House Size of *Euglena*

For purposes of comparison about how much logic is going on in an animal, let us assign a money amount. What if we arbitrarily assign one dollar for each nerve cell that a creature has? What would we assign to the *Euglena*? Technically it has no nerve cells. It doesn't have any cells other than itself. So we should assign $0 dollars. Even so, it clearly does things such as swimming toward light, though it does it in a fully haphazard primitive way, but it does do it. And it has other things it can do to, as far as how it interacts with its environment. Maybe we should assign it fifty cents. Or eighty-five cents. Maybe though it does more than one nerve cell could accomplish. But even one nerve cell is pretty sophisticated. Let's just assign one dollar to the *Euglena*. We will say that the *Euglena* is a one dollar fellow.

Well, that is our main point about the *Euglena*. But before moving on, we should note that in our native perspective the *Euglena* is terribly small. But that is not the only view into reality. The appearance of smallness is only the prejudice we inherit from our native physical being. From another perspective, one that is as real as our native one, the *Euglena* is "regular" size. And from still another perspective, the *Euglena* is immense, because it consists of 100,000 billion atoms, approximately. That's just one *Euglena*. And that is also a perspective as real as the one we natively have.

Bye, bye *Euglena*.

13 *Volcanoes, Euglenas, and Logic*

Logic external to mind

We have been looking at some of the logic going on in a *Euglena*, how the *Euglena* keeps twirling around as it swims, and how a red dot casts a shadow on its light sensitive tail thus changing the speed with which the tail lashes about, and how this causes the *Euglena* to swim toward light.

This logic has been going on long before there were humans on earth, possibly for a billion years before humans, since one-celled creatures generally go back toward the beginning of life on our planet. Here is the point. Many thinkers, and that includes philosophers, have been inclined to see logic as purely a creation of the human mind and therefore as something that "exists" (whatever that means) only in the human mind. But the logic of this clap-trap of a contraption little creature, the *Euglena*, has been going on for vast eons even before humans and human brains existed. Therefore, logic *exists independent of the human mind*; it is not purely a creation of the human mind. It exists *external to the mind* too.

The logic in the *Euglena* is as real as the *Euglena* itself, as real as the great storms of billions of years ago, as real as the frequent volcanoes in those early times of our mother planet. Just because there were no humans around to perceive those volcanoes did not mean that they did not exist. Obviously, volcanoes exist independent of the human mind, for science has volcanoes existing long before human brains, and this book, this journey, is from the perspective of science.

Detection devices indicate existence

As for whether something exists outside our thoughts, we might consider some kind of mechanical detection device. Science accepts the results of such devices. Thus, if we have a device that detects such and such, that means the such and such has existence out in the physical world, existence independent of the thoughts in our brain. Whether talking of volcanoes or *Euglenas* or storms, we could have a computer, attached to a machine of sensing devices, including perhaps some basic computer vision, which could detect and record the thing. It could record a volcano, it could record a storm, it could record the *Euglena*, and yes, it could record the logic of the *Euglena*, assuming that its technology was advanced enough, though its sensing would have to be sophisticated indeed. And so, this is another way to argue that the likes of storms, *Euglenas*, and volcanoes exist outside the human mind.

Definitional boundary problem

Some people may say that the logic going on in the *Euglena* is vague, and thus does not exist. For with most uses of the word "logic" in mathematics or formal philosophy, the word is referring typically to something not vague but very precise.

Yet many things have a vague definitional boundary and we hold them to exist. For instance consider a volcano. A volcano does not have absolutely clear boundaries as to where the volcano is situated, as to where the lava is required to be relative to the volcano, nor as to the precise heat the lava has to be at nor the speed it races at, nor the length of time an eruption lasts, nor the number of eruptions, nor the height of the cone, nor whether there is a precise cone and crater or only various blow-holes, nor the power nor continuity nor discontinuity of the eruptions, and on and on. None of these things are precise. Yes, a device that detected whether something was a volcano, along with the device's computer programs and sensing apparatus, would have to be pretty advanced to deal with these vagaries. But such a detection device is possible. And that means that volcanoes exist independent of humans or of the human mind: they exist out in that physical external world of science. And it would be the same with the *Euglena*, and it is especially the same with the logic in the *Euglena*. Computer and sensing devices could detect the kinds of *Euglena* logic that we have listed above. We could picture the device doing it somehow, and any difficulties of vagueness would be no greater than detecting whether a volcano existed at a certain location of land.

Hence a scientific instrument consisting of a computer and sensing devices could detect a volcano and also the logic in the *Euglena*. Hence both are real. Because they are detectable by a scientific instrument. Logic is as natural an existent as a volcano.

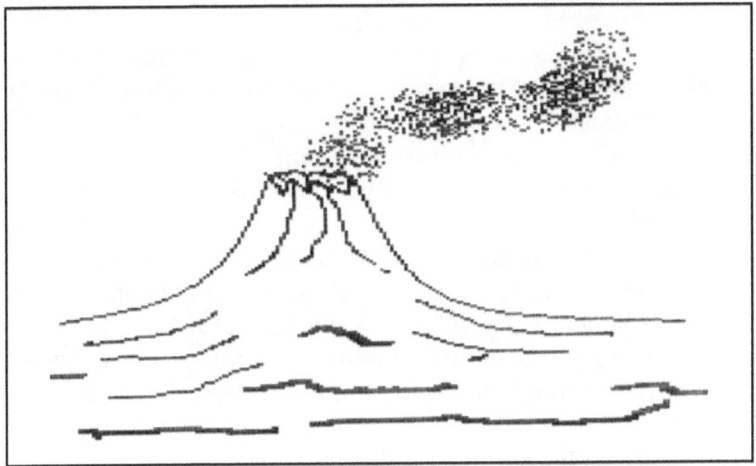

Figure 6 Volcano and Logic

Thinkers, including, philosophers, can get into strange states when it comes to issues of existence. Does the material world about us exist? Do trees exist? Do people exist, do automobiles exist. Does the half-mile long bridge you drive over to work every morning exist? These are all things in the material world. Some parts of philosophy have spent a great number of pages wondering about the existence of such. Maybe such things exist only in our minds. Now these issues are important, in their own way. There are some deep concerns about the ultimate nature of things, and that is what philosophy is trying to explore in these kinds of questions.

But this is a book from the perspective of science. We assume the existence of the material world with all the things in it. We assume the existence of all those things which are so important to us in our day to day lives. Incidentally if you drive to work across a bridge, perched over 200 feet of empty space above turbulent waters, is that bridge only in your mind? If so then you should have no qualms about driving off the cliff that is just to the right of the bridge. Because it's not real either. If the material world is not real, then let's see you go and drive off that cliff.

From the perspective of science, things like bridges, water, cars, people, volcanoes, as well as the whole material world, all exist.

One way to give definition to what it means to exist in the material world is to give a precise definition of a bridge, water, volcano, logic going on in a *Euglena*. One could do this by building a device that detected such things, but as with a volcano, so too with any of these, we encounter what might be called a *definitional boundary problem*: there are inherently vague areas of whether something is a volcano, or whether a bridge, or the logic going on in a *Euglena*.

But that doesn't stop us from saying that things like volcanoes or other things exist.

There may even be other problematic areas in addition to vague boundaries. But again, that is not reason to get muddled about their existence. The natural world is a messy place. But we still get by. It is important to us.

14 The Leech (Also, the Neuron and its Signals)

(Throughout this chapter, unless otherwise stated, all page references are to the book by Nicholls, Martin, Wallace, and Fuch's, *From Neuron to Brain, Fourth Edition*, 2001.)

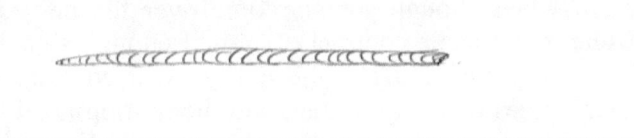

Figure 7 Leech

The leech is not a particularly favorite animal among humans. It burrows into our flesh and sucks the blood out. We can see the leech, unlike the *Euglena*, which was too small. Once the leech burrows down, it is harder to see. This worm-like creature can be about half an inch long, about a 1/30 inch in diameter, and consist of 120 or so little circular segments, the segments being pulled this way and that, by a number of muscle systems running this way and that, internal to the creature. The muscle system is controlled by a little nervous system running through the little creature.

This section of the book covers not a lot about the leech but lots about neurons in general, whether in the leech or human or cat or dog or insect, for the neuron is pretty much the same in all of them.

Nervous System of Humans and Other Creatures Too

Most cells in our nervous system just connect to other cells in the nervous system. All the nervous system cells are called nerve cells. Signals, which are a combination of

111

electrical and chemical, travel along all these interconnected nerve cells, sort of like a "hum" constantly going on throughout the nervous system. In terms of logic, the signal may be thought of as electrical, but the chemical process at every point along each nerve keeps the signal moving along – these chemical events take place with astounding speed.

Our whole brain and spinal column is all part of our nervous system; it is all nerve cells, in incredibly huge numbers, and it also consists of the nerves that go out, for instance, into you hands and fingers – those too are part of your nervous system. All muscles are attached to nerves, which, when enough signals come over the nerves to the muscle, the muscle contracts. Thus, if enough signals come into say a finger muscle, the muscle will contract and the finger moves this way or that, and the motion is all built on a system of bones as levers. Almost all the time, these signals originate in the big hum that is going on in our brain, with the brain sending out signals over pathways consisting of millions of nerves going out to muscles.

And it is not just to move a finger that a nerve fires a muscle. If you are holding something, the fingers need to maintain a certain pressure against the object in order to keep holding it in your paws – many muscles in most of your fingers will need to be constantly tensed in order to maintain the grasp of the object. All this is done by the brain continuously sending out signals along millions of nerve pathways, from nerve, to nerve, to nerve, down the top of the spinal column, across to the shoulder, down the arm, into the hand, and to the hundreds of muscles that must be kept flexed at the appropriate pressure for you to hold the object (actually, finger muscles are in the lower part of the arm and these pull cables attached to parts of the fingers).

Assuming that you are holding this book as you read, then, at this exact instant, your brain is sending out extensive barrages of electrical signals down these pathways across your back and then arms to hundreds of muscles for your fingers. In the reverse direction, now heading *toward* the brain, from pressure sensors in your fingers and from your eyes and so on, the brain is continually receiving signals from which it determines whether the book needs to be held a little more up or down, or to this side or that, and the like. So the brain is continually making elaborate calculations to adjust the pressures applied by your fingers so that the book will be kept in the "desired" position. And! Though almost all of these computations are done in the brain, they are done in primarily parts that are different than those that "do" our consciousness and awareness. In other words, though these finger muscle computation parts and our conscious awareness performing parts (those parts that are doing the reading and interpreting the sentences into ideas and processing those ideas – right now as you read this sentence) are both going on inside our skull, they are basically separate, and so we are unconscious of myriad computations going on to get our fingers to hold the book in the right way, even though those computations go on only inches or less from the part where consciousness is being done.

We have snuck in a few terms and ideas here. We spoke of parts of the brain performing consciousness. This will be explained much farther down the road

Incidentally, getting back to the human hand, did you imagine that the muscles operating the fingers where in the hand? There's not much room in the hand for muscles, not of the required strength. Mostly the muscles that operate your fingers and the hand as a whole are located way up toward the upper part of the forearm, and cables, like for a puppet, connect from those muscles up there, down the forearm, through the wrist, and to the bones in the hand and fingers. (Biologists don't call them cables – they call them "tendons" – but they are cables all the same – not made of steel – and when the muscles contract they pull the cable upwards, and the rest of the cable going all the way down to some bone in some finger is pulled up too, and the finger moves, or exerts pressure, as the case may be.)

You can see your cables pulling underneath the skin of your forearm if you turn your palm upwards, will that your fingers wiggle, and look at the forearm skin surface. It has to be night with only one light shining from the side across the skin surface. In the lower part of the arm you will see the cables pulling up against the skin. You can even see them a little under the palm.

So the muscles for the hand are way above the hand, and they couldn't fit in the hand even if the hand were empty inside. As a matter of fact, though, the inside is anything but empty. It is just jammed with stuff even more than a modern automobile under the hood. There is no room at all. It is crammed with cables, blood vessels, nerve cells, huge numbers of little bones, and a little bit of padding too, for the paw part.

This general picture of nerves and muscles and bones and brains applies to all animals that we can see. Whether cats and dogs or fish or leeches or humans or birds or snakes or bumble bees or ants (but not plants).

Incidentally, the bumble bees and ants and insects don't have the specialized skeletal system of bones that the others do, though they have nervous systems and muscles. They have sort of a skeleton, but it is made not of bones but of a substance called chitlin, and this skeleton is also primarily the surface of their body, so it is called an exoskeleton. But whether the skeleton is bones or chitlin, the logic of the skeleton is the same, to serve as something more solid, both to maintain the spatial shape of the being's physical form, and also for the muscles to anchor between and move one part of the skeleton with respect to another part – as in the bones in your finger moving with respect to bones in the rest of your hand and arm – or as when we walk – or as when bird's wing bones move with respect to the main body bones.

Back to Leech

The leech has 23 tiny little brains running up and down its whole body, all connected to each other. Though to call them brains is stretching the word. Except for the one at the very front, and the one at the very back, each brain, or *ganglion*, has about 400 nerve cells. The ganglion at the front and the one at the back are roughly maybe 10 times bigger. (Compare this to our own brain of a hundred thousand million nerve cells.) The word "ganglion", plural "ganglia", just means a cluster of nerve cells. The ganglia of the leech are so similar to each other that a professional can look in the microscope and detect the same nerve cells in the different ganglia. The expert can even detect the same nerve cells across different species of leeches.

There is a "chord" of a number of nerve fibers or connections running along the bottom of the leech, over which the nerve signals can pass between different ganglia. These bound-together connections are called a *nerve chord*. To be more accurate, the whole connected group of 23 ganglia joined over the nerve chord is one single brain. This grouping of 23 little clusters of nerve cells and connecting nerve chord are responsible for all of the leech's "movements, hesitations, avoidance, mating, feeding ..." (See page 293. Excellent diagrams of the physical leech structure and ganglia structure and individual neuron structure and kinds of nerve signal spike trains are on pages 294 – 300).

Neuron is another way of saying "nerve cell." Thus each ganglion consists of about 400 neurons.

Of the 400 nerve cells – neurons – in a ganglion, four (kinds) are known to fire whenever certain areas of skin in that section of the leech are being touched *lightly*. (The four skin areas for these neurons are the upper part of the 5 or so segments centered on the ganglion, the lower part, the upper right, and the lower left – professionals call these the *field* of the sensory neuron – the area on the skin that will actuate the neuron.) Whenever one of these *fields* is lightly touched a signal goes out from that particular neuron. Professionals call these four neurons *the P neurons*. There are a many other kinds of neurons too in these 400: for instance two *AE neurons* send signals to muscles in that area of the segments centered on the ganglion. Under a microscope, for experts, these neurons are "so recognizable" from each other "and so familiar from segment to segment, from specimen to specimen, from species to species ..." (*From Neuron to Brain*, page 293).

The Neuron Itself

Many, many, many neurons (nerve cells) in a general way look as drawn in this figure. You see lots of branches, and those are called the *processes*, also called *nerve fibers*. And all the branches are coming into or going out of something called the *cell body* – that's the thing that looks like a little blob or

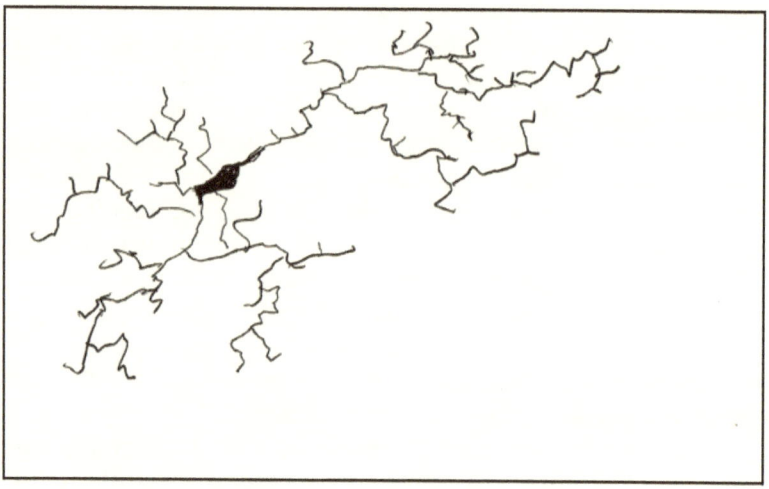

Figure 8 An Example of a Neuron or Nerve Cell

bump in the middle of the figure. For most neurons, there are certain branches over which signals come into that cell body, and there is a main branch leaving the cell body and then splitting up into smaller and smaller branches. (Incoming branches are technically called *dendrites* and the outgoing branches is called the *axon*.) The interesting thing, for most neurons, is that the cell body either *fires* or it doesn't: A signal either goes out completely, or not at all, on the branches leaving the cell body. There is no in-between. Furthermore, for most neurons, the signal that goes out from the body is *always* the same strength. What does vary is whether or not the body fires a signal in the first place, and sometimes how close the firings are to each other in a train of firings. But before getting into a little more of these details, let us look at the history of how humans came to be able to see the nervous system. It was not easy.

Starting in 1887, Ramon y Cajal, using the breakthrough Golgi staining technique, was able for the first time to make sense of nerve connections under the microscope. Previous to that, when people looked at the nervous system, under the microscope, it was a total mess, because no one could make out hardly anything in a dense massive confusion of lines and connections. As stated earlier on our journey, if you took the densest jungle on earth, and took all the vegetation therein, with all the vines and trees and branches and shrubs, and squished it all together, that is what it is in like in our brain, or in the nervous system of any creature. Only our brain is more vast and it is three-dimensional too. It was impossible to see anything under the microscope.

Golgi, not too financially well off, worked in his home and developed a unique staining technique of neurons so that only some neurons would stand out. Golgi's stain, for a reason still unknown, stains just a few neurons out of all the neurons around, but those that it does stain it stains all of down to the tip of every branch. So in effect, it selects just few neurons at random, and makes the whole neuron stand out from the jungle that it is embedded in. (*From Neuron to Brain*, page 5, last sentence of 2nd to last paragraph).

Golgi published his results in a short paper of 1873, but his report did not indicate much about the highly capricious technique nor did it contain any drawings, which would have shown the value in a concrete, stunning manner. Even so, a few people were trying the technique and Cajal discovered it from one of them. Here is Cajal's reaction in 1887 on viewing the results of the stain under a microscope.

"All was sharp as a sketch with Chinese ink on transparent Japan paper. And to think that that was the same tissue which when stained with carmine or logwood left the eye in a tangled thicket where sight may stare and grope for ever fruitlessly, baffled in its effort to unravel confusion and lost for ever in a twilit doubt. Here, on the contrary, all was clear and plain as a diagram. A look was enough. Dumbfounded, I could not take my eye from the microscope." (Shepherd *The Foundation of the Neuron Doctrine*, page 136-137. The 1887 date is in paragraph 2 of page 136. The 1873 date is in paragraph 1 page 84.) Such are great moments in history.

Cajal used this technique over a lifetime of famous work. This was the first time humans had such a perspective. Interestingly, one of the areas he worked on quite a bit was the retina of the cat.

Let us jump from over a century ago to our knowledge now. The retina is what is on the back of the eye, in cat, dog, human, and the like. This retina picks up the light falling on it and converts the light to electrical signals that are then sent into the brain. That is how we see. Incidentally, in humans the retina has about one hundred million sensory nerve cells in each eye – they're called rods and cones. After passing through only three levels of neurons or so, the hundred million signals are reduced to about one million nerve pathways, over which all the electrical spike trains are constantly moving along and out into that big hum in the brain. That is the beginning of everything that we see. The one million from each eye move off in two large cables of nerve fibers, the cables crisscrossing at about the center of our skull and continuing to the area of the brain in the back of the skull.

What is interesting about Cajal's observation of the cat retina, is that he drew little arrows along the nerve fibers. That's how the signals traveled from the sensory retina through the first few levels of the nerve cells. Cajal had no way of detecting any signal – he didn't even know what kind of signal it could be. Remember that the technology of that time was stunningly primitive and coarse compared to what we have now; as what we have now will compare to what we have a century from now. In 1906, Cajal and Golgi jointly received the Nobel prize in medicine for their work on the structure of the nervous system (though they never agreed over some of the interpretations of what they saw in the Golgi stained nerves).

And what about those signals?

The overwhelming majority of signals traveling through the nervous system are in the form of *spikes* – a spike is where the voltage spikes upward and immediately drops back down. It's a spike of voltage that travels along the nerve. That's from one firing of the cell body – remember the cell body, the little blob at the center of the drawing of the neuron. At any one point along a nerve fiber, the spike, as it goes zooming by, lasts maybe only 5 thousandths of a second (*From Neuron to Brain*, page 14). For instance, if you looked at a nerve cell, at some point on one of the dendrites, as a spike went by that point, the voltage would jump up then drop back down, in about 5/1000 = 1/200 second.

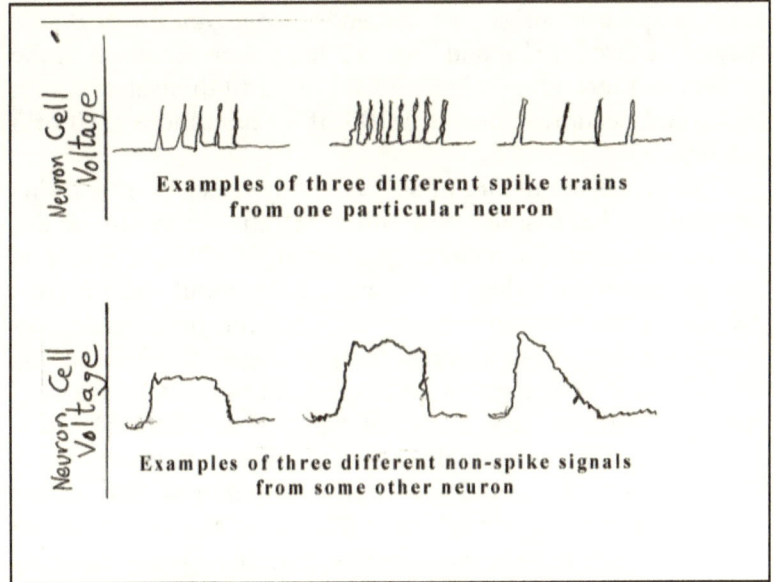

Examples of three different spike trains
from one particular neuron

Examples of three different non-spike signals
from some other neuron

Figure 9 The Two Types of Nerve Signals

So a spike comes into the nerve cell along a dendrite to the cell body. Various spikes may be zooming in along the dendrites to the cell body at any time, where they are in one way or another combined and influence the state of the cell body, and perhaps then the cell body fires, whereupon another signal – a spike – is sent out by the cell body, sent out along the axon, zooms along the axon, with the axon possibly splitting into many branches, till it gets to a place where the axon or branches "touch" the input dendrites of yet another cell, and the whole process continues with the next cell(s).

Thus, spikes are going in over the dendrites, coming up to and going into the body of the cell, where some kind of basic processing is taking place, and on that basis the cell body may or may not fire. If it doesn't fire, no spike at all goes out – no electrical signal at all goes out. If it does fire, then a spike is sent out on the axon(s) from the cell body.

Quite often there is a little *train* of spikes (see Figure 9 The Two Types of Nerve Signals). For informative graphs of real spike trains see for instance *From Neuron to Brain,* pages 14 through 15, and 295 to 296. Notice that each of the spikes in these graphs is about 80 mV (80 thousandths of a volt) and occupies about a 1/100 of a second to a 1/10 of a second.

So that is the nature of spikes – they occur in little trains or singly. That a spike will either be sent out totally or not at all was first discovered by Adrian in 1913. It really is rather fascinating that this occurs throughout the nervous system. (The phenomenon is called, appropriately, *all-or-nothing* firing. This all-or-nothing is *logically the same* as the binary 1 or 0 used in computers: in both there are only two possibilities, a 1 or 0, an "all" or a "nothing". It is interesting that nature "discovered" binary code long ago. Nevertheless, it should be kept in mind that this all-or-nothing is only one part of the logic at this level of the nervous system. Another part is that the spikes may be single or repetitive, they may be further apart or get very close together. This part of the logic is not two, it is not binary, it is a whole range of possibilities. Where there is a whole *continuous* range of possibilities when contrasted to a rigid, required choice of only two, then sometimes one uses the term *analogue* logic, as contrasted to *binary* logic.)

The spikes for a specific neuron can get closer together, or farther apart, but still *their height (voltage strength) stays the same.* Below we will look at some examples of this in the touch sensory neurons of the leech.

A few signals in the nervous system are not in the form of spikes, but rather show a signal simply by voltage change over a period of time, and the voltage change moves along the nerve for only a little distance because it dies out real fast (see Figure 9 The Two Types of Nerve Signals). The light sensory neurons of the retina send out signals like this.

Just in case you look into a neurobiology book, here is a paragraph of terminology. Another term for spikes is *action potential*. The kind of signals that are not spikes are called *local graded signals* because the intensity is shown by the amount of voltage (unlike spikes where something like greater intensity would likely be encoded – by nature – as more spikes crammed together in the same time frame, but still the spikes being of the same height). The signal can probably best be thought of as electrical, but in terms of the details, it is a combination of electrical and chemical, each feeding the other, as the signal moves down a dendrite or axon. Dendrites and axons are also called *processes* of the neuron. Of course, *neuron* and *nerve cell* mean the same thing. As axons go out to the next cell or cells, many times branching into smaller and smaller processes, at the end is a swelling or fanning out into something that looks like a drop piece of candy with a flat bottom, with the flat end up against a part of the next cell. This fanning out portion is called a *bouton*. Actually the bouton does not really touch the next cell. There is a very, very narrow region between it and the next cell called the *synaptic cleft*. The whole region consisting of more or less the flat bottom, the synapic cleft, and the adjoining side of the next cell, all this is referred to as a *synapse*, the idea being that a synapse is where a signal jumps from one cell to another. All this is incredibly small. It is small even compared to the nerve cell. Things happening just before the synaptic cleft are called *presynaptic*, and just after *postsynaptic*. The term *membrane* in these contexts usually refers to the surface of the cell, including the surface of the processes (dendrites and axons). This cell membrane is analogous to being like the "skin" of the cell, however unlike skin, it is selectively and dynamically porous to various molecules, and part of that dynamic porosity is involved in the chemical part of the signal moving along the processes. In all of the figures mentioned so far, the inside of the cell was negative with respect to the outside, about 60 mV (60 millivolts, or 60 thousandths of a volt) more negative on the inside of the cell membrane than on the outside. When the spike goes by, the voltage reverses, and the inside gets more positive, and in fact so positive, that the inside goes from being 60 mV less than the outside to 10 mV more for 1/200 of a second or for however long it takes the spike to go by. Incidentally the cell is always doing work to make sure that it's insides are 60 mV less than outside. (For different cells these numbers will be different.) The membrane on the bottom of the bouton (the candy drop looking tip) is called the *presynaptic membrane,* and across the synaptic cleft, on the other side will be the *postsynaptic membrane* of the other cell to which the signal will jump. What happens at the synaptic cleft involves little enclosed bubbles or sacks of extraordinarily small size, called *vesicles*, sometimes so small they carry only two molecules of for instance ACH (acetylcholine). As part of the signal getting to the presynaptic membranes and preparing to get across the synaptic cleft, the vesicles move to the presynaptic membrane, merge, and open to the outside, releasing the two ACH molecules, which move out into the cleft (Incidentally, when curare darts are fired into animals to paralyze them, the active ingredient in curare is ACH, and this messes up the transmission of nerve signals in the animal so that it becomes paralyzed.) In line with the hierarchy of the universe table presented earlier on our journey, all these happenings are extraordinarily small in space and in time — when viewed from our native perspective.

As stated above, the overwhelming number of signals in the brain are spikes. But the other kind of neuron, rather than sending out spikes, sends out a voltage that is a measure of the intensity of the signal. In spike trains, the intensity of the signal is *encoded* by having the spikes closer or farther apart, but the spikes themselves stay the same strength. However, for the other kind, there is no spike of voltage that jumps up and then right away falls back down. Instead, the voltage itself indicates the strength of the signal.

These non-spike signals travel hardly any distance at all before they die out, so the next neuron had better be pretty close. (On the other hand, spike signals can travel fast and far.)

There is some tendency for these non-spike encodings to occur in what are called *neuro-receptor cells*, alternately called *neruo-rceptors* or *neuro-sensors*. These are just the sensory cells scattered here and there over the body and sometimes inside us too. These are neurons, half of which is a sensory mechanism. For instance, the rods and cones in the retina at the back of our eyes are neuro-receptors. They are neurons, one side of which, in this case, detects light and/or color, and the other side of which is sending out a non-spike signal.

(In a neurobiology book, non-spike signals may be called *action graded potentials*, because the voltage of the signal is graded, unlike spikes where the voltage is the same. These non-spike signals may also be called *local graded potentials* or *localized graded potentials*, because the non-spike signal dies out so fast that it is only *local*.)

(Information on what kinds of signals there are in neurons can be found in most neurobiology books. In particular see *From Neuron to Brain*, pages 10 through 18, or for another example, pages 295 through 296).

Back to the Leech

As stated, each of the ganglia (other than the ones at the head and tail) are remarkably similar to each other and each has about 400 neurons. Some are neurons that get input from other neurons, process it, and send it onto to still other neurons. Some neurons get input from other neurons and send signals directly into muscles causing contraction of one of the groups of muscles in the 5 or so circular segments centered on the ganglion. And still others are sensory neurons. The leech has neurons sensitive to touch and light.

It's touch sensory receptor neurons pick up basically three kinds of touch in the 5 or so segments centered on the ganglion: light touch, stronger, and noxious (*From Neuron to Brain*, page 296), and these sensory touch neurons have been largely determined and can be spotted under a microscope by a knowledgeable person just from the location in the ganglion and the neuron's morphology (experts use the term "morphology" to mean form and structure).

Often researchers will label neurons by one or two letters. Thus the three kinds of touch neurons in the leech ganglion are labeled, T, P, and N neurons.

The T neuron will fire spikes if its skin field is lightly touched. The term "skin field" refers to the area of the skin that the neuron has dendrites to, and that is the area that can cause the neuron to fire. (See page 297 for example of a skin field of a P neuron.)

Actually, in each of the 21 ganglia one can detect six T touch sensory neurons, one detecting touch on the *top* of the 5 segments centered around the ganglion, and others on other places on the 5 segments. If one were to look at for instance at Figure 15.3, page 296 of *From Neuron to Brain,* one would see that this T neuron sends out a train of spikes when anything in its skin field is lightly touched. Notice how all the spikes are the same intensity – a jump from about minus 40 mV to plus 20 mV and immediately back down again.

What happens when the skin is touched, but more than lightly, say with 7 grams of force? Again, the T neuron fires a train of spikes, but this time there is a brief train at the start of the touch, and nothing then even though the touch continues. When the touch stops, there is again a short train of spikes. And that's it for the 7 gram touch.

Notice that no matter if it is a light touch or a 7 gram touch, the strengths of the spikes are still basically the same.

You might ask why is there that gap in the T spikes with the 7 gram touch: spikes at the start and at the end but nothing in between. Who knows? Maybe the researchers have some explanation. Maybe they don't. Maybe in two hundred years they will more deeply understand things and then they will have a fairly good explanation; maybe it will be that this general kind of neuron tends to respond in this way in all kinds of animals, or maybe it will be, well, just pretty much randomly that's how that neuron turned out in the leech species, or maybe there will be some other explanation.

Let us move onto the P neurons. They detect harder touch. They fire not at all with a light skin touch. However they do have the regular continuous spike train for the 7 gram touch. It is basically the same spike train as for a really hard 21 gram touch – the 21 gram touch is called the "noxious" touch. It's the one that triggers the stereotyped U-shaped bending of the leech away from the noxious touch.

The N neurons fire basically only for the noxious 21-gram touch. (We call it "noxious" because 21 grams is pretty hard for the little leech.) When a noxious touch occurs on one side of the body, the N and P neurons pick that up and send spike trains into the little hum (400 neurons) going on in the ganglion, and eventually signals are sent to the ganglia on both sides and eventually those signals trigger muscle neurons (also called motor neurons), M, that cause muscle contractions that cause the body to bend in the characteristic U-shape away form the noxious stimulus. That's how the little leech does its U-bend.

Notice in the pictures in *From Neuron to Brain*, that the spike trains for a particular neuron are all or nothing, and when they do occur the spike intensities are all about the same. For T neurons they are about 60 mV spikes, while for the N and P neurons they are all about 85 mV spikes.

(mV is a thousandths of a volt; so 60mV is 60 thousandths of a volt, or a hundred times less than the electrical pressure delivered by a 6-volt battery.)

That's the leech. We will now got down to some detail of the nervous system, some idea of what neurons are and how they operate, and a rough idea of some of the logic going on in a leech.

By the way, remember how a *Euglena* was a $1 house, roughly in terms of the logic that went on inside it. How big is the leech if we assign $1 to each neuron? Given that the inner 21 ganglia each have about 400 neurons, and the first and last ganglion (head and tail ganglia) might have 10 or so times as many neurons, the leech is, roughly speaking, a $15,000 house.

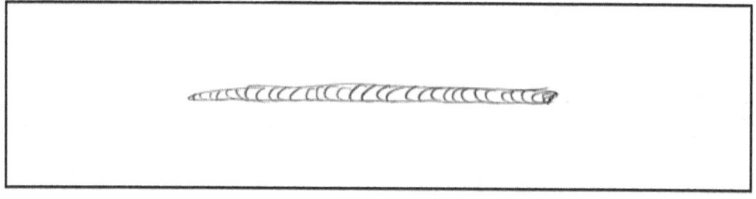

Figure 10 *Euglena* - a $1 house. Leech - a $15,000 house

Meaning of "Neuron Code" – Encodings

Back in the desert, I unexplainably found myself sitting with some other hikers on a log in a dusty little opening, but with pleasant Creosotes and Joshua trees about. The hikers seemed to be as uncertain as to how they got there as I was. Well, that is not quite true. I had a suspicion.

We talked about neurons – nerve cells – and their signals, and the leech.

"Hey I noticed in the diagrams you have the word 'code'," said one of the hikers named Jack.

"What?"

"I mean here, you have 'T Neuron firing pattern' and then you have 'code'. What's the 'code' stuff?"

"Oh, that just refers to, uhh, actually that's a good question. It doesn't mean too much. But it is emphasizing an idea – to see this from the perspective of an idea. How are the signals that are coming out of neuron, or going into it, *coding* the information? Remember, all signals are spike trains, whether it's the neurons in back of your retina neurons picking up a pleasant tree, or your ears picking up the sound of the wind, or the touch sensation in your finger tips as you hold this book. All of these are spike trains (once we get a neuron or two away from the sensory neuron itself). *Coding* refers to how the information is encoded in the spike train. The diagrams show that. That's why I wrote 'code'. It does not mean much."

"Well, I think it means quite a bit," Jack retorted.

"Well, uh, maybe you are right. Sensory input photon receptor rod and cone neurons in our retina send out signals coding for the light or photons that are hitting them. Some sensory neurons in our skin have various codings for heat or cold. Sensory neurons in our ear (cochlea) send out coding information about what kind of sound they are picking up. And who knows, maybe deep in our brain somewhere is a neuron that fires when our grandmother is in the same room we are in – whatever kind of signal that neuron sent out would be referred to as the code in this neuron that expresses that your grandmother is in the same room.

"Both us and leeches have sensory neurons that detect touch. Or you could look at the kind of signal a muscle neuron puts out to a muscle as it causes it to contract.

"All these, in different creatures, and in different situations, are all spike trains – the height of the spike is the same no matter what – what varies is lots of spikes being closer together or not so close together in a spike train – and also what varies is that sometimes only one spike is sent out. For instance, one particular kind of a spike train from one particular neuron might indicate that a certain shade of green is being picked up in a certain portion of our eye (Don't forget though that there are tens of millions of spike trains ultimately originating from our retinas.) For another particular neuron, a certain kind of spike train might mean some aspect of the music we are listening to, and yes that is also nothing more than a spike train. Another spike train might be one of thousands of trains coming in from touch receptors in our fingers and hands, due to holding a book. For some other neuron, a certain kind of spike train might be associated with a particular feeling of happiness. *The point is all these are just spike trains, but they encode various meanings.* The way that information is coded into spike trains at a particular neuron is spoken of as the *encoding* of information at that neuron."

"So," Jack interrupted, "let's summarize this. All the computations the brain does, of any sort, are technically done on and with spikes and spike trains. But in terms of the larger picture, all the computations are done on meanings going into a neuron and are producing meanings coming out of a neuron. We speak of those meanings as being encoded in the spike trains. All our feelings, all our actions, all our moods, all of those are also meanings, and they are also encoded in terms of great numbers of spike trains."

"Yes. All of that is in solely in terms of spike trains, a vast ocean of spike trains, in our head. A vast ocean of meanings too."

That was the end of our little discussion.

(For examples and advanced statistical analysis of encoded meanings, see endnote 5 on the book *Spikes* by Rieke et al.)

15 Magnetism and Polarization in Ants and Bees

Some ants have regions of eye facets that are polarized and thereby the ants get information about what direction they are walking in with respect to the sun. Furthermore, though the ants nervous system is extremely tiny, nevertheless it uses this polarization information to keep track of where it is with respect to the nest, so that once it finds food, it can head straight back home, in spite of the zig-zag, exploratory path it took to find the food. And all this is in the tiny ant's brain (Nicholls et al, *From Neuron to Brain, Fourth Edition*, 2001, page 306).

Bees detect and use the earth's magnetic field for their navigation (Nicholls et al, *From Neuron to Brain, Fourth Edition*, 2001, page 312).

Thus, insects, although they have a very little brain, especially the ant, sometimes possess neural sensory and navigational systems that we are missing. It is interesting to think about all the behaviors that are packed into the ant's brain, and how small that brain is, being only a minor portion of the ant's body, which is already quite small, especially for the smaller species of ants. Some of the ants behaviors: The ant detects ants of its own nest versus others, has a capacity to fight other ants, to feed and take care of the young, to search for food, and it has a range of behaviors for evaluating the amount of food, its desirability, how movable it is, and techniques for getting it back to the nest. It has a range of behaviors to arrange grains of dirt or of almost anything. It responds to water and so on. All of this is in an incredibly little brain. When we look at all the ant can do, and then compare the size of its brain to ours, is it any wonder that we are such special creatures on earth?

16 Zero Crossings

The senses of some creatures, and the nerves behind those senses, are quite extraordinary, at least as they appear in our native time perspective. Certain creatures are so amazing in that their nervous system detects every up and down cycle of a sound wave impacting their ear. In some cases it is not their ear being impacted. Later, when we look at electric fish, we will see that they have points scattered over much of their body and that these detect waves of electrical voltage (pressure) in the water impacting the point. Each point detects every up and down cycle. The nerve signal sent out contains an encoding of the time of impact. Even if the waves are impacting at 100 or 200 or even 500 times a second, every one of those times of impact is being sent out in the nerve signal.

The brains of owls use this sort of capability to determine the direction a noise is coming from. A noise might be coming from an edible mouse, which for a hungry owl is important. The direction of the noise affects the difference in impact times between the two ears. The owl's brain compares the impact times coming in from the two ears and from that computes the direction. Good news for the owl, bad for the mouse. But the mouse has its techniques for survival too. More information on this auditory processing in the owl brain can be found for instance in the section on owls, and in particular on "phase locking" on page 139, of the book by Peter Simmons and David Young, *Nerve Cells and Animal Behaviour, Second Edition*. This book is fairly easily read by the non-specialist.
From our native perspective of time, it seems astonishing that the network of neurons is making comparisons of the signals arriving from the two ears possibly hundreds of times a second. But from a more helpful time perspective, all the signals would be moving around slowly, leisurely, through the neurons; and now the

comparisons between the signals from the two ears would be slow and straightforward.

As we will see for the electric fish, it is not two ears, but many points all over the body, and conceptually, comparisons are being made between any two points. Well, probably more or less. The comparison is sophisticated. As stated earlier, the waves are not pressure waves in the air, but waves of electrical pressure in the water. We poorly know all the sorts of things the fish might be detecting with this information.

Nervous Systems in General

For one of these points on the skin of an electric fish, the situation is simple. The neuron sends out a spike at the very start of every up and down cycle of the voltage waves impacting the skin of the fish.

In the electric fish, in this book, we will call this spike the zero-crossing spike. We call it this solely because at the start of an up and down cycle, the wave is at its zero point – the point that is half way between the crest at the top and the trough at the bottom. We use this terminology only in the electric-fish chapter coming up. (We leave it unspecified in our discussion, whether the signal is emitted only during the beginning of an up then down cycle, or whether it is emitted once on the up cycle and again as it crosses downward past the zero point.)

Notice that the zero crossing, the point in time when an up and down cycle starts, is something going on in the pressure wave *outside* the fish. It occurs in the wave whether or not there is even a fish there to detect it.

Here are a few technical points for those readers who wish to dig into some books on these matters and on the spikes associated with them. As stated earlier, the way the time of the up and down cycle is encoded in the spikes coming out of the neuron is not always direct. Sometimes there are *latency* times before the neuron fires. Indeed, there are a variety of latency time factors in the electric fish. For some species, the encoding is more complicated than for the electrical fish. For instance in the owl (page 139 of Simmons and Young's book above), several kinds of information are encoded in the signal coming out of the neuron, one of those being the times of the up and down cycle. These issues are sometimes put under the term "phase locking."

This completes the material on zero crossings. We now move on to electrical fish, where eventually we will see the sophistication of the logic of zero crossings in the environment outside the fish, and we will see how that logic is duplicated in the fish's brain.

17 Electric Fish

One Big Good Example

The leech was not the most pleasant of creatures and it had only about fifteen thousand neurons anyway. Good bye leech. Now we move onto something much livelier, a fish, and an electric one at that. These fish are about a foot long, maybe cute, maybe not, depending on how you react to their appearance. The important point is the immensely greater logic these fish have when compared to the creatures we have looked at so far.

Figure 11 An Electric Fish (*Eigenmannia*)

The motivation for our looking at this fish comes from the perspective of science. We want to look out in the world at examples of brains, and we want to look hard, thoughtfully, and carefully, because reality offers more than we can create on our own in our minds. The piece of logic that we will look at in these fish is the most sophisticated example on our journey, and indeed, it is among the most understood of all the more advanced logics that scientists currently understand in detail in the brains of any creature. The logic not only has sophistication, but scientists have traced it down to the various parts of the fish's brain, and even down to the signals going on in individual neurons. That is why we look at this particular logic in this fish. In accord with our approach from the perspective of science, this is just the kind of logic we want to hone in on, because for our times, it is advanced and known in much detail.

It should be emphasized that this piece of logic is not the only logic going on in the fish's brain. Not by any means!

Furthermore, even though there are gaps in our understanding of even this logic, we must still honor those who have expended effort over decades in order to obtain the understanding that we now have. Tomorrow we will go further, but we must always remember the people who have worked so hard to bring us the awareness we have today.

So let us begin.

The E-fish Account

Electrical waves and a sort of seeing with them

If you were to look casually at one of these fish, you would notice nothing special, at least nothing like what we are going to observe. A great deal is happening that we are able to see only because researchers have worked so hard to see it.

By the way, there are a variety of electric fish, but only some kinds have the logic we will be looking at. The most studied such fish is the *Eigenmannia* (Heiligenberg, *Neural Nets in Electric Fish*, 1991). Perhaps one might call this fish the e-fish, as short for electric fish or for *Eignemannia* fish.

The e-fish is continually sending out into its watery environment waves of electrical impulse (waves of electrical pressure) at about 200 to 500 times a second (Heiligenberg, *Neural Nets in Electric Fish*, page 10). The fish uses these waves to get information about what is going on its environment, and even when it is night, or the water is muddy, the fish can still "see" some of the aspects of the area around it. This gives the fish a definite evolutionary advantage in terms of information about inanimate and animate objects about it, including presumably those it might want to eat and those that might want to eat it.

Most of the fish's long tail is continually sending out the electrical waves.

By using electrical sensors scattered over its body, the fish is constantly monitoring the electric waves as they fall back on its body, and, by analyzing how the waves are changing, the brain of the fish is able to get information about what objects are in the water around it (objects in the water around it affect the electric waves). The fish may be able to use this kind of electric vision within only a few fish lengths. But this is of value.

In fact, included in this information is motion itself because that also affects the electrical waves. The objects could be rocks on the lake floor, rocky surfaces to the side or around the fish, twigs drifting in the water, leaves, other fish, or food. We have no idea of what it must be like for an electrical fish to see electrically.

This kind of seeing is called *electrolocation*. This term should be compared with that for bat sensing, where bats send out high-pitched shrieks and determine the location of things around them by analyzing the echo – such sensing is called *echolocation*; or the term can be compared with one or our senses, the two sensory devices in our face that we call "eyes," and which detect photons flying into them, with the brain using this information to derive other information about the location of objects around us, and so this kind of sensing can be called *photonlocation*

How awesome is the world, or creation, or nature, or whatever we want to call it. This "electrical vision" is one of the many perspectives that *our* neurosensory system does not pick up.

Interference Due to Neighbor. Definition of JAR

One of the problems for the fish, with this method of "electrical vision," occurs when a neighboring fish happens to be emitting a signal close to the same frequency. Then the electrical pressure waves being emitted by the two fish combine in a way that messes up the ability of either of them to "electrically see."

Here's the neat part. *The nervous system of each fish detects this bad situation and the two fish shift their frequencies away from each other*! This is the piece of logic that we will be looking at. Neurobiologists call this piece of logic the *JAR*, the *Jamming Avoidance Response*.

This is the sophisticated logic that we are going to look at. As already stated, it is by no means the only logic going on in the brain of the fish. But it is a logic that is both advanced and remarkably well understood, in terms of our current level of understanding.

Detecting Neighbor Interference. Two Electrical Eyes

The following way is my own way of explaining how the fish does this.

Here is our fish, and not too far away is a neighbor, and our fish's nervous system wants to know if the neighbor's frequency is higher or lower. Let us mentally jump from this to ourselves with our two eyes. When we look at something, we see it as slightly different with each eye. Areas in our brain, but outside of our consciousness, use those slight differences to determine the three-dimensionality of what we are looking at, for instance, whether something is 8 feet away or 12 feet away. And the result of that determination is what we perceive as the three-dimensionality around us. If you look at the room about you right now, keeping your head still, and opening and closing alternate eyes, you will see objects jumping around with respect to each other. That simply shows that each eye is getting a slightly different picture of things because each eye is in a slightly different location. That is the little differences that we are talking about that the brain uses to determine the three-dimensionality, and that determination is then passed on to your consciousness. Now the technical details of how this works for light and distance are extremely different than for electrical waves in water and frequency. So we do not want to look at the details of our own eye perception. But the idea of two eyes seeing slightly different things and using that slight difference – that idea is the same for us and for the electric eyes of the electrical fish.

The fish looks at its neighbor with two "electrical" eyes - not the two eyes in its face, no no, but we will come to that later. With two "electrical" eyes, the fish looks at its neighbor. And electrically it sees something slightly different with the two eyes. And just as we humans put the two views together from our two eyes in order to get three-dimensionality, the electrical fish's brain puts the two views together to get whether the neighbor has a higher or lower frequency.

We will first describe how this is done for two "electrical" eyes. Then we will add some further details.

The "Electric Eye" - Very Basic Description

Each electrical eye is strikingly simple compared to our own two facial eyes that detect light, and unless you know what to look for you probably would not even notice anything special on the skin of the fish, where these eyes are scattered all over the skin. Each electrical eye detects just a measly two things. On one nerve fiber going out from the eye is the *strength* of the electrical waves in the water at the present point in time. On the other is a signal sent out on every zero crossing of the waves (zero crossings were described before this section on e-fish – every time one of these electrical pressure waves or voltage goes from low pressure to high, that is called a zero crossing, and the electric eye detects that and sends out a spike on its nerve fiber, which then enters into the little hum that is the fish's little brain – we leave it unspecified whether there is also a signal sent out whenever the electrical pressure is going in the reverse direction from high to low). If the fish's tail is generating electrical waves at 100 cycles per second, then those waves move throughout the local water, and that includes the waves hitting the electrical eyes scattered over the skin of the fish. But those waves will not hit the eyes at exactly 100 zero crossings per second. Various kinds of objects near the fish can affect the evenness of the number of zero crossings hitting the eye per unit of time, and that is one of the things the fish's nervous system analyzes in determining information about what objects are around it. The eye itself does no such analysis. All it does is send out on one of its fibers a spike every time there is a zero crossing.

Those two nerve signals are the only thing that this simple little electrical eye sends out:

The pressure (that is, strength or voltage) of the waves <u>and</u> the zero crossings.

Technical Explanation (The Logic of the JAR)

This bit of logic of the electric fish is fascinating. So get ready for a ride.

Two-eye Case (Dot on the Paper)

(This is presumably a purely mathematical result about the physical nature of the environment. All one needs is *two* locations in the water (where the two eyes are), and a third place where the neighbor is, and one set of electric waves being generated near the eyes, and the other set being generated where the neighbor is. We will describe one "oval" for each eye location with respect to the other eye, and the sum of all ovals will indicate whether the neighbor frequency is higher or lower.)

Take a sheet of paper out. Draw a line down the middle,

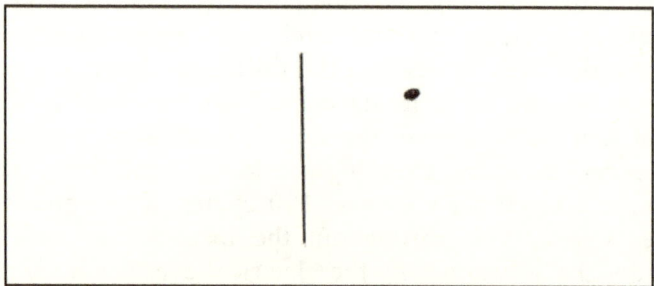

Figure 12 Line down Middle of Paper

We are going to describe a dot that moves around on the sheet of paper. The interesting thing is this. After you read the description of how the dot moves around, of course the logic of how the dot moves around will be in your thoughts. *But that same logic in your thoughts is what is wired into the fish's brain.*

The sheet of paper will be for one eye, with respect to just one particular other eye. (Eventually we will look at all eyes with respect to almost all other eyes, but that will be much later).

Now let us get back to just one sheet of paper, with respect to just one other eye. We will call the eye, "A". And the eye it is with respect to, "B". Here is a picture. We draw them as eyes, but as stated earlier they really are not at all like the two eyes on our face. You likely could not even see them if you looked on the body of the fish.

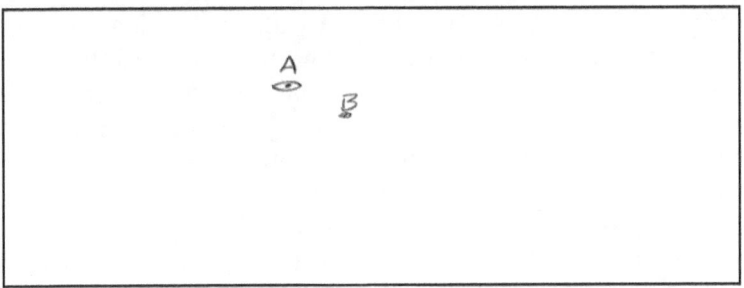

Figure 13 Two Electric Eyes on the Fish, Eye A and Eye B (B is a little farther away) (Drawn as if regular eyes, but are not)

Here is how the dot moves around on the paper. The strength or amplitude of the waves (how big are the waves in each tiny stretch of time) – this amplitude will be how high the dot will be from the bottom of the paper. Since the strength *at* the eye will be changing, the dot will be moving up and down. As the strength is getting greater, the dot will be moving up, and when the strength is lessening, the dot will be moving down.

However, in addition to moving up and down on the basis of wave strength, the dot will also be moving at the same time to the right and left, based on the zero crossings. Remember those zero crossings – when the wave is exactly at the half-way mark between maximum pressure and minimum pressure – the sensory neuron emits a signal.

Because the two eyes of the fish are in different places on the skin, it turns out that a nearby fish affects the two places differently, with the zero crossings at A eventually drifting ahead of those at B, and then after a while drifting in reverse so that they are behind those of B, and after a while ahead again, and on and on, back and forth, ahead and behind ahead and behind those of B.

(Of course, from our native perspective, all these things are happening extremely fast, with for instance maybe 50 to a 100 zero crossings a second. But in this situation our native perception is not good because it gives us the impression that everything is an impossibly fast whirl of events. It isn't. That view is only from our native perspective. If we look at these waves from a different time perspective, there is no wild whirl at all, but instead one zero crossing calmly happening, and then after a little time passes, another zero crossing, and so one. From this perspective everything happens slowly and easily and with time for the nervous system to do work on the signals. And this perspective gives us a more truthful view of reality in this situation. And remember that this perspective is fully true reality as much as our native perspective, except that our native one gives a blurred confusion in this case.)

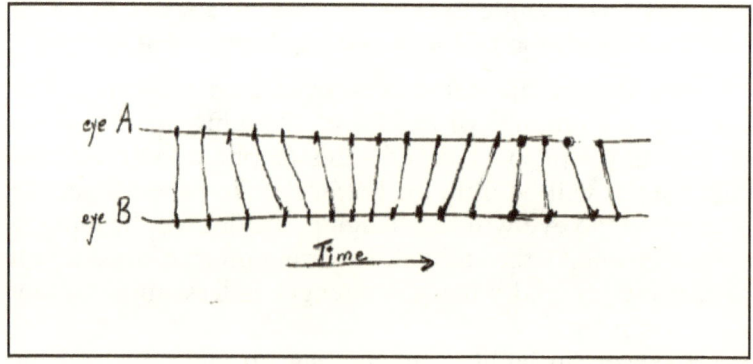

Figure 14 A-B tick marks (zero crossings)

In the above figure, time goes from left to right. The tick marks show when the zero crossings are occurring at A and at B. The zero crossings (on the left) are simultaneous, but then those at B slowly pull ahead, then slowly fall back to where they are the same, then continue to fall further back so that B's tick marks are behind A's, then slowly those at B start to pull ahead again, and so on.

As stated earlier, how high or low the dot is on the sheet of paper will indicate how strong (voltage) the waves are. The next figure shows four examples of dots on different places of the paper (we also throw in the zero-crossing information for the dots).

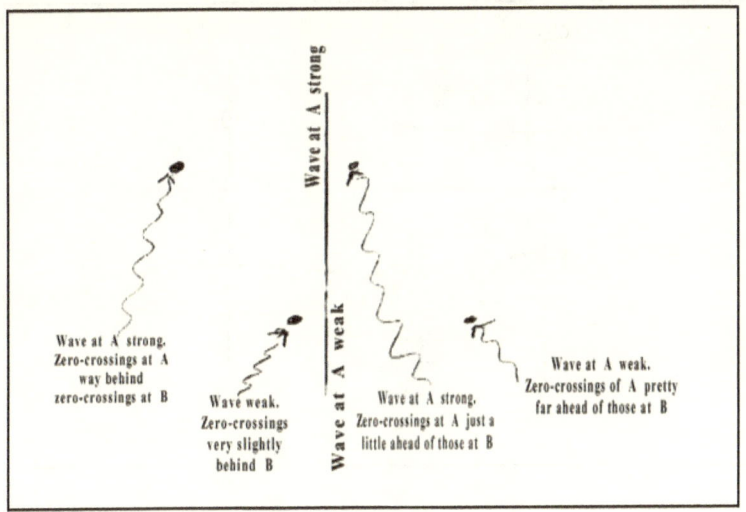

Figure 15 Example of Four Dots on Sheet of Paper

We now come to the final part. If the electric waves of a nearby fish are interfering with the waves being generated by our fish, it turns out that the dot moves around in an oval-like pattern, which can be going clockwise or counterclockwise. Why? It just inheres in the nature of how the two wave sources (from the fish and its neighbor) combine.

(The size and shape of the oval and whether the dot moves clockwise or counterclockwise will depend on the orientation of A and B with respect to the neighbor, and also will depend on whether the neighbor's frequency is higher or lower).

And now we get to how to determine whether the neighbor's frequency is higher or lower.

Add up the area in the oval for A (with respect to B), and B (with respect to A), with clockwise counting plus, and counterclockwise counting negative. The sum tells whether the neighbor's frequency is higher (sum is plus) or lower (sum is negative). Why? Again, it just inheres in the nature of how the two wave sources combine.

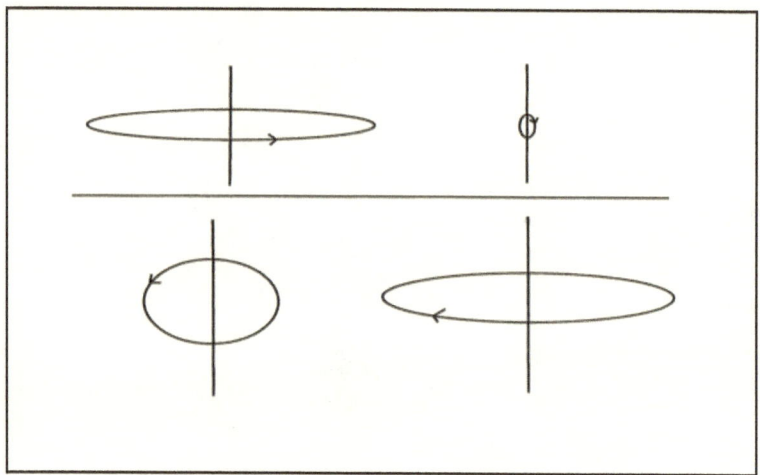

Figure 16 Each Example is Pair of Ovals, Which are Conceptually Added Up

Is that amazing? Yes. In practice, the ovals may be all kinds of different sizes, and the two may both be going clockwise, or both counter-clockwise, or one may be going one way and the other the other way. But in all cases, if you add up the two ovals, you will get that a positive sum means that the neighbor has higher frequency, and a negative sum means the neighbor has lower frequency.

I believe this has nothing to do with whether any fishes are even in the water. If you have two locations (where the eyes would be) and a strong electrical signal close to them, and a neighboring signal further away, then this technique of adding the circles up tells one whether the neighbor's frequency is higher or lower. (There may be restrictions such as for example that the two frequencies cannot be too far apart, otherwise the method breaks down.)

Story closed!

Would that were so. In science, just as in the story of our daily lives, and in fish brains too, the story is hardly ever closed, and it is not quite closed here either.

Multiple Eyes

As stated earlier, there are lots of electrical eyes all over the fish's body. And the fish looks at its neighbor, so to speak, with almost every possible *pair* of these electrical eyes. In terms of the perspective of electrical engineering, if multiple signals for the same thing are coming in, and if those signals are averaged, that reduces the error and also increases the sensitivity of the detection process. It is the same with all these pairs of eyes. The fish's brain averages out the results of all the pairs and comes up with a sum that is remarkably free of error and amazingly sensitive to the smallest difference in the neighbor's frequency from its own. Perhaps it is the same as if we had eyes scattered over most of our body, and you were outside on an extremely windy, dusty day. Suppose your body used every possible pair of eyes, and then averaged it all out. The average would get rid of almost all the obscuring blowing debris, and further you would be able to see three-dimensionality of the scene with ripping precision. It is the same with all the fish's electrical eyes combining their information.

Encodings and Where in the Brain is JAR Logic

Just where are the parts of the logics of this JAR logic going on, where are these neurons in the brain of the electric fish, what are some of the neurons like, what is some information about the encodings of meaning in their spike trains of electrical signals? This is somewhat technical; this material is for those who want to know in more concrete detail this particular reality. (All page and figure references in this section are to Heiligenberg's book, *Neural Nets in Electric Fish*.)

The brain parts mentioned below are shown in several perspectives of the brain on page 20: (A) A side view of brain and spinal column of fish (head of fish would be toward left, tail toward right); (B) Looking forward onto the fish, a slice of the brain about halfway between front and back (*midbrain*); (C) Looking forward onto the fish, a slice toward the back of the brain (*hindbrain*).

The overall logic of the JAR in the electric fish brain is basically the sum strength of circles or ovals going clockwise or counter-clockwise, combined with several averaging processes.

The "computation" of the size of ovals does not occur localized to one region of the brain, but rather one part of what is needed occurs in one region, and that feeds signals to another brain region where further things happen to the signals, and that region sends signals to other regions and so on. The primary regions where the significant work is being done to compute the size of the ovals are these.

ALLG – "anterior lateral line nerve ganglion" – actually just a relay area – The electrosensory neurons over the skin of the fish all come to the ALLG, and the ALLG relays signals on to the ELL.

ELL – "electrosensory lateral line lobe" (The ALLG is at the bottom of the hindbrain, the ELL is a little above that, with one part on the far right, the other far left, side.)

TS – "torus semicircularis".(The TS is a larger area, centered left-right, and further up in the midbrain. The TO, the Tectum Opticum, the visual area, mentioned later below, pretty much surrounds the TS.)

The ALLG and ELL are way in the back of the brain, the ALLG being at the bottom and the ELL being half-way above toward the top of the brain. The TS is way toward the front of the brain.

Not discussed here are several regions (NE, PPn, and Pn) that are small, and through which the signals pass through from the TS in the front and then to the back and on out to the spinal chord, where they fire the electric wave generator of the fish. These areas are primarily summing and averaging areas.

The electric sensors on the skin are called T and P electrosensory receptors. (These are also called tuberous receptors. There are other receptors for lower frequency but these are not used in the JAR logic, and are called ampullary receptors.)

Electro-sensory neurons T send out the zero crossings from that point on the skin. The encoding is straightforward: a spike is sent out at a zero crossing. (In practice there is a small latency, a short time between the zero crossing and the spike.)

Electro-sensory neurons P send out the strength or amplitude of the electric waves at that point on the skin. The encoding is in terms of spikes. The greater the strength of the waves at any point in time, the more likely the neuron will send out a spike. Thus, as an average, the number of spikes being sent out in a unit of time indicates the amplitude of the waves. This is how the amplitude is encoded in the firings of P. (Figure 4.2, page 54)

The T and P neurons go to the ALLG, which just relays the T and P signals to three kinds of neurons in the ELL.

One or more T outputs (zero crossing's) from a field on the skin go to a single spherical cell in the ELL, which in turn sends a single summarizing signal on to lamina 6 of the TS (fig 4.9).

It is in lamina 6 that a comparison is being done between the zero crossings of all possible (or at least between many) pairs of areas on the skin. This is done by the use of two kinds of cells in lamina 6, giant cells and small cells. The T (zero crossing) from a single area on the skin goes to a single giant cell and to many small cells (all being "somatotopically" ordered in much the same spatial layout as they are on the skin). The giant cells relay this information by sending it out on "extensive axonal arbors" (page 85) to small cells from many different skin areas.

Each of these small cells is computing the difference of zero crossings of two different areas on the skin (fig 4.17 page 84). It is here then, in lamina 6 of the TS, that zero crossings from each area of the body are being compared with the zero crossings of all or many other areas of the body.

The encoding of the output of these small cells is intermittent spikes, with the rate being determined by the amount of difference between the two zero crossings coming into them.

Let us back up now to the P cells. Each P electrosensory neuron signal (strength or amplitude) goes to two different ELL cells, a basilar pyramidal cell (called an "E-unit") which detects and fires when the amplitude is increasing, and to a non-basilar pyramidal cell (called "I-unit") that fires when the amplitude is decreasing. Both the E-unit and I-unit send signals to laminas 3, 5, 7, and 8 of the TS (fig 4.9, page 70).

In lamina 8 of the TS, there are neurons that are firing for when the point is in different portions of the oval (fig 4.27 page 99). For instance,

> There is one physiologically distinct class of neurons that is firing for that part of a particular oval when both the amplitude is increasing and the difference in zero crossings is increasing.
>
> There is a different physiological class of neurons that is firing for that part of a particular oval when the amplitude is increasing while the difference in the zero crossings is decreasing.
>
> There is still another physiologically distinct class of neurons that is firing for that part of a particular oval when the amplitude is decreasing while the difference in zero crossings is increasing.
>
> And finally, there is still another different physiological class of neurons that is firing for that part of a particular oval when both the amplitude is

decreasing and the difference in the zero crossings is decreasing.

These four physiologically distinct classes of neurons are called: E-advance, E-delay, I-advance, I-delay, with "advance" meaning the phase (the zero crossings) difference between the two skin areas is increasing, "delay" meaning the reverse, and E meaning the amplitude is increasing, and I the reverse (page 99).

Moving out of the TS and backwards toward the spinal column, through the small areas NE, PPn, and Pn , there are neurons whose firing distinguishes whether the neighbor fish is emitting waves at a higher or lower frequency, depending on the direction of the neighbor, and for other neurons, independent of direction of the neighbor (fig 4.38, page 118). There are sections in these regions of nerves where presumably huge amounts of such information are being averaged together. Then, at the end, finally, the signals the neurons are firing are quite accurately attuned to whether the neighbor has a slightly higher frequency or slightly lower frequency.

Pointers for reading Heiligenberg Electric Fish Book

For those who want to dig into even more details in Heiligenberg's book on electric fish, here is some information.

The horizontal axis of some graphs in the book will be labeled with something like "HB – HA" or "phase at B – phase at A". This means the zero crossing at B minus the corresponding zero crossing at A. A and B are two different areas (fields) on the skin of the fish. To state this more carefully, this is the time at which the zero crossing at A occurred minus the time at which the corresponding zero crossing at B occurred. This is the amount of time that the zero crossings at B are ahead or behind those arriving at A.

The vertical axis of some graphs will be labeled with something like |S|A. This is the strength (amplitude measured as voltage) of the *total* electrical wave impinging at place A on the skin of the fish. The book does not say "total". I use the word simply to emphasize that we have an electrical wave produced by the fish (often referred to in the book as S1) and another wave produced by the neighbor fish (this wave referred to as S2). These two waves *combine* to produce a resultant "total" wave. The amplitude of this resultant "total" wave, at area A on the surface of the fish, is referred to as |S|A. The resultant "total" wave may also be written as S1 + S2. It is thought of as being at some place on the surface of the fish (although S1 + S2 actually exists everywhere in the water, not just on the surface of the fish). On the surface of the fish, S1 will be much stronger than S2, since any place on the fish is so much closer to the fish's own electric generator in its tail than to the neighbor's electric generator.

The phrase "more caudal" refers to being toward the rear of the fish.

Figure 4.38 on page 119 of Heiligenberg's book shows the firings of two neurons that quite accurately determine whether the neighbor is emitting waves at a slightly higher or lower frequency. One neuron accurately fires when the neighbor is in any of the four directions shown. Another neuron works for only two directions. At the end of each arrow is a picture, and the top half shows the situation for Df > 0 (meaning the neighbor is emitting higher frequency), the lower half for Df < 0 (meaning neighbor is emitting lower frequency). Each half of the picture shows the amount of neuron firing at the various places in the beat cycle.

Cube of Logic: The Right Spatial-Time Picture

In our mind, when we picture the JAR logic going on in the fish, there might be a tendency to think of just *one* electric eye on the skin, sending some nerve signal to some point, and from there another solitary nerve signal goes out, which give rise to another signal going somewhere else, and so on, forming a thread. If your way of picturing logic going on in the fish filters out multiplicity, then this thread is a valid perspective. Otherwise, one should picture the JAR logic as, conceptually, somewhat like a whole cube of logic, *all* points of which are *all* the time active.

Electric eyes (electro-receptors) *all* over the fish's skin are all producing, *all* the time, their zero-crossing signals and wave strength (amplitude) signals. All these constantly active electric eyes may be thought of as on the left face of a cube (we will imagine the cube as sliced into slabs going from left to right.)

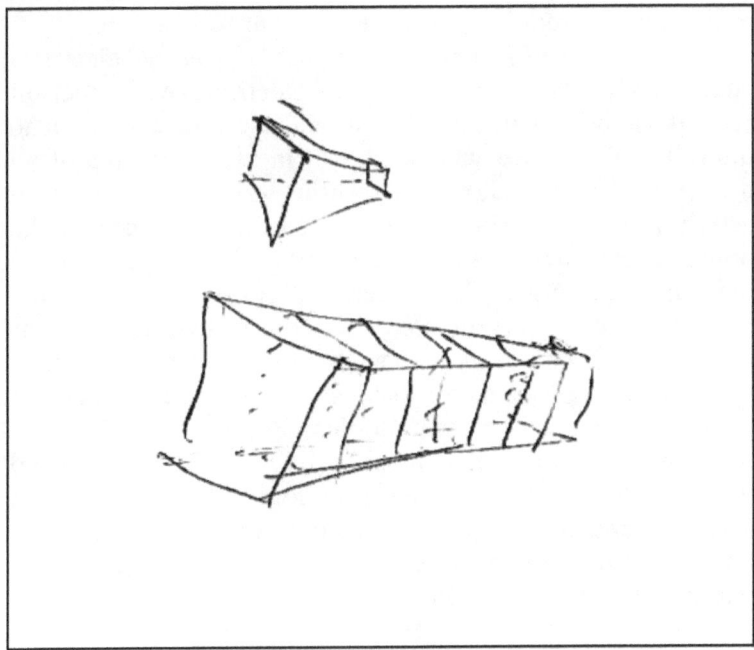

Figure 17 Cube of Logic

All the points in a slice and all the slices are all constantly active and filled up with signals. Each slice is like a mirror reflecting an image. The left face of the cube – that slice – is a mirror consisting of all the electric eyes: they receive status about the electrical waves in the watery environment, then each eye sends out signals about that status, the signals being the eye's particular reflection of the environment. All the eyes together are a mirror on the left side of the cube, a mirror that reflects to its right a particular image of what it received from its left. Each slice is like that, reflecting to the right an image of what it received from its left. The cube consists of this kind of bunch of mirrors placed next to each other. Note that a mirror that is reflecting an image from another mirror is reflecting an image of ultimately what the first mirror is reflecting. Therefore, all the slices in the cube, each being a mirror, are reflecting their own particular image of what is to the left of the cube. The whole cube is like a three-dimensional mirror, reflecting different images of the original source.

If a change occurs in the fish's watery electrical environment, then a whole area of electric eyes on the left face of the cube will be affected, and the image sent out to the right of this area will be shown in the same area of all the mirrors to the right. The whole is one solid mass of activity. As for the actual location of these slices in the brain, several may lie next to each other in one area, several more may lie in a different area, but conceptually they are like this cube. (Neurobiologists will use words such as "map" and "topo" to refer to such an area in the brain, thus speaking of the area as for instance a "somatotopic map" or a "somatopic map" or a "tonotopic map".)

Many logics in brains must be like a cube of logic, and that includes parts of our sentience logic. We seem to have a strong tendency to want to see our sentience as being at one physical place, or at least logically being inherently at one physical point at each point in time. But like with the cube, such a part of sentience logic is taking place *inherently at a whole range of places at once*. Nevertheless, the tendency to picture sentience in this way might derive from a truth that we do not yet scientifically understand.

This section spoke in a way that is approximate, conceptual, and the last sentence is speculative.

House value of electric fish

Any thinking person must be in awe that a klutzy little fish has this much logic going on in its brain. And this is just one of many logics going on. What must that say about the rest of the fish, and the rest of animals, of creation, on Earth?

Remember that the *Euglena* was a $1 house, and the leech a $15,000 house. How big is the electric fish? It's quite a step up from the $15,000 leech house. It is a twelve-hundred-fifty-million-dollar wealthy compound. That is a big difference in size.

To get a better feel for this, suppose that we consider mansions, each mansion costing five million dollars. Suppose that a city block is a tenth of a mile by a tenth of a mile (typically the long side of a city block is a tenth of a mile – we will assume both sides of the block are a tenth of a mile). We will assume that ten of the 5-million-dollar mansions are placed on a city block. (There would be more than this is in crowded parts of major cities – but otherwise this is reasonable if there is to be some grounds for each mansion, and it still is in a city.) The electric fish would occupy 25 city blocks of 5-million-dollar mansions. That's a quarter of a square mile.

Electric fish house: quarter of a square mile of 5-million-dollar mansions (250 5-million-dollar mansions. That is quite a few).

Different Explanations

It should be kept in mind that neurobiologists, physicists, mathematicians, and logicians all likely have somewhat different ideas as to what constitutes a complete explanation. There are even cases where the fish's brain itself may do the "wrong" thing. All this is part of the imperfectness of the real world. As the decades go by, no doubt more complete models of all sorts for all of these areas will be developed.

18 Closing Points

Nice If Purely Math Theorem was General

It would be nice if there were a theorem of mathematics and physics, the theorem capturing all the relevant aspects of the interaction of the electrical water waves being emitted by the fish and its neighbor. Perhaps there is a theorem about all kinds of environmental situations in a very general logical setting and with all kinds of factors (differential nature of waves, averaging at various points, places where the method breaks down - yes there are some situations of wave parameters where the method above will not work). Let the theorems be phrased in ways readily usable by people studying such things, useable without having to be excessively knowledgeable in mathematics. As for generality, a good example is Byron and Fuller, *Mathematics of Classical and Quantum Physics*, chapter 5 on Complete Orthonormal Sets of Functions.

Existence before humans

For millions of years these electric fish swam through the waters of our planet and surely for millions of years they had the JAR piece of logic going on in their head. To reiterate an earlier theme, this fish existed long before us humans, and so the logic going on in its neurons in its brain cannot be solely ideas in our minds. In fact, even the mathematical and engineering knowledge needed for us humans to understand the logic were not known till the last two or three centuries. And the fish has been around far longer than that

But all of this is by way of saying, again, that the idea from philosophy that logic is all in our minds is thus proved false. *Logic exists independent of the human mind. Logic is not a creation of the mind* any more than are the other real-world observations of science (Remember that in this book we take the physical world of science as a given).

The same point was made by looking at the *Euglena*, in its one-dollar house of logic, or at the leech, in its fifteen-thousand-dollar house of logic. The logic, existing millions of years before humans, must therefore exist independent of the human mind.

Volcano as Illustration of Some Aspects of Logic

We saw in the various creatures, that the logic going on in them has some vague aspects. The logic might be seen as going on in a place in slightly different ways, there might be boundary situations where it is not clear whether the logic is going on, and so on.

We pointed out how this was not reason for saying logic does not exist. For many physical things also have the same problem of having a vague definition of whether they exist, and if they do exist, precisely where are they and how much are they there.

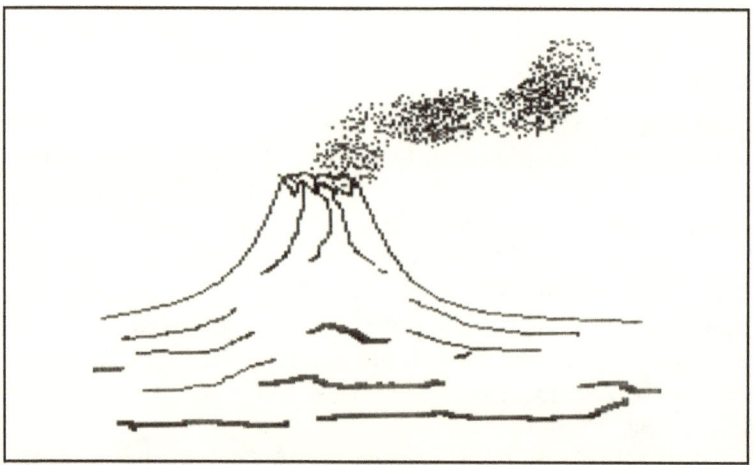

Figure 18 Volcano

If one attempts to give a definition of a volcano, a definition that could be used in a mechanical way by a mechanical device, one sees what kind of definitional boundary problem there is. A volcano is – there exists some dirt and rocks that make up one part of a side, and more dirt and rocks that make up another side, and so on, and these sides are arranged somewhat like a cone, with properties that would basically limit molten or heavy liquids to using primarily usually one central channel, a channel being an area of non-matter, well at least no molten matter – maybe – oh and air and gas is OK in this property just as long as it is not hard – well oh it can be hard before the volcano erupts, but during the eruption, no the channel – well there might be little channels more than one – hmm I guess there can be more than one large channel – but there shouldn't be too many, and definitely there should be an elevated heap of matter (dirt and rocks, well yes of course there can be some trees rivers and so forth, but I mean generally, whatever 'generally' means) that is somewhat constraining the molten stuff. Not only that, what is the matter or material the volcano is made out of is vague, how far does the cone matter extend from the volcano – well you look at a picture you can tell – we can't be exact but in 'general' – well don't get too technical. And as for that molten stuff – the lava – it is a part of the volcano as it erupts – one would think so – but after it gets far away and turns to hard cold rock – no absolutely not – so don't talk about a volcano as if it is characterized by the matter composing it. As to exactly what dirt or rocks compose the volcano, what matter that composes it, it is vague. All these things of what a volcano is, are properties. Yet the volcano is quite real. It is a hard physical thing. Philosophically some would say a volcano or tree – its existence – is a naïveté of the scientific perspective. Some would say it is not naive. But from the perspective of science, a volcano is real, it exists, and it existence does not require the existence of us humans because volcanoes existed long before us.

Since this book is a journey form the perspective of science, the definitional boundary problem cannot be used as an argument against the existence of volcanoes nor against the existence of logic.

Science starts at hard, physical reality but does not stay there

People could question hard reality. We do not. Because that is one of the starting points of this book. If you want to question it, there are many other books out there that do. If you wonder about that, keep in mind that all of science and engineering start from these assumptions. (It should be emphasized that they *start* from these assumptions – that does not mean they are limited in some naive way to only what is hard physical stuff – as was pointed out with gravity and with light waves – science *starts* from the hard physical stuff, then branches out according to hard logic.)

Definition by pointing

In Part 2 of the book or journey, we have defined logic by pointing to it, repeatedly, in different situations. We take external physical reality to be primordial. What exists in it may be beyond our current understanding. Logic going on is there. It would certainly be nice if we could define this more precisely. But our state of understanding is not at that phase. All we can do at this stage in the journey is to be like some ancient physician, point in certain directions, and say logic is there, logic is going on there. Two hundred years ago we could not even see the logic going on in the nervous systems in this external physical reality. (Maybe three hundred years ago we barely had *ideas* of this 'logic' as in 'logic going on' – and it is computers today that have helped us apprehend a little more the idea of 'logic' – but we still are not all the way there).

So! We point. We point at the logics going on in the *Euglena* and other animals.

Pointing is a logically legitimate method of definition. Because of our state of understanding, the pointing may not be as precise as pointing at something which we do understand, like a chair or a volcano. But it is all we have at present. And too, the pointing that has taken place so far in this book has given a feeling, or as logicians and mathematicians say, a set of intuitions, about this idea of logic. This set of intuitions so far obtained may not be enough to give a feeling of full, solid grasp, but it does start to give a picture of the pieces of logic.

The reason this logic is important is because we are to be found in it. That is where sentience is.

When we attempt to look at this "thing" through the concepts of engineering and mathematics and science, it seems surprising that all this is going on in reality, whether or not we see how. In fact, as stated, the logic of the JAR in the brain of the e-fish went on out in reality for hundreds of thousands of years before we humans were even able to understand it, and it went on even before we humans existed.

It is only in the last few decades that we have the technological ability to look somewhat inside the fish's brain at the neuron level and to see some of these logics. There is much else in reality that we just do not have the technology at present to see, but as time goes on we will be able to see more and more.

Amount of logic in human

As we close Part 2 of the journey, it is natural to ask how much logic is going on in our own brain. We said about $1 dollar of logic was going on in the *Euglena*, about $15,000 of logic in the leech, and that the electric fish with a $1,250 million was a quarter of a square mile of five-million dollar mansions.

And *our* brain? The logic in our brains would correspond to a hundred square miles of five-million dollar mansions. That is how many neurons we have, working away, right at this very moment.

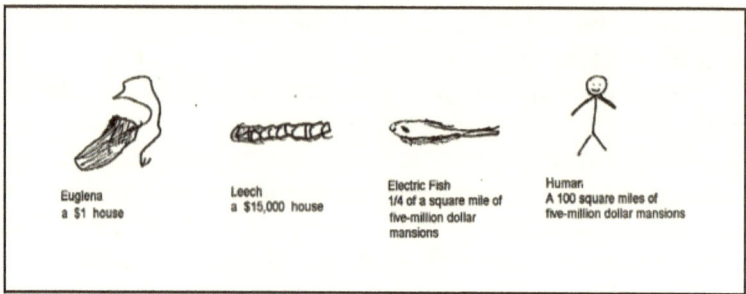

Euglena
a $1 house

Leech
a $15,000 house

Electric Fish
1/4 of a square mile of
five-million dollar
mansions

Human
A 100 square miles of
five-million dollar mansions

Figure 19 The Human House

Even the small nervous system of the approximately 15,000 neurons in the leech is far from being fully comprehended by science. As for the big-brained creatures, especially the humans, the sheer immensity is the underlying challenge of trying to understand what kinds of things might be going on in such a brain, including that special feature, sentience (or relational aspects of soul).

Structure of the Logics Going on in the Brain

In a discussion with the god, I found that a certain issue will come up again later in the journey. It is important. It clarifies things. Oh well. That's what the god said, so I guess I would have found it out eventually anyway.

The discussion took a while, so I will just condense the results.

The JAR logic described earlier in this chapter on the electric fish is not just at one point or one area in the fish's brain. It went on in a region that consisted of one area toward the lower back of the brain, another area that was in the upper back, another in the middle front, and another in the further back, and so on. In other words, the region where even *one* logic is going on can be scattered across several areas of the brain.

Although we did not give them official names, there were several logics that made up the JAR logic. There was the logic dealing with changing strength of the electrical water wave signal on a field of the skin. And parts of these logics were logics for the changing strength of the signal at on point on the skin. There were logics that handled rising or falling differences in phases between the skin areas in a pair of skin areas, and there were logics that summarized that for several pairs of skin areas. There were logics that handled all combinations of falling/rising strength of the signal at an area of the skin combined with the falling/rising of difference in phases between two areas in a pair of areas. And there were logics that summed up these logics.

Not mentioned earlier, the electric fish sees a combined, "overlaid" picture, from its regular eyes combined with that from its electrosensory receptor eyes scattered over its body. For sure, this is interesting, to us humans, because we don't have that kind of vision. From a technical perspective of the brain activity, this comes about because the TS (Torus Semicirularis) has topographic projections to the TO (Tectum Opticum) (Heiligenberg's book, figures 4.1, page 52, and 4.16, page 81). This means that some of the logic sent to, and therefore part of, the JAR is also being sent to the visual area of the brain.

As stated, for sure, this is interesting. But our interest here is that there is logic in the TS that is being used both in the fish's visual logic *and* in the JAR logic. It is part of both logics. The general principle is that a single logic can be part of several other very different logics.

Now let us put all this together. There are logics going on in the brain. A logic hardly ever goes on at a point in the brain. It goes on in an area or region. More than that, the region can be in several areas not even next to each other. (This principle came from observing the JAR logic, where the going on of even this one logic took place in several widely separated areas of the fish's brain.)

These logics have a kind of structure. Logics are parts of other logics. They are composed out of other logics. One logic can be part of several very different logics.

Finally, although there is messiness in the details, and there are other issues of vagueness too, nevertheless there is much logical cohesion of activities in the brain – and sufficiently strong occurrence also there of meaning from the environment – that naturally pulls the activities into what we are calling logics in the brain. Talking even more philosophically, briefly, after rightly abstracting, we speak of these as *logics going on* in the brain.

Clearly, these ideas extend to any brain, not just those of electric fish.

(For further comments, as well as a stronger version of the above statement, see endnote 4.)

I told the god that some of these notions were abstract. The god continued gazing toward the horizon.

All this is part of the structure of the logics going on in the brain.

References and Numbers

See endnote 6 for books on electric fish, endnote 7 for books on neurobiology and history, and endnote 8 for a delightful book on animal behaviors in terms of presenting the logic of the behavior.

As for the number of atoms in the single-celled *Euglena*, I took that to be roughly the same as the average number of atoms in a cell in the human body. (See the appendix section on "Numbers".)

Part 3 Onwards

19 The I

Talk with the God

"You've gone through two parts of the journey. You've done what a scientist would do. And that is good."

"Yes," I replied.

"You went into brains, well some beyond that, and eventually looked at what was going on there, and from a larger perspective. Maybe now go into a good selection of all the things that comprise human consciousness. But look at the structure of their logics in the brain. To extent possible. Pure logical form of their going-on's in the physical – brain. Some of."

"Alright," I said weakly with uncertainty. "Oh, I see. The first part was general background, the second was the animals and doing what science does. Now it is onwards, and start to look further at the real material of the journey."

The god was silent. I was not really sure what we had just discussed. Eventually however, I wrote the following report.

A Perspective of Logics in the Brain

For anything that we are aware of, there is a certain logic, going on in the brain, that is the awareness itself. Maybe this statement is obvious to the reader. Maybe it isn't. If it doesn't feel obvious, then think for example on the JAR logic in the electric fish's brain. While that isn't consciousness, nevertheless it is the same issue of logics in the brain. The JAR is an example of a logic going on in a brain. To be sure, there are logics that are a part of the JAR logic. And indeed there may be logic in the fish's brain that the JAR logic is a part of. And there are certainly logics that are pretty much or wholly different than the JAR logic. And there are parts of the JAR logic (the parts that feed into the visual system and into the remainder of the JAR logic) that are also parts of other logics than the JAR, in addition to being part of the JAR logic. Nevertheless, one may take the JAR logic as an example of a logic going on in a brain.

It is in *this sense* that we mean, "For every awareness itself that we have, there is a certain logic going on in the brain." From the scientific perspective, every awareness that we have must literally be something going on in the brain. Further, that something must have some kind of logical form, must have meaning, and that form is what we take to be the logic that is going on in our brain. There are also many things, or brain activities, that are not awareness, but likewise have a cohesion of logical form and meaning, and they also are logics going on. But all these logics, whether or not "part of" awareness, go on in our brain just as the JAR logic goes on in the electric fish's brain. It might take one or five or ten centuries before science advances enough to trace out all these logics. But they are there.

Thus when you become aware of a C-major chord being played based on the middle C of a piano, or when you have the awareness of red (the awareness that is the seeing the color red somewhere in the field of your vision), there is a certain logic going on in the brain, which logic is *exactly the experience itself of hearing the piano chord, or of seeing the red* somewhere in the field of vision. Or suppose that you have a feeling of worry over the doctor telling you that your parent, not only having bone problems in their legs, is now showing signs in the blood of developing gout. This experience of gnawing worry, the experience itself, is a specific concrete logic going on in you brain.

To be sure there is a huge variation on any of these, for instance on awarenesses of hearing the C-major chord on a piano. There are different C-major chords depending on which C of the piano the chord is based. Musically, there are several "inversions" of the C-major chord. Even for chords based on the center C, the piano could be nearby or far away, it could be in tune or out of tune, the chord might be played all at once, or one key at a time in succession, the piano might be crystal clear, or have a muddied sound, and on and on. All these examples of awareness in our mind are logics, literally in the brain, that surely are quite similar, and yet have some differences between them. Likewise, there are the different ways in which we can have the awareness of seeing the color red. These are different but similar logics in the brain. And whether one *views them* as different logics, or variations of the same logic, regardless, they are logics going on in the brain.

Understanding how all these logics relate to each other might be called understanding the structure of the logics in the brain.

(Certainly a complete, total, detailed understanding of *how* a logic *goes on* in a brain would include such things as which neurons fired, in what kind of order, and if relevant, if the neurons themselves get into certain relevant states. As with all understanding involving vast numbers of things, in this case vast numbers of neurons, the understanding may involve instead of a specific neuron a specific category of neurons, or a statistical category, as well as statistical aspects of signals flowing from one category to another, of state information about neurons, and so on. At this point, we will not look off in these directions. Here we are only interested in the simple, clear, true perspective that there are logics going on in the brain for any awareness that we have.)

(There is another direction which we do not want to move off into at this point. But we are constrained to address the idea because it occurs to many people: "When I have a feeling or awareness, why does that have to be something in the brain?" A good answer, from the scientific perspective, is this. Every awareness that you have, you could talk about. In fact, generally, the more knowledgeable you were, the more ways you could potentially talk about an awareness. If we move in and look closer at the brain, we would see the signals going to your vocal chords, and to the muscles in your mouth, that phrase the various syllables that come out of your mouth. Those syllables are everything that you say, and the *signals* that generate the muscle movements that generate those syllables come *solely* from the big hum in the brain. This is the point of the scientific perspective. Those signals do not come from nowhere. They come from somewhere, and where they come from is the big hum in the brain. Everything you say or could say, under any circumstance, comes from the logic going on in the physical brain. Thus, all of our awareness is certain logics in the brain. Sigh. As if this long parenthetical paragraph distraction weren't long enough, the god suggested that I add a further point: we should not think that the logic inside the brain is separate from the logic outside the brain. As for myself, I don't want to think further what this could mean because we must get on with our journey: let it be stated that for any awareness itself, there is a certain logic going on in the brain.)

When it comes to a new area of understanding, the value of knowing something in the concrete is immense. There is no denying how valuable it would be to know, in concrete detail, the logics in the human brain, but surely such scientific understanding, with the needed technology, must be centuries off into the future. In an upcoming chapter on "Machines" we will look at the development of such technology, up to the time when it is fully capable of seeing these logics in their full specific concrete detail. That will be an awesome day. For the current chapter, we basically assume the results of such future science and technology. This allows us to think of the sentience in a more concrete way. We imagine that we have technology that can literally look into the brain and see these logics going on, and see the various aspects of these logics.

Touching Left-hand Middle Finger to Shirt

Suppose that you touch the middle finger of your left hand to your shirt. Maybe you wiggle it around, ever so lightly, to get a better feel for the texture of the cloth. You could do that right now. At that moment, you have an experience, in your awareness, of the touch or feel of the shirt. We will now look at the flow of signals through the neurons, from the finger tips up to the points far in you brain where the logic of the experience itself takes place.

When your finger touched the shirt, hundreds of finger-tip touch receptor neurons fired signals. These signals are in the form of multitudes of electrical spike trains, each starting from a touch receptor neuron, then jumping to another neuron and moving through it, then jumping to the next, and so on, all beginning at the tip of the finger. Let us follow the spike trains as they move through one neuron after another. They eventually get to the area of the first knuckle, and through. Then to the second knuckle, the third knuckle, across the sweep of the hand, through the wrist, onward to the elbow area, then upward to the left shoulder. Since the shoulder, at the top of the back, is not far from the spinal column, the spike trains travel a short distance across to that immense trunk of nerves (this great trunk is packed with neurons over which spike trains are constantly traveling between the brain and all parts of your body – and "constantly" means right now – it is just that this is invisible to our native perception). Then up the spinal column the spike trains go, whereupon enter they the big hum, the human brain, of about a hundred thousand million neurons, each basically like a tree with no up or down, a tree with tens or hundreds or thousands or tens of thousands of branches going out to other neurons. Before entering the big hum, there is probably little processing of the spike trains, except for some summation, coalescing, error reduction and the like, that kind of logic being basic.

Transition to getting into the brain

But it's when the signals start entering further layers of the brain that the logic is no longer basically a simply sending of a signal. Things start happening with the signal, in the e-fish, and in us too. While we know quite a bit about the logic of the electric fish processing the specific logic of the JAR, we know virtually nothing about what is happening to those signals of finger-tip touch after they enter the brain. However just because we currently know virtually nothing does not mean it is not there, right now, functioning, in full concrete detail. Several centuries from now, our technology will be able to see it.

Even so, we can easily presume this logic transforming into deeper and deeper forms (that is to say, more and more processed forms) in the brain, until it eventually becomes the logic of our awareness, that logic the activation of which is the very awareness of the experience itself.

Lots in brain not in consciousness

This might be a good time to stress that there are broad amounts of logic going on in the brain which are outside of awareness (sentience). There may be processing in parts of the brain that contribute to the speed of your heart. Typically, unless we do perhaps Yoga, we have no awareness of any part of that processing, and so that logic is outside the logic of our sentience. Likewise any processing in the brain that guides your eternal, unconscious in and outs of breathing, all that processing or logic is outside the processing or logic for sentience. And such is also the case with the considerable and advanced logic that keeps us from falling over as we walk along the ground, balancing our body over one swinging pole (one leg) after the other pole (leg). Many logics are outside sentience.

Lots in consciousness

Just as there is a lot of logic going on in the brain that is outside consciousness, so too there is a fair amount of logic going on that is part of sentience. By definition of sentience, all such logics are intimately identified with the feeling or experience itself. At the moment that you touch your shirt with that third finger, you may also be aware of being in a room that is too warm, and the logic of experiencing that will be going on in your sentience, along with the logic of experiencing the touch of your shirt. You may not have eaten much for lunch, and dinner time is approaching and you experience a mild feeling of hunger. The logic that is the experience itself of hunger would also then be going on in your sentience at the same time as that of logic of the experience of the touch of the shirt. And so too with the experience of seeing whatever you were seeing at that moment, hearing whatever you were hearing, as well as many other touch sensations at many places of the body.

We might call these logics the *perceptual space* of sentience – that part of sentience which is more directly related to our senses.

Parts of the touch train outside sentience.

What parts of our finger touch train are outside sentience? Somewhere in the brain are determinations as to how smooth or rough the shirt surface was, and this is computed in some way just from the volleys of spike trains. These determinations are certainly not in our awareness. Rather, we are simply "instantly" aware of the results of the determination, the results being some measure of the roughness or smoothness of the texture. Surely there are also other determinations being computed from the volley of signals, the carrying out of these determinations also being outside our sentience.

Part inside sentience. More definition of issues of sentience logic.

As for the touch logic that is in sentience, at our time in human history we have not a clue. Our technology is too primitive at present. Even our current understanding cannot conceive of how something like touch, or our experience of red – the experience itself – is nothing more than a logic. Nevertheless there are some important things that we can say about logic, in the brain, that is in sentience logic versus logic not in sentience.

Consider the spike trains going on for the determination (computations) of the roughness or smoothness of the shirt touch. As said, we have no awareness of those computations. We just suddenly feel the texture. If in that part of the brain where these determinations took place, one did not allow the signals to move outward to other parts of the brain, then the person would never experience the touch – they would never feel the shirt. It is similar with the situation where the spike trains would never be allowed to move up beyond say the shoulder. There may well be some processing of averaging and error removal below the shoulder. But if the spike trains were never allowed to move further up from the shoulder into the rest of the nervous system, the person would feel nothing.

On the other had, if a logic is in the larger logic of sentience, and is going on, then the person experiences the touch of the shirt. And they experience the touch of the shirt even if no other signals are allowed to move onward. And they experience the touch of the shirt even if the signals were artificially created for instance during surgery (in the far enough future, there will be various non-surgical techniques for direct firing of a large number of arbitrarily selected neurons throughout the brain). The person will feel the shirt touch even though there are no spike trains coming up from the shoulder.

Issues with this definition

There is a problem with this way of determining what logic is in sentience and what is not. The problem has to do with blocking further propagation of the spike trains. Consider that logic going on in the brain which is the experience itself of touch from the finger tip. Suppose that researchers artificially activate that logic (by some way of activating various key neurons), and so the person has the experience of touching something. Suppose however that we truly block any signals from leaving that area. Then no signals would get to anything like memory or to anything like perhaps a pre-speech area. So presumably the person would experience the touch, but immediately after have no memory of it, nor would they be able to speak about it at the time because no signals ever went to anything like a pre-speech area. The signals moving up from the finger tips to the brain are basically a linear train moving upwards, a train that can be physically, and logically, stopped at a certain point. However, inside the big hum in the brain, this linearity often disappears; we no longer have this physical characteristic to guide our understanding.

In spite of this problem, I have faith that someday the kinds of logic going on in the brain will be worked out enough that the above will serve as a basis for a more complete definition that will indicate precisely what logics are experiences themselves and which are not, that is, what logics are in the logic of sentience and which are not.

Intersection

There is a further way that we might suppose the logics in the brain are often organized with regard to sentience. This is based on looking at the common parts of our experiences. We have already looked at touching our finger (the middle finger of the left hand) to our shirt and the logic connected to the resulting awareness in the brain. We now look at a number of other awarenesses, or experiences, and what they all have in common.

Forefinger Instead

Let us look at an experience or awareness that is almost the same. Suppose that you touch your shirt instead with the forefinger.

Again, we will suppose that you ever so slightly wiggle the forefinger against the shirt to get a better feel for the cloth texture. The multitude of spike trains coming out the finger tip is probably quite similar to the situation before, where we touched the shirt with our mid finger. But now the trains of electrical spikes go out over different tracts of neuron pathways, certainly through a different finger, and quite possibly the neuron path is different up until it gets into the big hum.

But somewhere further into the big hum, the logics produced by the two different fingers, if only we could see them, become the same logic, but just have some different "parameters" indicating which finger this block of touch information is from, as well as perhaps some other small component of the volley of signals, indicating the sensitivity of the finger area (if perhaps different touch areas have different sensitivities). But apart from these few small components, the majority of the signals (being the physical basis of the logic) would be information about the kind of touch being received – softness, roughness, variations, some standard corrugation patterns, and so on. The volley, and possible offshoots, moves to deeper and deeper parts of the paths, some going to sentience, some not.

Third and Forefinger. Sophistication of Brain Logics

Instead of comparing the touch from one finger with that from another, what if we look at both fingers touching the shirt at the same time. Suppose that you touch the same area of the shirt with two fingers at once, again slightly wiggling them around to get a better feel. This will illustrate some of the more sophisticated aspects of the logic that goes on in the brain.

Two volleys, one from each finger, of hundreds or thousands of spike trains would set off into the nervous system, and would probably remain separate until they go up into the brain. Somewhere, thereafter, they would become one volley, one logic, the combined sensation of the two fingers, the combined information. But a component of that one logic would indicate that that one and same logic is occurring simultaneously from two different fingers, and possibly the component would further indicate which two skin areas – which two fingers – are the source, and also might indicate other information such as relative amount of softness, corrugation, intensity, and the like, that there is between the two simultaneous areas of touch. In other words, the component might have some sophisticated "quantifier" information.

Overview from Touch

It is important not to get buried in the technicalities of the logics. This chapter is about sentience, experience itself. The purpose of the above, and of the whole chapter, is to approach closer to the sentience logics (even though we cannot at present imagine how a logic going on could be sentience). For instance, the computations of characteristics of texture on touching the shirt – the results of those computations are in the logic which is the very experience itself in our awareness. We know there is such a component in this logic that is the experience itself, because we experience the texture through the touch.

A theme running through this whole chapter is that there are actually these logics in the brain that are the awareness itself, they *are* awareness, they are one and the same as awareness – at least they are, when you grant the scientific perspective that everything we are is in the (physical) brain – and after all this book is from the scientific perspective. Nevertheless, as the journey proceeds, the god told me, these ideas will be filled out a little more, not contradicted, but filled out in a direction that gives a perspective different from the one we might have at this point. Nevertheless, even later on in the journey, the idea will remain that there are logics, in the brain, absolutely identified with sentience.

Music via Violin and Piano

Of course the feeling of touch is not the only thing we are aware of. If a violinist pulls the bow evenly across the same note, let us say the A flat below middle C, we can then look at the cochlea, just behind the eardrum, which contains the neuroreceptors for sound. It's 15,500 hairs detect different frequencies, and cause neurons attached to the hairs to fire, and that is the start of the logic going on in our nervous system. If we could see enough, we could follow the going on of this logic as it moves further and further into the brain.

Suppose now that instead of the violin, a pianist hits the A flat key below middle C. Again, the hearing logic going on in the brain, starts at the neurons connected to the 31,000 hairs of the two cochlea. The strongest neuron firings will be for the 440 cycles, for that is frequency of a pure A flat below middle C. But there will be many other firings, of much less strength, among them of quire varying intensities, for the striking of a piano key carries not only the primary frequency but a rich set of harmonics, generated by the complex interaction of the hammer, the set of strings for the note, and primarily by the backboard of the piano. Further, the harmonics of a struck piano key are nothing compared to the richness of the overtones and harmonics of the violin note.

The path of the going-on logic starts at the neurons coming out of the cochlea, firing not only the basic 440 cycles per second for the theoretic pure A flat, but also firings for a multitude of overtones and harmonics – and off it all goes into the brain. Actually, it doesn't have far to go since the little cochlea is pretty much in the brain already (unlike the finger touches which had to travel a long distance through the fingers, the arms, shoulder, and back of neck).

There must be a number of different measures computed on the volleys of these signals, for we have an incredibly rich awareness of the aspects of the qualities and harmonies of these sounds. Since not even the most professional of musicians or musical instrument technicians or instrument theoreticians have near a full understanding of the relation of these harmonics and overtones, it is not possible at present to say much even about the frequencies in the outside world. Nevertheless, the brain is processing and picking up on rainbows of aspects, in the logic going on, probably over and over as it goes deeper, moving into the area of beauty within our sentience, and all the joy and emotions that music seems to miraculously bring. It is all there, in excruciating detail, in the brain, in the areas of the brain. The shear computational magnitude of this is a wonder that will someday be revealed across regions of the dense jungle that is the brain.

And if that computation and going on in the brain is for a single note, it goes on many times over for a chord whose notes are played sequentially or, also, all at once. Somewhere in that long extended dense forest that is the brain, one pattern, one connection, one relation after another, is being drawn out in the logic that is going, throughout, all over these areas of the brain, a kaleidoscope of explosive activity all over regions. And yet that is not all, for it is not just the notes of the violins and pianos and other instruments at an instant in time, but throughout time all are playing avalanches of frequencies and overtones, all together, all individually, and there are other regions in the brain comparing the results of earlier regions with the results over time and coming up with characteristics of the *form through time*. And the brain builds hierarchies of analysis on analysis of the forms in the music. All this is in the brain. (A book which analyzes some of the hierarchic form of music is Cooper and Meyer, *The Rhythmic Structure of Music*. There are logics in the brain that do all this and more, and centuries from now, researchers will concretely see these logics in the brain.)

And somehow all the activities from all these regions eventually are moving into the areas where the logic of consciousness and sentience is going on, whether or not of the emotions of music, and this is itself the being of the most extensive emotions and feelings that we experience, to the greatest heights of heaven of feeling and happiness and sadness, to the turbulence of hell, and the gamut of most other emotions, from the intense psychological feeling of sexuality of the young, to our joys of beauty and food, all in a form of us, from the view of kaleidoscopic activity of the valleys in our mind, forms embedded and compared with forms, both instantaneously and through time, separately and with each other, awesome regions of the forest extending over horizons, all alive with thriving activity. Centuries hence, with the help of countless supercomputers, we will have analyzed and understood all of this, in terms of logics, and in terms of incredibly concrete detail, of what is going on in the hundred thousand million neurons of our brain.

Vision via Shape of Small Table. Space.

Suppose that you are looking in the direction of a small circular table, and that you are focusing on the shape of the table. The path of the ensuing logic starts at the back of each eye, a hundred million sensory neurons, rods and cones, each firing a signal that encodes information about the intensity and, if a cone, the color at that point. Already within one or two levels of neurons within the eye, the one hundred million signals are processed at a simple level, and the result appears encoded in signals going out on 1 million nerve fibers, from each eye, which go off into the big hum. Like the cochlea, the eye is pretty closely surrounded by the brain. Those 1 million nerve fibers from each eye pretty much form two large cables of nerves that go diagonally toward the middle of the head, crisscross, and continue toward the back of the skull, where presumably the first more substantial processing of logic takes place. As you notice the roundness of the table top, somewhere in the wide variety of processing done to those 20 million points of continuous input, is the detection and analysis for features and monitoring of the perimeter or shape of the table. Somewhere, somehow, a volley of logic is going out encoding the shape of the table top, whose primary characteristic emphasized for you is the roundness. Eventually the volley is entering into the sentience area where it becomes your experience of the shape of the table top.

Or perhaps the table is smallish and square with rounded corners, and again we suppose that you are concentrating on the shape. Eventually the logic going on gives rise to logic going on in your sentience, that is the experience itself, of the shape.

We tend to think of our sensation or awareness, experienced on seeing shapes, as geometry inherent in external reality. But if one looks at minimalist axiom systems that characterize geometry, whether a Cartesian version or a minimalist non-numerical Schwabhäuser version, a great deal of the properties that we are instantly and inherently aware of when we see a shape requires a great deal of derivation and computation from the underlying axioms. Likewise, there must be many other properties, of these shapes, that do not instantly and inherently appear in our awareness. Thus, our awareness of even pure shapes has relations to only a selection of only some of the logic that always goes on with the shape. As for the experience itself, that which we have when our eyes pick up on the shape, that experience may not be as much the collection of geometric properties of external reality as we imagine. Our sensation of perceiving such pure shapes may be as "indescribable" as is any sensation or feeling, as is the experience we have of seeing red. We should not be misled by the fact that we can relate it to geometric characteristics out in the external world. The sensation itself, like that for seeing the color red, derives in some way from our own particular evolutionary history. More than that, we should not forget that the scientific conception of space initially derives from such internal experiences, and the conception may still have aspects of those. The goal of science is to create a perspective of space that consistently includes all other perspectives of space, but we should not assume that science has achieved that goal yet.

Intersection of Vision, Music, and Touch

There will be logics that go on for touch, and logics that go on for music, and those for vision. Some of the logics that go on for touch will also go on for music, and presumably those are closer to the logics of sentience itself. Similarly some logics will occur both for vision and music, or for vision and touch, and so on. Those logics that occur during touch, and during vision, and during music, could be presumed to be closest of all to being those of pure sentience itself.

(Here is a technical aside. As illustrated by the JAR logic of the electric fish, logics do not have a simple correspondence to the notion of locality in the brain: A single logic may be scattered through different regions of the brain. And there can be complexities such as parts of it being shared with other logics. This correspondence can be expected to be even more challenging when considering the structure of a large number of logics. Nevertheless, the intersection of logics, in the brain, due to different areas of external experience, such as vision, music, and touch, can be expected to show up to a rough degree in the intersection of areas of brain activity – for instance in amount of blood flow as shown by a PET scan Done check spelling. Thus, areas of the brain that are active no matter whether the person is experiencing music, or vision, or touch, could be areas that support some of the logics closer to those of pure sentience itself.)

Feeling of Emotional Pain. Will.

There are other logics that we sometimes speak of as a feeling, for instance the feeling of emotional pain or of concern. For instance, if one's relative had indications of illness, one of the pieces of logic going on in oneself will be the experience of worry, and that piece of logic will be in the sentience logic. It seems hard to imagine that the logics of experiencing emotions would be the same as the logics of experiencing various visual sensations, or sound, or touch and the like, but as far as being logics, and as far as the logics going on via electrical spike trains, in this sense, all the logics are alike.

Returning to the logics themselves, we might ask, as we did above, what is the logic that is common to all these, to the experience of music, generally to the experience of sound, to shapes in vision, generally to vision, to touch, and to emotions. It would be sentience but it would be not the experience of vision, not the aspects detected in sound, not sensations, not emotions. If there is logic in the sentience that is not part of any of these, could this be called the "pure I"? And who knows, maybe logic going on in the pure I without any other of the logic would not be sentience would not be awareness. Maybe one needs the other experiencing to go along with it for us to be conscious or aware or sentience or experience being? But these are issues for the far future.

Another kind of thing that "enters" our awareness are issues of will. We raise our arm, we pick an apple, raise it to our mouth, and bite. These are acts of will, and we have a particular kind of awareness of willing such. There is some question as to whether our acts of will arise in sentience or unconsciously outside. But regardless, acts of will have a unique kind of awareness for us. This kind of logic, that is the experience itself of will, is going on in the brain, and again we can look at what is in common with it and all the above types of awareness, and in doing so, we might contemplate whether what is in common to all is the "I".

Awareness of Thinking

There is another kind of awareness, or experience, that we have in relation to thinking, what we describe in our speech as "*I am thinking of such and such*". This awareness too is a certain kind of logic going on in the brain, not just the thought, but the logic that is going on that is our awareness of the thought. One component of the experience logic would be characterizations of the kind of thought, for after all, when we have the experience that we are thinking, we also have some general awareness of the structure of thought taking place.

Intersection of All

Once again we might wonder what is the logic that is common to all these logics – the experience of sound, of vision, of emotions, of will, of thought. This would be the logic that always goes on when we have any kind of awareness. It must be an amazing kind of logic. This would be the "I".

Someday, with the help of gargantuan supercomputers, we will see all this in detail. Perhaps it will be a hundred to a thousand years till we do. But we will.

How will technology see sufficiently into the brain, that jungle, so extensive and complex? How will science see even to the logic of single neurons? The second chapter after this will look toward this daunting task, through centuries of the future, as to how the human race will achieve the understanding of the logic of sentience.

20 Devices to See Logic

Neurosurgery

It is centuries in the future. A door opens, whereupon a neurosurgery specialist beckons us into one of the rooms where a patient lies on an operating table. Thus starts the presentation and our visit. Looking close up at the opened human brain, we see naught but a bunch of gray slime. However, we should not be distracted by what our native perspective so emotionally pushes at us. Instead, we must somehow take our perspective on what we are natively blind to, *the logic going on in this slime*. That is everything; the slime is nothing more than its physical substrate.

But what *is* logic? Logic is the same sort of thing that meaning is. The going on of a logic in some region of the brain is the going on of meaning in that region. Indeed, even at a single nerve cell, neurobiologists speak of the encoding in the signals. This is the encoding of meaning in the signals – the encoding is the way in which the signals represent meaning. If we look at how the animal acts in different situations, if we look at how the creature may have come to have those actions through the process of evolution, and if we look at how all that comes to be going on in the creature, it is always the meaning that is encoded in the signals in the brain. It is not the signals themselves, it is not even the specific encodings, for they could be anything (as long as they are capable), but the *meanings* in the encodings, the meanings that are coming into the brain, and that are going out of the brain. The meanings are everything. The signals are just a way to carry the meanings, they are just a physical basis for carrying on the meanings. The different interactions between signals are really interactions of meaning. For instance, when we look at the Jam Avoidance Response (JAR) logic, we see different parts of the logic going on in different regions of the brain, and in all those parts, all the signals are nothing more that trains of electrical spikes moving over nerve fibers. But it is the meaning encoded in those spike trains that is everything. At

183

some neurons, the meaning encoded in the signals is that of the amount the amplitude (of the electrical waves in water) is rising or falling, at other places it is that of the amount of difference between wave phases at two different places on the skin that is encoded, at other places the meaning in the spike trains is the amount the amplitude is falling combined with the amount the phase differences are increasing, and so on. Likewise, when neurons combine spike trains from several places, and in turn output a new spike train, what is important is that the input meanings are being combined to form the output meaning. (Individual neurons can sometimes be sloppy in their encodings but, when averaged over a number of neurons, quite accurate.)

The Devices

Our surgery specialist of the future brings us over to a device to look into the gooey mess. The device can show so many perspectives which people back in the 21st century would never be able to see.

At first, the device displays nothing more that what our native perception shows, a slimy grayish mass. There is a button. We push it. Now the device shows multitudes of little greenly-lit strings flitting about all through the slime, in spaghetti-like tangles. These are the nerve signals, which our native vision is blind to. There are dials on the device to slow them down or speed them up. In their regular speed, they are way too fast for us to see. There are other dials to look closer in or farther away, or to move to this area or that of the brain. If we hit the button again, the little green strings disappear leaving only the gooey mess visible. The button shifts back and forth between the two perspectives. We push it again. Little green strings are back.

The perspective in which we see the little green strings moving around is a picture of the logic going on. Still, we would have to know a great deal about how those strings were moving around before we were able to understand that picture, before we could understand what meanings we were looking at. Nevertheless, if one takes that into account, we would indeed be seeing the logics going on, as the little green strings wound there way this jagged way, and that jagged way, through that huge dense forest of the brain.

We see a different button and push it. The jagged paths that the strings move in straighten out! Now they are mostly perpendicular to each other. It is still the same green little strings moving around, yet here they move in something that looks like a complex engineering diagram, with straight lines in all different directions. We use the dial to look further away, then we move closer in.

"My gosh," you think, "how much easier to see where these signals are going. Well, sure it is still pretty much a mess, but nowhere the mess it was, where everything was going around like spaghetti all twisted and chopped up. Here I don't have to trace a particular signal winding all over the place through the spaghetti. Now it goes directly in a straight line to one place, maybe turns a perpendicular corner, goes to another place, and so on."

The surgeon in the lab comments to you, "Naturally, the spatiality you see can't be exactly the same as in the brain. The points," – the surgeon points to the screen – "don't match up to quite where they are," – the surgeon points to the brain the device is looking into – "but surprisingly they're not too far off. But the logic! It is exactly the same! Just converted to a more orderly presentation. Nothing added, nothing taken away. You are seeing exactly the same logic, just presented better." (For discussion of this last point, see endnote 9.)

Now we notice even more buttons on the device. One of them changes the little moving green strings to a pinkish red. Another control changes them to a yellow against a dark blue background. Still another causes two colors, brown and red, of threads to move over the engineering diagram. "I wonder what that means!" but you don't have the time to find out. An earlier button causes the brown and red threads, still there, to revert from the engineering diagram back to moving through the spaghetti mess. Only now there are two colors winding all through the spaghetti slime. You push another button and pull back in shock. Everything has gone back to the neat, orderly engineering diagram, but now the diagram consists of fixed lines of hundreds of different colors, with the little strings of color moving along them, though the strings are a different color than the line they are on. "Wow, but too complex," you think, "What is this other button?"

This one is fascinating. Instead of little strings moving along the lines, now the whole line lights up for a second, with a different color. The whole engineering diagram becomes one of various line segments lighting up and then going off. "Interesting," you think.

All these are perspectives of the full logic, provided you watch long enough and you understand enough. This is different from the perspective of the dining table, mentioned some time ago on our journey, where the top perspective showed only some of the table and the underneath showed only some. Our perspectives through this device show the full logic (provided we watch long enough and understand enough). You are seeing the actual logic going on in the brain on the operating table, only shown from different perspectives. You play away with the fascinating device. One control even displays information about the condition or state of individual neurons and then displays whole logics using the perspective of conditions and states. But you move on.

Hours have passed. You are getting rather proficient with all the buttons, dials, and computer controls. You have discovered a variety of computer screens. On one of them one may set ranges of all kinds of controls and options. You have moved far beyond having the device show just little signals moving around.

You push a button. "Choose a logic," the screen says. You press "help," and the screen says, "There are currently 312,849 meaningfully, separable logics in this X954.2 region of this particular human brain (using the delta-5 separation scheme, version April, 2541, with usual values for major alpha divisional parameterizations)."

Suddenly the screen changes the "312,849" to "312,855", and then quickly jumps it up yet two more.

Now as you use the device to see into the brain being operated on, you see not little green lights moving along fixed paths but flashing squares and circles, with the flashing moving from left to right through the figures, but at uneven speeds, sometimes stopping at one place, sometimes continuing all across, often briefly changing colors in places, though sometimes nothing at all happening to the squares and circles – nothing lit up or moving. This is a very different perspective.

The bottom of the screens says, "Logic L9134A-B45GH82-1.429, near pre-visual inner level 7 awareness logic area HJ-9." You hit the "more" button, and then decide to click "choice 4," not having much idea of what the choices mean anyway. On the screen appears, "There are 78 copies of such logic (separable with beta = 5.3). Not fully understood at present. Logic depth 429 (unusually large, but possibly not for logics that get closer to the more concreto-sec attempts of the sentience, which is still basically non-resolved / non-understood even at level 1 of logic structure resolution.) Zero evidence of retinotopic spatiality."

Suddenly some of the middle squares and circles light up with a few different colors, but equally suddenly they all become quiet. Something just went on in that logic. What was it? You glance at the patient lying on the operating table, wondering who is this person, this being, this sentience, this logic going on. Did it just have some experience? "Behold ourselves," you think.

Back on the screen, after having selected choice 4, several further choices appeared. One says "evolutionary." Hitting it, the screen adds at the bottom the following additional information for this Logic L9134A-B45GH82-1.429. "There is evidence that structure of this brain logic is derivable from evolutionary logic going on in the environment from 400 to 350 million years ago (see evolutionary logics G983-5865,6,7 (environment logic dynamics K (subtransfer 5 range)), also possible relation to 1.5 billion years ago (environment logic dynamics, posited

P41))." You see that most of the identifying numbers can be clicked on to get a description of that logic.

Interesting, you think. There are logics going on not only in the brain, but also across vast stretches of evolutionary time, and these evolutionary logics gave rise to most of the fundamental logics in the brain; the device can show perspectives on these evolutionary logics too, just by clicking their ID number.

The evolutionary logics and these brain logics are more connected than the previous sentence indicates. But it is time to leave this room.

The surgeon specialist signals us to move on, and we pass through a door into the next room. Lying on the operating table is a foot-long *Eigenmannia*, an electric fish. Here is something we know about, at least a little, and we eagerly place the device over parts of the gooeyness that is the fish's brain. What logic is going on here?

In this world of the far future where so much more is known, we select JAR. JAR stands for Jam Avoidance Response logic. This is the logic we looked at back in Part 2 of the book. Seeing the look on our face, the surgeon says, "don't worry; you don't have to remember the details of the JAR logic". Relieved, we randomly select on the screen sublogic JAR-Z1, and the following appears. At the bottom of the screen is "JAR-Z1. Logic of primary initiation of zero-crossing information." Laid out over the top of the screen are hundreds of points with little green strings of signals periodically moving out in great waves from them, and then wind all over the place, spaghetti like. Well, not quite that bad. But sort of spaghetti like. Having got better in using the machine, you play around with some buttons and pretty quickly bring up a completely different perspective of this zero-crossing logic.

There is a single point on the screen. There is a big line going into the point, and another line, on the other side, coming out. Although there are various distracting terms like "description JAR-Z1-a5 Input" on the screen, you know enough to know what the screen is saying.

This large point on the screen is a sensory neuron, which means that one side of it is exposed to the environment, getting information about that environment, and the other side is producing electrical spike trains that encode the meaning. In this case the aspect of the environment being focused on is the electric waves in the water. The output of the point is when the zero crossings of the waves are occurring. Actually, the output is a spike train, and that spike train encodes a meaning, and that meaning is "when the zero crossings are occurring."

Pushing some buttons quickly – for now you are a whiz, at least in some ways – you simplify the screen image so that it only shows meaning. On one side, going into the point, is written "meaning: electric waves in the water". On the other side is the output, written on the screen as "meaning: when zero crossings are occurring." "Now this picture is nice and simple!" you say with a self-congratulatory pat, for it took you a while to get to this level of proficiency.

What we have done is to get the device to show a category of neuron as one single large point on the screen. On an input line to the neuron is the meaning of the environment, on an output line is the meaning that is encoded in the spike train signal and that is output from the neuron.

You push some more buttons for some details – not too many details – just some. The screen shows that on this particular fish there are 312 such neurons, 150 for the right side of the body of the fish, 162 on the left. On typical fish of this sex and subspecies there are 320 such neurons, 160 on each side of fish. There is some additional information on the standard deviation of these numbers for the typical such fish. Oh what is this down in the corner of screen? Do I really want to know where these 312 neurons go? Fast peek. "ALLG – anterior lateral line nerve ganglion" and there is a little ALLG in the brain. "Enough," you say.

Pressing a button for more information shows that this JAR-Z1 sublogic was the identical logic used in all kinds of logics outside of the Jam Avoidance Logic, no doubt because they were all computed in one place in the brain and then sent off to the various other places. And the screen lists *all* the other logics (places).

Of course there were closely related logics, like JAR-Z-Summary1, which took several "meaning: when zero crossings are occurring at this sensory neuron," and summed them up to a more solid, "meaning: when zero crossings are occurring in this little field area on the skin." As usual, one could push buttons to see the details of how these meanings were encoded in the spike trains. You push some. What a horrid mess of graphs and statistical data, you think. The main idea of all of this is that it is the meanings that are being worked on, combined, reduced, separated out, built up, and so on; it is the meanings that evolution is working with, and on.

And there was a screen for logic JAR-AMP-INC2. As listed on the screen, the input is "meaning: strength of the water waves at a certain area on the skin". The output is "meaning: how fast is the strength increasing."

There was a screen for JAR-I-Delay4. The meaning is shown as "how fast is the strength decreasing *and* that the zero crossings of the comparison neuron are falling behind the zero crossings of this neuron." This logic has a lot of complicated additional details. If one wanted to see them. There are complicated histogram "graphs" that give more information about the encodings. There are statistics about how some of these logics tended to be "directional" in that the interfering waves produced strongest results when the neighbor fish was in certain directions; and also the screen showed in what percentage of such neurons direction did not matter. And so on for quite an amount of detail. If one wanted to see it.

Then there was an interesting logic, JAR-OVERALL – and you select it. What appeared this time was a different way of showing a logic. The language was not graphical, with points or lines or strings of figures. The language on the screen is totally made out of letters, digits, and punctuation. Here is what was came up on the screen.

```
(JAR-OVERALL   (section 1  (stats 24.3   41.2 . . .) Group
      (  (stats 31.4  56.7 . . . )   JAR-I-Delay4 )
      (  (stats . . .            )   JAR-E-Advance4)
      . . .
      degree-of-presence .85
   . . .
```

And this was only the beginning of what appears on the screen. The "stats" have something to do with how many times the next item occurred, and what were its variation characteristics and so on.

What was interesting was the "way" in which this logic was . . . uh. Well, the device simply said that this particular JAR-OVERALL was occurring to degree .96 (which was a pretty high degree) in the fish lying on the table. There were slightly different JAR-OVERALL's that were going on to degree .91 and one even to degree .89.

This is to say that there are many slight variations of this particular logic, and they are all going on to some degree in this particular physical system of matter in time and space, in this case, in a particular stretch of region in the brain matter of the electrical fish. These variations are because physical reality is a somewhat messy place, especially the brain with all those bushy tree-like neurons leaning up against each other all over the place. For most of the important logics in a brain, they are going on fairly solidly, at least in most fish of the species. But it is never a perfect affair. On one hand no logics ever go on perfectly; on the other, there are many slight variations of a logic, all of them also going on to degrees in the same area.

(Here is a technical aside. This issue of variations of a particular logic, all going on at once, is separate from the issue of various idealizations being approximations, similar to the way in which a number with more and more decimal places approximates a "real" length out in external reality. Except with numbers it is a simple one-dimensional affair, whereas with logics, there may be no limit to the dimensionality. Indeed, these two issues appear even in the way in which a larger logic is structured into smaller part-logics. When one considers the great issues of the metaphysics and ontology of the universe, it really is odd when you think about it. But there it is. That is just how it is.)

It should be noted that whether one has little green worms of light crawling around on the screen, or whether all kinds of letters, names, numbers, and parentheses, are ranging over the display, still, in all cases, the viewer must "understand" enough of what is appearing on the screen to know the meaning being referred to. What is on the screen invokes a meaning in our mind, and that is the meaning the device is saying is going on in the brain the device is looking into. In other words, what appears on the screen is not the meaning itself. In all cases, our mind must impute – read – the meaning from the lines, points, letters, words, and so on, that are on the screen.

More on the Structure of Logics

One way or another, a device that displays logics in the brain is displaying a perspective on the structure of logics going on. The idea of the *Structure of Logics* was introduced toward the end of Part 2, in the section by that name. Back in that section, we said that a logic goes on in a *region* and *not at point*. Why did we say that? After all, with the JAR logic in the brain of the electric fish, the electric spike trains output by a specific neuron at a point had a pretty specific meaning or logic. The answer is that the *full* computation for the meaning that was output at an individual neuron was computed *in a region consisting of a number of neurons across the brain*. The JAR logic, as well as the logics that are pieces of the JAR logic, take place in *regions* of the fish's brain; they do not take place at a single point. Similarly we may consider the logic in the human brain that is sentience. Clearly it is not at a single point in our head but in a whole region. (If one wishes, this last sentence may be stated more carefully. The logic that goes on in the human brain that is sentience clearly goes on not at a point but in a region.)

A second question or issue is this. Can one say more about how to distinguish logics that are a part of sentience logic versus those that are not? For instance, a logic that is an experience we are having will clearly be "in" sentience, or a logic that tells us the conscious parts of a thought we are having will clearly be "in" the sentience logic. On the other hand, most of the logic and its signals dealing with our leg muscles to keep our body balanced as we walk are not "in" sentience (and there are quite a few such signals indeed, for we are balancing our whole body on two alternating moving poles, our legs, as we propel the whole body through space). None of these balancing signals are in our consciousness at all; their logics are not "in" the sentience logic. (Incidentally, a scientist or mathematician or thinker might, in an academic sense, raise the question, why might there not be a second sentience logic in our head for which all these other logics and signals are "in" that sentience; there is a second "being in our head" that is conscious of all these other things. We reject that there is any second sentience logic going on in the head.)

Logics and signals for balancing are not the only ones not "in" consciousness (sentience). Consider the zoomoscope and the circuitry in the brain that computes how large we perceive objects as. Because of the way this circuitry works, it assigns a smaller size to objects when we look at them through the zoomoscope. The only signals in this circuitry that are relevant to sentience are the output signals indicating how large the object has been computed to be, for as long as you maintain these signals, the person will perceive the object as unchanging in size. But if you artificially vary these signals, even though the internal signals of the circuitry do not change, the person will perceive the object as having different size. Therefore the signals "within" this circuitry are not "in" sentience logic. This is one way of determining what signals and logics are not "in" sentience. Likely in the 25th century we will have found more ways of determining what logics and signals are not in sentience. Far enough in the future, we will use a range of techniques to delineate the structure of logics in the brain, to determine what logic is in sentience and what is not in sentience. This will be one of the fundamental steps in determining what is sentience logic.

A third question arises concerning the nature of sentience logics. Can we even say that sentience logic has the characteristic that other logics (e.g. experiences, thoughts, balancing logics, and so on) are "in" or "not in" it? Yet it is not a matter of whether it has such a characteristic. It is a matter of whether it can be displayed from such a perspective. Clearly a device could display it in this way. On a large screen might be a large circle representing sentience. Logics such as experiences or thoughts and the like would appear in the circle if the person (the sentience) was having that experience or thought or the like. In other words, a sentience logic can be seen from the perspective of other things (experiences and so forth) being in or not being in it. That is one of the types of perspective on a sentience logic.

(The area of human study called *formal logic* looks at "logic", but it is not the logic talked about in this book, though there are similarities. As stated in Part 2, "On Animals," the logic discussed in this book existed before humans and before human thought. Now there is something in formal logic called *first order logics*. If characteristics of sentience logic where brought into formal logic, it would require stepping outside first order logics, or otherwise using some of the standard apparatus that is used to model higher order logics while staying within first order. For a sentience logic may have other logics "in" it, as described above. Capturing the idea of this "in" in its purest and most general form would require higher order logics, assuming we stay within the standard categories in formal logic.)

Cavalieri

When we were in the operating rooms playing with the display device, there had been some discussion about how one display showed the same logic as the previous display. The little green strings moving along the engineering diagram showed the same logic as the little green wiggly lines moving in paths wiggling all over the place. Mathematically, this could be rephrased by saying that the second display was a transformation of the first, and that it was a transformation that preserved the logic being so displayed.

There are many transformations that apply generally to almost any display and they transform it into a new display, so that a even a series of such transformations might be applied, one after another, but since none changed the underlying logic being displayed but only the way it was displayed, the final result would still be a genuine display of the original logic.

I would imagine some of the transformations could be fairly remarkable. At this point, the god said to mention that some are "time-space transformations". The god said that the best way to imagine a whole bunch of these was to keep in mind that, mathematically, a two-dimensional display, such as you see on a computer screen, can be mathematically thought of as a three-dimensional object – two of the dimensions being the spatial dimensions of the height and width of the screen while the third dimension would be time. Now there are a whole bunch of ways that you can rotate a three-dimensional object and each of these is a transformation, most of which interchange time and space in various ways. All these would produce a new display of the same logic. I told the god that I was not completely sure but it seemed right. The god answered by placing the following thought in my mind. Suppose that the display shows a dot moving from left to fight. A time space transformation of that would show a display where there was nothing and then suddenly a whole line all at once appeared from left to right and then disappeared. That would be one time-space transformation of the moving dot display. This is only one of the unlimited number of transformations that could be applied.

I thought the god's point interesting and the god said that time was an important issue, especially in regard to something the god referred to as logics of evolution being "isomorphicly imaged in the brain". All I could do was shake my head. My own thought was on transformations of the sort that take a whole surface of points and transform them into another surface of points – well, lines, surfaces, volumes, whatever. One word relevant to such transformations is "Fourier" or "Fourier transforms". Other words and terms might be "orthonormal sets", "Legendre polynomials", "spherical harmonics", "Hermite polynomials", of which the Fourier transforms are often taken as the prototype (see for instance chapter 5 of Byron and Fuller *Mathematics of Classical and Quantum Physics*). These are connected with transformations that take one surface and transform it into another surface. Patterns, and hence logics, not visible on the first surface, may become visible in the transformed surface. Patterns, and hence logics, not visible to the human native perspective in the first display can become visible in the transformed display, and that is what device displays of logics are about: the device shows us logics that we are unable to see directly.

There was a sixteenth century Italian mathematician, Bonaventura Francesco Cavalieri, 1598 – 1647, who developed a "theory of indivisibles," presented in his *Geometria indivisibilis continuorum nova* of 1635 (endnote 10). The method seems pretty obvious now, but at the time lead to new, simple, and fast ways of computing the volumes of certain geometric objects. It was a primitive version of the calculus, which was not around then. The technique came to be known as Cavalieri's principle. It stated that two objects have the same volume if all the slices through one, parallel to some plane, had the same area as where the plane was slicing through the other object.

Although Cavalieri's principle is not quite analogous to the repeated transformations above, there are similarities. The sameness of a sequence of areas is similar in a general way to the sameness of the pure logic form or structure underlying the sequence of transformations. Also, just as the Cavalieri principle was an early aspect of a part of the calculus, so too the above sequencing of transformations might be an early aspect of eventually getting a more solid hold on sentience logics. So maybe we might refer to the technique of repeated transformations of a display as a *Cavalieri sequence*. A *Cavalieri sequence*, as used here, is a sequence of display transformations each of which preserves the underlying logic so that the final display, no matter how different it appears from the first, is still of the same logic.

We are Pure Logic, a Secular Miracle

We are not in this mucous-like matter of the brain. We are in the huge logic going on in that wet gray matter. Now logic is a … well, very abstract thing.

We, our being, is in pure logic. All of our innermost experiences, our innermost being and consciousness itself, is nothing but the most abstract of form, is logic, is pure logic, and in this case, this pure logic happens to be going on in regions of this moist gray matter. Our essence is in logic, which is implemented purely in forms of the dynamics of systems of matter. This is special, unusual, odd. And the more carefully one looks at it, the stranger it gets. It is a kind of secular miracle.

(The term "secular miracle" is used because this is a book from the perspective of science. A religious audience might be happy with the word "miracle" without the "secular". That we are purely in logic is a miracle. On the other hand, scientists, whether or not they are religious, stress the scientific perspective in their work, the perspective of "no gods," and the term "miracle" in a general way comes too close to gods. They might prefer the same sentiment be expressed thus: That we are in logic and that we are in the forms of dynamic matter is quite amazing. Yet, for the non-scientist, the term "amazing" misses the mark. The term "secular miracle" indicates that this really is a miracle, when you think about it, but it is a miracle of the fabric of the universe, and such it is whether or not one is referring to a specific religion.)

Close

Just as, say, an amoeba is a logic going on in a mass of matter, is a logic going on in a cloud of maybe a 100 trillion atoms, so too are the logics displayed by the devices in this chapter. These logics are also going on in matter, and the devices in this chapter make the logics visible to our native perspective. Just as the logic going on in the cloud of atoms is an amoeba, and just as the microscope device presents a visual version of that logic to our native perspective, a version that our native perspective is unable to see on its own, so too are the logics in this chapter going on in matter, and so too does the device in this chapter present visual versions of that logic to our native perspective, versions that our native perspective would not be able to see on its own.

21 Machines

Introduction

This chapter is about machines and their growth in the future. By the term "machine" we certainly include computers and computer-like devices. We look at future history, at what kind of machines and science will be used, and at how we will get there from the present day. This chapter is about machines helping us dig into sentience in order to understand and solve this ancient mystery.

Current Technology Primitive for our Purposes. PET Scans and so forth.

Modern technology such at CAT scans, PET scans, and MR have opened up seeing into the brain as never before. Even today these areas continue to advance. "CAT scan" stands for "Computed Axial Tomography" scan. This takes x-rays of many slices of a part of the body, from different angles, and a computer puts them together for a three-dimensional view. The CAT scan allows a researcher to look at the soft tissue, say of the brain, then at the bone or skull, and then at the blood vessels. All of this is in the CAT scan and can be displayed by the computer.

"PET scan" stands for "Positron Emission Tomography" scan. A positron is the antiparticle of an electron. Positrons are sent into the brain, or into whatever is being scanned, and eventually near an electron, they and the electron are annihilated. The annihilation produces two high-energy gamma-rays (photons) moving in opposite directions, which detectors spaced around the person record and figure out. Doing this from various angles, a three-dimensional picture is determined and built up in a computer.

The great advantage of the PET scan is that it allows for the determination of a range of metabolism functions. Thus, when the person is given some mental task, the PET scan shows the amount of activity in the various regions of the brain.

In PET scans, CAT scans, and MR, "tomography" refers generally to getting a two-dimensional image from one angle, then repeating this for many angles around the object, then using these to construct a three-dimensional perspective of what is being looked at. This requires a fair amount of computer processing based upon a fair amount of geometry related to tomographic issues.

These perspectives are marvels compared to what was available in the past. The computer can display various slices from the three-dimensional structure, or even structures within what is being looked at, and can optionally color-code all in various ways. Even so, as impressive as these are, we must remember that the picture is from an extremely big perspective.

The total representation inside the computer is composed of a huge number of points, but each of the most detailed possible points still blends together the activity of thousands of neurons, and possibly much more than that. In other words, the activity of thousands and thousands of neurons shows up only as the smallest possible point that you could see on such a computer display. A lot of logic can go on in a group of thousands of neurons. None of that will be shown.

Consider an ant. If you want, you may even consider an ant of one of the smaller species. They are quite small indeed. And then consider all the activities which that ant can perform. It can detect various aromas, and it can then shift its behavior into motion toward the aroma, and then at the object of aroma it checks again with direct touch with its antennae. The logics in the nervous system then initiate actions to determine if this is food. All this is in the neurons of this tiny thing. How small are neurons and how much can go on in a so small a place. If the logic that determines whether this is food so determines it, the ant goes into a variety of exploratory efforts to see if the food can be moved, broken into pieces, or absorbed. Further, such logic contains smaller pieces of logic which determine if the ant has tried enough to move the object. And there are other behaviors related to struggling to get food back to the nest, as well as for finding the way back to the nest. (There are some ants that use the polarizing angle of sunlight combined with the direction in which they are walking to constantly maintain a mental map of where they are in relation to the nest. Thus they can head back in a direct line to the nest, no matter how much they have roamed in zigzag directions since leaving it – Nicholls et al *From Neuron to Brain, Fourth Edition*, 2001, page 308.)

There is logic that determines what happens when it encounters another ant, whether to fight or to exchange information with the antennae. And there are logics that carry out activities from rolling or carrying little specks of dirt into mounds, feeding babies inside the nest, and so on. Darwin remarked on all the capabilities in such a tiny creature . So much in so small a space.

How big would the ant's nervous system be in a PET scan. Like us, maybe nine tenths of the inside of the ant is used for other than neurons, used for muscles, esophagus, stomach, heart, and other organs. Would its whole nervous system even fill the smallest point on the display of a PET scan? Yet, consider all the logics that go on in what shows up as the smallest dot on the computer display.

It will be a long time before our machines allows us to easily see the detailed logic going on in an ant. It may be centuries to see such in the massive human brain.

By analogy we might consider Craig Ventner's work creating algorithms and computer programs to finally analyze human DNA. Vast powerful computers and algorithms were needed to determine the 3.2 billion letters in the human DNA (alternately called the human genome, human chromosomes, or blueprint for the human). Yet, where there are 3.2 billion letters in the DNA, there are 100 billion neurons in the brain, neurons that one day will need to be all tracked and analyzed. Where there are only four letters total in the whole alphabet used in DNA, there are as many kinds of neurons as there might be configurations of branches on large trees. Our future machines must analyze all this. Where the letters of DNA are in effect one huge long string of letters, the 100 billion neurons are not in such a simple relation to each other. Instead, neurons interconnect to many other neurons in a complex three-dimensional space. Future computers must deal with this. Where a letter is just a letter, the state of neuron may undergo some shifting, which will affect how it responds to incoming signals. Future computers must handle this too.

The problem of figuring out even the lowest level of the brain in a more detailed manner is greater by a number of orders of magnitude than figuring out the human genome. And so will be the required computers, their algorithms, their data bases, and the like.

Phases and Exponential Proof

Let us sketch the future phases in the development of our machines, from now to the point where the mystery of sentience is finally solved.

Phase 1. Sensing and monitoring

In this phase of history, scientists and technicians will develop ways to place, in effect, sensors inside the brain. To really have a basis to understand in detail what is going on in the brain, eventually such sensors must be placed at almost every neuron. Even though this is utterly impossible with today's technology, researchers will start out small, and over years and decades and centuries they will extend their capabilities.

One scenario for doing this might be nanotechnology (for instance, see Scientific American, September, 2001). This is the science, technology, and development of extremely small devices. The science is new and as time goes on, the devices will get smaller, and of more kinds and more intricacies. Eventually such devices may be produced in huge numbers, might eventually be as small as huge molecules, and might be used in combination with other techniques to see into the brain.

Another scenario might be the development of a kind of small molecule that would be injected in the blood or simply added to the food of the creature. To get into the workings of the brain, these molecules would have to be small to get past the blood-brain barrier (Nicholls et al *From Neuron to Brain, Fourth Edition*, 2001, pages 150-153). The situation might be adjusted so that only a few molecules would be incorporated into any neuron. Somehow, individual molecules, or any pair of molecules, might have some unique aspect. Further, we might suppose that with a certain kind of probing by some kind of electromagnetic waves, each such molecule echoed both its unique identification as well as a measure of the electrical neuron current going passed it at the time. Such a mechanism could form the basis for a technology which with ever greater advances could report on the electrical signals in the brain, possibly using a complex relaying system by larger molecules moving along with the blood.

All these directions might well be combined with future developments of the current PET scan, CAT scan, and ER technologies. And people will think of other methods too. Human ingenuity never ceases.

Much needs to be accomplished. We are looking over major spans of time for phase 1. Early steps might be injections of small amounts of fluid into brains of animals, perhaps very simple animals, and at this point science would focus just on getting the information from sensors in the brain to detectors outside. Further early steps would be a microscopic determination of when and where in the brain the signals originated.

Ultimately, to have a really good, solid vision into the brain, we need to detect the firings of every neuron. Can this ever be achieved?

Exponential proof for all phases

Over time, every five, ten or so years, there would be some improvements in the techniques for this first phase, or indeed, for any of the phases we will describe. The improvements will be of various kinds: in our understanding of various issues, in the algorithms our computers use, and in all kinds of aspects used in the process.

Can full monitoring of the whole human brain, down to the level of neurons, be done within seven hundred years, even though there are perhaps a hundred thousand million such nerve cells? We will argue yes, because technological progress is, on average, exponential, if not faster.

If we start to successfully monitor a small section of the human brain with injected molecular sensors in say the year 2030, then probably in ten years, 2040, the monitoring capability will be double. And in a further ten years, a decade, in 2050, it may well be double again. After every ten years, it more or less doubles.

Let us follow how this doubling every ten years would work.

In the year 2040, we have double what we had in 2030.

In the year 2050, we have double even that, which is 4 times what we had in 2030.

In the year 2060, we have double that, which is 8 times what we had in 2030.

In the year 2070, we have double that, which is 16 times what we had in 2030.

In the year 2080, we have double that, which is 32 times what we had in 2030.

In the year 2090, we have double that, which is 64 times what we had in 2030.

In the year 2100, we have double that, which is 128 times what we had in 2030.

If one continues this sequence to the year 2200, it will be 131,072 times what we had in 2030. In the year 2300, it will be about 134 million times more than in 2030!

In 2400, this goes to 137 billion times more.

And in 2500, it is 140 trillion times the capability that we had in 2030! Since the brain has only about a hundred billion neurons, this will more than do the job of full monitoring of every neuron in the brain.

Is this estimate into the future realistic? Again, surely, ever being able to monitor over a hundred billion seems not possible. But the above increase in numbers is the nature of exponential growth (as well as, indeed of compound interest) over long enough times of centuries. Perhaps not every ten years will the neuron monitoring become twice as good. Some decades, it may almost stay the same; other decades it may double twice. In a span of centuries we may also expect longer periods of not much growth, but just as surely there will be periods of greater growth. Further, every now and then, every fifty or a hundred years, we can expect some great breakthrough, which will expand our ability to monitor the neurons in the brain by 10 fold or more. So yes, this computation is approximately correct, within a rough estimate. If we had almost miraculous luck, we might monitor the whole brain by 2090, though more likely by around 2500 AD, and with much greater sureness at least by 2700 or 2800 AD.

Exactly the same sort of reasoning applies to each of the next four phases too. On average, every certain period of time, there is doubling of what can be done. Sometimes the doubling will occur more often, sometimes less, sometimes there will be breakthroughs that jump progress quite a bit. Sometimes there may be difficult periods on earth where progress will be almost non-existent for a few decades. But, on average, we have the doubling phenomenon described above. For instance, after 50 doubling periods, we have an improvement of about 1000 trillion times.

So it would seem that all five of the phases that we describe could be completed by the year 3000. It would certainly seem they could not be completely any sooner than 2100. When it happens, we will have a detailed understanding of the logics (meanings) going on in the brain. This will include a resolution of the mystery of sentience.

Phase 2. Elementary recording of all data

In the second phase, machines will help us further. The first phase was to monitor more and more the firing of individual neurons. In the second phase, the machines will keep track of generally where neurons are in the brain, what neurons fired in what circumstances, and the machines will also assist the researcher search the thousands of millions of neurons for certain kinds of patterns that researchers may want to investigate for whatever reason, and the machines will also allow the researchers to keep track of researchers' developing ideas for various neurons and areas of neurons, from the microscopic to the non-microscopic. This will be no small task, considering the hundred thousand million neurons in the brain, each with perhaps tens or hundreds or in some cases even thousands of substantial connections to other neurons in various directions. Eventually, as part of all other work, data records and their relations and circumstances must be kept and organized in a way to be quickly accessible, for all of the neurons in that crammed, magnificent tropical forest that is the brain. To have machines (computers) large enough, capable enough, is a dismaying task. But broken into enough pieces, and given enough time, it will come about. Earlier, we went through an "exponential growth" argument as to why such amazing technology will be created; for whether it is in one century or ten, it will come about.

Phase 3. Machines determine basic characteristics of the structure of logics

In the previous phase human researchers and thinkers will have rapid access to these records of combinations and relations and various situations. More rapidly than ever before, *researchers* will be able to test hypotheses about what kind of logic might be going on in say brain area HO89.ab-387. Research will be faster than ever before. But with a hundred thousand million neurons, the number of logics going, and their structure, will be beyond our capacity even if every human on earth were put to work as a researcher.

The machines will start to do more of what the researchers were doing. At first, in only simple ways will researchers begin to think of whole structural categories of some aspects of logic, and then they will be able to turn the machines toward searching the vast jungle for any of these aspects, and having the machines record the results as to where the logics might be going on. This will be phase 3.

In phase 3 computers start to carry out elementary analysis of all this data, of all these signals, analysis leading to the beginning of basic characteristics of almost all the awesome number of logics going on. Over time, from experience with getting the machines to do these tasks, humans will figure out ever more advanced activities for the machines to check, and humans will also figure out ever more advanced theories of what it is that needs to be checked. These theories will be implemented in the machines.

This third phase will start soon after the second phase of recording data becomes substantial. Thereafter all three phases will develop naturally synergistically. Each advance in sensing technology will produce new levels of incoming data, and that will necessitate more work in the second phase to figure out how to record all this new detail of data, which in turn will advance the third phase of the machines figuring out and recording the basics of the structure of logics.

Phase 4. Machines analyze lots of the logic in the brain

In phase 4, our machines are going to analyze lots of parts of the brain. And not just simple analysis. We have to deduce the logics going on – logics related to non-sentience and sentience. Society will no doubt have interest in the astounding algorithms discovered to be going on in the brain, and will start to use them in all kinds of applied and theoretical fields. But those algorithms will also be part of a foundation for the sentience logics. All these directions of research will all come together in various ways the details of which we cannot currently foresee. The algorithms will be beautiful, of quite remarkable power, and built into the brain by millions, sometime by hundreds of millions of years of evolution (a million is a thousand millennia).

But it seems naive to think that researchers will be able to complete their job with only this "shallow" assist from machines. Much of the logic in the brain surely is deep, and deep logic takes much more time for a researcher to analyze. Again, considering the vastness of the big hum, there simply are not enough researchers to spend a hundred thousand years carrying out the task of working on say each group of a few million neurons (There could be about 20,000 groups each of 5 million neurons – and 5 million neurons can develop some pretty sophisticated logics).

The machines must be programmed with the abilities of researchers tyring to figure out what logics are going on in a neural system. This is the beginning of phase 5.

Phase 5. Machines figuring out and understanding for us

In this phase, we, and our machines, will go even further. The machines, our assistants, will themselves start to come up with new concepts and understanding.

We will start to develop concrete theories of what is the nature of explanation and understanding. Simultaneously, from our investigation of the brain, we will get insights and information and algorithms as to how the brain handles explanation and understanding and concepts and how it comes up with new explanations and new understanding and new concepts. Then we can put these algorithms in our machines, at first in simple experimental ways and then in ever more developed forms.

Humans will do this to extend understanding of many things, not just related to those of the brain and sentience. But for our goal, it is so that the machines can assist in understanding of sentience.

This may be required, for perhaps no system, human or otherwise, can fully understand itself. Perhaps humans simply cannot fully understand their own sentience logic. It may not be a matter of time or effort. It may be that we do not have the mental equipment to fully understand our mental equipment. Perhaps there is a characteristic of computational complexity to the effect that to understand a system of complexity level x requires a system of level x+2. If so, then the machines will be required to go far deeper than possibly we ever will be able to.

That might be important, for our inability to look into brains to the required depth may be one of the keys to our inability to understand sentience.

Does all this seem far off in the future? It is. We are talking of likely hundreds of years up to a thousand years. We predict that the mystery of sentience will be scientifically completely resolved within a thousand years.

The Unfolding Development

How will all this unfold into the future? The phases above merely describe functional divisions of human machine activity. But how might our other efforts unfold as we resolve this great mystery?

Even today, in the lowest level visual processes of frogs, cats, and dogs, we have quite a bit of the basic logic that goes on. We are starting to get some other areas of logic, for instance in speech recognition in humans. What about the larger future picture?

Figuring out logics

One of the earlier tasks of researchers, and then of machines, will be to get a fair idea of what groups of nerve cells there are and which groups are basically connected to which.

Researchers will work on getting an idea in a general way of the amount of connection between the connected groups, somewhat like the connections between various groups of cells in the electric fish for the JAR avoidance behavior. At the same time, we need to slowly unravel what each cell in a group is doing in regards to the cells it is connected to, in regard to logic events elsewhere in the brain (where there is a correlation), and with respect to events in the environment around the individual. All cells are dependent on the cells they're connected to, but we do not always need to understand how those connections affect the given cell in order to understand the logic at the given cell. In fact, in our earlier examination of the electric fish, and its JAR, it seems that much logic of cells may have been figured out by looking not at adjacent cell behavior but at the environment; for instance cells in certain areas of certain structures of the brain were computing phase differentials between various pairs of points on the fish's skin, others amplitude differential between pairs, and others the integration for various pairs seemed to be taking place. Researchers determined the behavior of these cells not so much by looking at the behavior of connecting cells as by looking at the environment taking place outside the fish during various cell signaling. Nonetheless, there were other cells where the behavior was deduced from the cells particular role in a small group of cells. The cells that negated the signal of another cell is such a case. Here the researcher determined the behavior primarily from adjacent cells.

Researchers will use these ideas, and many others still un-thought of, to analyze the hundred thousand million or so neurons in the human brain. Researchers will get machines to use these ideas too.

Homologous structures and logic, other species, will help

Researchers will not only be looking at brains of humans, but of all kinds of creatures, for much understanding is gained by analogy. In the field of biology, *homologous structures* are surprisingly similar structures that occur across species, and such also occurs across brain structures of different animals. These structures assist experts with insights and global understanding across a range of species.

Far in the future we can expect that once a new piece of information is known about a homologous structure, with a certain degree of sureness, that information will automatically go out to the thousands of computers around the world which are doing work that can benefit from that information, and the new information will automatically be incorporated in the analyses and searching of each computer.

Such rapid, ongoing, synergistic update between computers all around the world will be quite more extensive than in our times.

Future computers of this sort will be more like neural chips

As implied throughout the above, future computers will be based on neural chips. Long before they understand all the details of all the brain, scientists will have a fair amount of understanding what is going on in the brain regarding specific aspects. They will be putting these logics in the brain onto neural chips. Of course at first the logic going on these chips will be only primitive compared to the human brain. But as stated, over time it will get better.

As they discover how the brain does things, researchers will naturally turn their attention to whether there are better ways. For instance, mentioned toward the end of this chapter is an item that possibly we, our human brain, can only keep 5 plus or minus 3 things in our awareness at once. Any more, and we forget one of them. When future researchers discover the relevant algorithms in our brain, and when they move these to neural chips, they would certainly consider extending this 5 plus or minus 3 to something larger, perhaps 7 plus or minus 3 (if it seems possible, for there may be inherent combinatoric limits). This would be done even though the first logics put on the chip were a simple attempt to duplicate what went on in the brain. And the same with concepts, once we see what looks like the handling of concepts in the brains of humans and other species. Even though our knowledge of these things will be relatively primitive at first, over time we will see certain ways that the handling of concepts could be made stronger than what goes on in humans, and the researchers will explore putting that too in the logic on neural chips. Even as today our machines that do physical labor can surpass human strength in many situations, so too in the future, eventually these neural chip machines will surpass human strength of thinking in many situations. And just as with machines that do physical labor, where over time they are able to do more and more things that humans used to do, yet still there are situations where they cannot do what humans do, so too will it be with these neural net machines.

Three contributors

There are three main contributors to the overall task of understanding sentience. Humans (naturally), machines (obviously), and the brain itself (surprisingly) as a golden repository of information. When we move into the more advanced states of the task, we will find in the brain all those incredible algorithms that make human ability to analyze the world so magnificent, as shown by the accomplishments of our scientists, experts, artists, and thinkers. In this sense the brain itself, as an object of study, will contribute these advanced algorithms to us, algorithms which we would never figure out on our own, and which we will need to assist our more advanced thinking. As we start to put those discovered algorithms on our computers, we will crystallize them a bit, massage them, and in those ways where it can be done, extend their ability, sometimes vastly so. Much farther down the road, we may need the computers to do some pretty deep analysis as to what is going on with this thing we call sentience logic.

Self-reporting

Somewhere in the future, for a century or two, self-reporting will be an indispensable.

Until we know an immense amount more, and until our technology is incredibly more advanced, we will need self-reporting because at present we ourselves are the only ones who can say what we are seeing or hearing or feeling, particularly if we are self-reporting after some structure or process or signaling has been temporarily altered in our brain.

After some technological alteration of signals in our brain, we might say, "this color, I simply have never seen anything like it. It's definitely a color though. Not like that last thing that wasn't quite a color. This is definitely a color. Frankly, I have to say I horribly dislike the color."

Or. "Oh wow. No, I know it's a circle. But yet, it looks like a triangle." The experimenter asks questions but gets nowhere as to what such a statement means. "I can't explain it. I know it's a circle, but it looks like a triangle."

Or. "No, this is not a color. Hah hah. No. It's a shape. But I simply can't 'see' it that well. I know 'see' isn't the right word, because it's all in my mind. But it's sort of like seeing, except I can't see it."

Or. "Astounding! Wow!" We shake our head repeatedly. "Oh Wow! I can't describe it. No, it's a shape. But, but ... oh Wow." The researcher asks, "does it have corners?" We pause, then reply, "corners?, hmm, let me, let me think." The researcher asks, "What do you mean 'think'? Does it have corners or is it like a smooth bending shape." We are both intense and confused. And after a while, we can only shake our head, and more than that, we go back to repeatedly saying, "Oh Wow," not caring that the researcher is even there.

Other specific brain signal alterations will cause the subject to become happy, or miserable, or dangerous, or thoughtful.

Certainly not right away, but eventually technology will become sufficiently able to see into and understand parts of the brain so that researchers will be able to read "internal speech" (Internal speech is when we partly phrase our thoughts to ourselves, though we do not say anything out loud; we just think words. Internal speech is more elliptic than imagined, but even so, at some distant time in the future, researchers will begin to be able to access internal speech, and they will do so better and better over time.) This internal speech, automatically recorded by machines, will be an incredibly valuable tool to researchers. It will be like an ongoing self-reporting during waking hours. And these telegraphically, internally speechified thoughts, informationally recorded, will present a great deal of information about our sentience, when correlated with other signaling throughout the brain. It will be an extremely valuable source of information for the researchers' great task of explaining sentience in terms of the full logic that is going on in the dynamics of electrical signals.

As presented here, in this story of unfolding future development, it might seem as if the logics and the correspondence to self-reporting are relatively simple (as long as one ignores the prodigious technology to see into the brain). But this is not so. There are the hundred thousand million neurons in the human brain. This is the brain that has produced every work of literature, art, science, construction, from those of Leonardo daVinci, to Michael Faraday, to Isaac Newton, to Shakespeare, to Eifel, and while all these are exceedingly deep, even the simplest of our own vision, whenever our eyes are open, is sending information continuously on two million channels into the big hum. Some of these logics will be in the mystery of sentience itself, including exactly the logic that is the experience itself when a human sees the color red. So deep may be these things, that only our machines will be able to understand them.

(Technically speaking, self-reporting is not needed to solve the sentience problem. With incredible additional effort we could work out the structure of logics without such help. But this might add two to twenty centuries to the task. Clearly researchers will use self-reporting, especially because it will give us so many enjoyable insights during the early phases.

On the other hand, it might turn out that the combinatorics of the logics both in the brain and as they extend through evolutionary time outside the brain are such that it is inherently impossible to directly deduce the structure of the logics, no matter how much machine power is applied; while self-reporting might be able to leapfrog across these combinatoric barriers.)

What Understanding is, and Future Machines.

When the light goes on, when we say, "Oh, now I understand," what is it that has happened inside our head? It means the pieces have fallen in place. And just what are those pieces and in what way are they now "in place"? To start with, the pieces are some kind of subsystems in our thought processes, with the subsystems having relations of various kinds to each other.

There are actual and potential relations between these pieces, and there are also various kinds of actual and potential relations to other pieces that the logics in the brain are categorizing as already understood for purposes of the current task.

To understand something, just means that the brain has to find a set of relations all of the right sort; it looks for these from among potential relations. Today researchers cannot say what is the nature of those relations in our head. Nevertheless our brain in fact does work with and recognize the right kind, and when all the relations are of the right sort, then that is when the feeling of "ah, now I see" is generated and is sent into our consciousness logic. Which, by the way, indicates that understanding is one of the feelings that our awareness can have. In the case of this feeling, there also enters into awareness, knowledge of some kind of overall framework of the essence of the "understanding," and, in the case of suddenly understanding something that we have not been able to for quite a while, there further enters into consciousness a reference to the piece of the puzzle that finally made the whole thing fall into place.

(It is likely that logicians could, or possibly in some cases already have, constructed formal systems to partially capture – i.e. to partially express in a formal language – these relations and subsystems. It is also likely that the complexity and richness of the relations in our head far surpass such formal relations. Furthermore, the methods our brain uses to try to come up with the right kind of relations for all the subsystems to fit together will of course be awesome. And some day our researchers will be able to see enough into the brain to be able to start seeing what these relations are and what techniques – i.e. algorithms – the brain uses to try and make all the pieces fall in place. Then the researchers will start to place those techniques – algorithms – in our machines.)

Whereas logicians may create formal systems in which something is either totally understood or totally not understood, with no possibilities in between, the brain has a whole range of kinds and degrees of understanding. Indeed, humans in different activities in our social system use the word "understand" differently. I believe that Alfred North Whitehead said something to the effect that we advance by how many operations in our mind we can manipulate without understanding them.

Probably the basic aspect in the brain regarding the logic of understanding is that understanding is relative to a "ground" set of concepts or subsystems which we may assume as given for the kind of task at hand. The rigorous scientist or philosopher or theologian or researcher will have a much more insistent categorizer for what may be assumed as given, but we all have our rough ideas of what, in various situations, is taken as already known.

When we have the feeling that we have understood something, it means that processes in the brain have reduced it to this "ground", even though the brain shifts that ground depending on the area of thought.

In the far future our machines will increasingly perform these processes of understanding, eventually far surpassing us in some ways. They may even give us concepts and ideas which we are unable to fully grasp, but using them we will be able to see far indeed. And we can do and enjoy this, greatly too.

We might worry about such machines giving us information and telling us what is the truth. But it will be no different than today when experts give us information or tell us what is truth. Today, science, technology, and all areas of expertise have become so extensive that we rarely fully understand any of the things told us. We could not understand it even if we wanted to, unless, first, we spent years of studying background ideas, and second, we obtained detailed results and proofs from the humans who made the statements. Today even scientists in one area of science often cannot fully judge statements made in a nearby area of science. It will be the same in the future with machines. The only difference is that it will be humans *and* machines that will be the gateway to pronouncements about reality. For most people this will feel no different than today, and even for most (human) scientists it will be no different. They will not be able see the full details of why something is true. In all these cases, truth boils down to judging the validity of a *source* of information.

Will we be able to trust machines as a source of statements about reality? Yes. In fact, we will be able to trust them more than we can trust the human experts of today. Statements that such machine experts make will be provable, just as is a theorem in geometry in high school math, or in other areas of math. The future machine will produce a proof of any statements that it makes. And any humans who want to, can check the proof, either in whole, or randomly select areas, and check that each step in that area follows legally from some earlier steps. More importantly, there will exist any number of independently written computer programs that will automatically check that each step in the proof is valid. So the statements produced by these future machines will be quite certain, indeed, quite more certain than the statements made today by scientists and experts.

These are important issues because some of these proofs may not be understandable, in the large, to humans. Further, they might have hundreds of thousands of steps and could never be fully checked by one human. Yet in all cases, the proofs can be ascertained as correct, and with a greater level of confidence than can a few hundred-step proof today. However, the gist of the proof, the idea behind the proof, may simply be inherently beyond human understanding.

One final comment on these future proofs is in order. To some degree there are different kinds of proof. This is true even in mathematics, where for instance, from the perspective of Brouwer's intuitionism, only very "constructivist" proofs are allowed (and one doesn't even have to go all the way to intuitionism in order to become involved with kinds of proof, but can simply question certain kinds of proof by contradiction, which easily fall into using non-existent objects). When we move from mathematics to science, the kinds of proofs no doubt get even larger, where for instance one kind of proof involves "thought experiments," while another involves them not at all. And as we move through various areas of human investigation, there are likely many other categories or classifications of kinds of proof. The above process of computer generation of proofs, as well as independent computer checking of proofs, will flag proofs as to what category they are. The certainty of such flagging we have not at all today. In fact, *all our past* scientific concepts and ideas will be carefully analyzed by these advanced machines, which will check and classify them much more fully than ever we did or today can.

Could we someday have neural net machines work to understand things? Could these machines accomplish understanding and then give that to us? Yes. How?

Some day far in the future, a group of researchers will say what they want the neural net computer to do. In the above case, it would be to accomplish understanding as that process is described above. So this group of researchers will say this is what we want from the computer in terms of the relations we are interested in and the "sort of"s involved and what the computer is to do with these. They will know nothing about how in the world you could get computers to mimic such relations and so on. They just postulate what they want. They will know almost nothing about the remarkable mathematical algorithms that can be made to achieve sometimes what seems like the impossible in terms of combinatorics.

As for the above phrase "sort of", it can correspond to a loose similarity of mathematical logical structure with the indicated area, and there are algorithms to work with this aspect too; also, over time the *researchers* might be able to flesh in the "sort of"'s, and that too would be added to the statement of the given that they want the computer to mimic, even though, again, they have not the foggiest how the computer could accomplish any of this.

Then another group of researchers enters the picture. They receive this description of what the first group wants, what the first group postulates as a given, the relations and the like the computer is to mimic. This new group's skill is in coming up with amazing combinatoric algorithms, for instance dynamic programming algorithms. These people also have an elementary knowledge of how relations and the like are to be mimicked on the computer. This group comes up with a set of algorithms that will get the neural net computer to accomplish what the first group wants. (Nature, through evolution, must already have elements of such amazing combinatoric algorithms in the brain. The fact that significant parts of the inner most constituents of the sentience logic must surely also utilize such algorithms, and hence that it naively seems that some of our awareness is "in" such which seem so logically distant from what they are accomplishing, would seem to be further evidence for the substantiality of logic.)

Finally, just as today, a third group of experts will make sure that the whole thing comes together and runs in the right way on the neural net computer. Whether this third group will keep their name they have today, "computer programmers", is not known.

Example Capabilities of the Brain

The remainder of this chapter describes some of the amazing processes that go on in the brain. This material, in detail, is not needed for the rest of the book. However it does show in a general way just how surprising is the logical depth that goes on in the gray mass between our ears. As if without effort, the results of these processes appear in our sentience. So we presume there is no effort involved.

Some day, when scientists begin to see how the logical processes described below are carried out, they can move those too partially into our machines. Over time they, the scientists and the machines, will see better and understand more. Thus will develop a synergism between the machines seeing further, leading to more powerful algorithms being put in the machines, so that they see even better, and so on.

The processes presented below, though outside consciousness, compute results that are sent immediately into consciousness; except for the section on biting your tongue.

Quantificational issues/every cat gets a saucer of water

A significant part of our analytical thinking has to do with what might be called quantificational issues.

Dear reader, please read the following paragraph.

"Jerrold, who lives down the street, has quite a few cats in his house, because he loves them so much. What's interesting is that twice a day, because he loves them, he puts a separate saucer of water, on the living room floor, for each cat. 'How do the cats know which is their saucer?' you might wonder. Well, they don't. But Jerrold puts a separate saucer for each cat anyway."

Have you read the paragraph? You would be surprised at the depth of logic that just went on in your brain – captured in an astounding number of signals flying around carrying on that logic. We will look at just one piece of logic, and that will be the meaning of the phrase, "a saucer for each cat." Now this phrase sounds deceptively simple, but that is only because the power of your brain has done so much work that is outside your consciousness or awareness or I-ness. In your consciousness, there has been no awareness of this huge work.

So what about this seemingly simple phrase, "a saucer for each cat"? You look at the phrase, and say, "OK, each of his cats has a saucer of water. That's what's in my mind, a *picture*, with each cat having a saucer of water." I ask you, "please tell me how many cats are there?" You respond, "you didn't say how many cats there are." I say, "then how can you picture each cat with a saucer of water? If you don't know the number of cats, there's no place in your brain where you are picturing the cats and each with a sauce of water in front of it. If you don't know the number of cats, you simply cannot picture them."

What your brain constructed as it read the paragraph about Jerrold was something more a like a generic cat, or a generic set of cats, and your brain also constructed some kind of logic which captured the idea that the generic cat is associated with a generic saucer. In other words, what was constructed, somewhere in your brain, a few seconds ago, was something considerably more abstract than a simple, direct picture of some cats; for in this picture there were generic aspects, and all of that was constructed and remembered, in logics that took place somewhere between your two ears a few seconds ago. In some form it still resides there (though it is now cluttered with the several additional paragraphs since then, paragraphs trying to analyze what went on in your brain when you read the Jerrold paragraph – and the meaning of these additional paragraphs is also in your brain, in still other logics).

Returning to the original paragraph about Jerrold, the point is the meaning there is not at all simple and concrete. There are some surprisingly deep aspects of a *quantificational* nature. But as you raced over the paragraph, you had not the slightest idea of what advanced processes really did flow through your brain, with speed, to build the meaning of that paragraph. This is an example of the remarkable things, the wonderful power of your brain at this very second, and of the concrete logic, about abstract thoughts, going on right now in various places inside your head – such an intimate part of you.

Before moving on to the next example, let us note two important points.

First of all, there is sophisticated depth of analytical logic and concept going on inside our brain, but outside our consciousness. Indeed, we simply were not aware of the abstractness of a seemingly simple phrase like "a saucer for each cat," until it was brought to our attention.

The other point returns to the subject of this section, that the machines of technology will eventually be able to see into the brain. Someday, with the help of millions of nanotechnology molecular sensors floating through the brain, researches will actually be able to watch the regions in your brain where this quantificational meaning of "a saucer for each cat" is created, with the sophisticated labelings of something like "generic cat", "generic saucer" or some such thing. And indeed, the "some such thing" is what will be so valuable. We will see how the brain handles such sophisticated generic issues of meaning. And we all also see how the brain is able to quickly draw conclusions about Jerrold's cats, even though it has no idea of the actual number of cats. Researchers will see the logic the brain uses for grasping deductions about this sort of thing.

Also importantly, they will be able to take that logic and put it in electronic chips for neural nets. They will be able to duplicate these wonderful kinds of logic that the brain uses for these sophisticated concepts and for the sophisticated deductions the brain does on this kind of information. So we humans can start to have neural net chips that are carrying on this kind of processing the *way* the brain does.

And this is just one of the kinds of things researchers will discover in the brain. Just one among thousands and thousands of techniques. The investigation of the brain will be a diamond mine that we will have never seen the likes of before, and we will never see the likes of again, in terms of quantity and sophistication of logics going on.

(Computer programmers, computer scientists, and even cognitive scientists, use the term "algorithm" as a word that roughly matches up the logic itself of the going on of the piece of the logic. They use terms such as "the representation" to indicate something like the logic of the meaning of say "a saucer for each cat." Using these words, the above would be rephrased as follows. The investigation of the brain will be a diamond mine of representations and algorithms, the likes of which, in terms of beauty and depth and sophistication and power, we will never have seen before and we will never see again. There is so much in the brain that it will dwarf the algorithms and representations that computer people have come up with so far. Likewise, after we have got these algorithms and representations from the brain, we will never again, in all time, come across such a quantity of new material. Even if in some far future date we come across creatures on another world who are vastly advanced compared to us, and they give to our researchers what all the logic in their brains does (more likely their machines would give this to our machines), that new information would not compare to the first diamond mine.

Quantificational Issues/"close file"

Here is another example of quantificational meaning, a situation that takes place outside our consciousness, but where clearly some logic of a deeper sort is going on.

Suppose that a person is learning to use one of the standard word processing programs, and that whenever they are done editing a file, they hit the "file exit" and the program closes down (possibly after asking them if they want to save the file). Let us suppose the person has used the program for some time and has always used it in this manner.

One day a friend is visiting, and both happen to be in front of the machine, which has the word processing program running on it. They are exploring the menu, and come across the item "file close".

"What is this 'file close'?"

"That closes the file."

"I thought that's what 'file exit' did. I've always used 'file exit'. When I'm done editing a file, I just hit 'file exit'."

"'File close" is for when you have more than one file you're working on."

"*More* than one file? What do you mean. I edit the file I'm working on, and when I'm done I exit."

"Ah, but if you wanted to, you could be working on more than one file, all at the same time, all in one program, but jumping back and forth between the files whenever you wanted to, jumping back and forth to wherever you left off in the last file."

"Oh?" the person says with a look of thoughtfulness spreading across their eyes. They never knew that they could be editing more than one file at once.

At this point, let us look into the brain of this person who just said "oh?". Always up till now, somewhere in that big hum between the person's ears, was a collection of neurons in which was going on the logic that indicated this program was for editing a file. During the time the person said, "oh?", there was significant shifting in the logic, shifting to a more sophisticated quantificational meaning, where now, the program dealt with a bunch of files. How many? Unknown, in some sense maybe vaguely pictured, or presumed to be pictured, an indefinite number of files, and something equivalent to a generic file with the last position that was being edited; and also a logic has been added that as one jumped to any of these "generic" files, the word processor went to this "last position" associated with this "generic" file. Since in the generic situation the person does not know the exact number of files, all this is created in their head in a generic form. Someday we will look into the brain and see this all this shifting as the person says, "oh?"

As for the thoughtful look that spread across the person's face when they discovered that they could work with several files and what that meant, that particular facial expression is generated by the human brain when there is a significant and "interesting shift" in a logic, which is what happened in this case. Yet the facial expression lasts for a second or two even though the brain had already determined at the start of the expression that an "interesting shift" occurred. Why does the seemingly vacant expression continue for a second or two after that? One might be struck by the vacant expression being the same as if some of the deeper visual processing areas were temporarily pulled away from regular vision in order to take a look at this "interesting shift" in logic. Remember, none of this has to be in our consciousness. However, one part that is in consciousness is that unique feeling, the experience itself, that we have when such a blank look has spread across our face. That feeling was generated by a part of the brain (and future researches will see also these logics in the brain) and "sent" to consciousness as an experience; its function is to mark time in our awareness because some "interesting stuff" is going on, so "wait a second." (Can all this be pictured in this way, by a device displaying logics, as logics "entering" "consciousness" and so on? If the display device is set up to display the logics in this way, then, yes. If the display device is set up to display the logics in some other way, then, no, it will not be displayed in this way.)

Once again, we may look into the future when researchers, using an injection of sensors, or using some advanced technology, will be able to look into the brain where this is happening, and see the before logic, when the person thought that the program worked with one file at a time, and the after logic, when the person found out that the program worked with any number of files at once. As in the earlier example, the researchers will see the kind of logic that is for the simple case, and the kind of logic the brain used for the more complicated, quantificational case. Indeed, the researchers also will see the logic used to switch from the simpler logic to the quantificational one. And as before, they can use this information to program electronic neural chips so as to be able to handle such issues of quantificational meaning and reasoning, according to the same logic used in the human brain. So they will have discovered another gem in the diamond mine.

Keep in mind that these logics, whatever their logical closeness to each other, could potentially be scattered in pieces throughout the brain, as discussed in the section, "Structure of Logics in the Fish's Brain". Yet devices that display the logics could show them in anyway that is appropriate.

Propositional Attitudes

Of course there are many things that go on in the brain that have nothing to do with quantificational meanings, though like such, they are more easily talked about through the device of language.

There are a range of "states" of sentience that philosophers call *propositional attitudes*. Here we are interested in them not for their being "states" of sentience, but for their showing how much the brain does. (For an overview of propositional attitude states, and the surrounding issues, see Lowe's *An Introduction to the Philosophy of Mind*, 2000, starting page 40.)

Consider for instance the statement: Ranor fears that it will drizzle tomorrow. Ranor is said to be in a state of fearing. Fearing what? Fearing that it will drizzle tomorrow. Ranor has the *attitude* of fear toward the *proposition* that it will drizzle tomorrow. Ranor could have other attitudes; for instance, Ranor might hope that it will drizzle tomorrow, or desire, or believe, or even conceivably intend that it will drizzle tomorrow. And it is not just that it will drizzle tomorrow, for Ranor, a human sentience being, could have all kinds of attitudes toward all kinds of propositions. Ranor might desire, fear, hope, believe, intend that the bill be paid on time, that the light be fixed, that there is an umbrella in the trunk, that the dinner is good, and so on.

Clearly there is something going on in the logic of sentience in each of these cases that is characteristic for that attitude and proposition. Again, someday, researchers *will* be able to see exactly the brain logic that is going on that characterizes each of these propositional attitudes. As usual, since we currently have so little idea of what that logic is, we simply cannot realistically imagine what aspects of the logic would characterize these fears, intents, desires, hopes, and the like, in the sentience logic.

These propositional attitudes illustrate even more things that go on in the brain. Separate from whether you yourself have some fear or desire or hope toward some proposition, you can read about someone having such an attitude. Without conscious work, your brain immediately constructs some kind of meaning of what you read, and sends a reduced version of the meaning to the sentience, which plays its own role with that reduced meaning. As usual the full meaning has much that is needed to perform the fast deductions that the brain might be called up to make in various circumstances. The reduced meaning sent to the sentience has everything for how the sentience (meaning you) might feel or understand or experience what you have just read.

If one assumes that the brain handles the propositional attitudes in too mechanical a manner, consider reading the sentence "Ranor fears that he fears going to the film tomorrow." If the brain handled each attitude (each of the two "fear"s in the sentence) in a mechanical way, it would seem it would handle the second "fear" as fast as the first. But it does not, as you can see from recalling how you read the sentence the first time. Additionally, the brain sends something like a feeling to the sentience logic which causes the sentience to experience something like difficulty or concern in trying to understand this sentence. If the brain processed each "fear" in the same mechanical manner, it would seem that it would charge through both of the "fear"'s without difficulty. The fact that it doesn't shows that it is trying to find a deeper meaning to the two "fear"'s but can't do it "instantly."

Quantifier meaning of 'the'

The word "the", which occurs in many languages, has a quantifier-like meaning, indicating that in the context under consideration there is only one such object. For instance, you might read in a book, "the first time they saw the car, they fell in love with it." The "the" in "the car" has the meaning that there is one car, here not specified, as being singled out of a collection of possible cars, and further, future references to this car will use phrases such as "the car" and "it". Logic indicating rather sophisticated meaning must be set up somewhere in the brain for all this to work. It must be sophisticated indeed, considering that all of what the sentence is "talking about" might be generic – when we read the sentence, we might not know anything about the car or cars it was selected from other than that the sentence said it. The sentence might be from a novel, in which case nothing in it is real. And the brain automatically is keeping tabs of that fact too. The brain is keeping tabs of all kinds of things right next to consciousness.

Some day researchers will see the logic going on, down to individual nerve cells, for all these things that go on in the brain. Maybe seven centuries from now? Imagine.

Human Language Issues in General

In general we see how much the brain does regarding language. The above are only three examples. Languages often have hundreds of thousands of words, and native speakers of a language may know a third to almost all the words in a language. No theory yet has been able to explain language satisfactorily in spite of thousands of years of trying and in spite of today's computers. It is a sign of how much goes on in the hundred thousand million neurons within one foot of either ear, or eye, that the brain can process combinations of so many words, assign meanings as descriptions of the world and its thoughts and feelings, and carry out deductions on the basis of the "meanings" of those combinations of words. Some day experts will see in concrete detail all this right before their eyes. Neither the reader nor the author will be around at that point. But we can dream of an amazing future.

A lesson to be drawn from this is that just because a system is mechanical, such as is the brain, or just because it is created via the process of evolution, one should not mistake just how deep is the logic that goes on therein. It may be so deep that we will never understand it fully, even though some day our machines will.

All this is going on right now, in its full glory in the most intimate part of you, even though it take a thousand years before science is able to see it.

5 plus or minus 3.

There was once, I believe, a famous paper called "5 plus or minus 3". This research paper summarized many studies that showed humans could keep at most 5 plus or minus 3 "things" in their mind at once. Any more, and they would forget some of them.

Some day researchers will be able to look and see why this is so. They will see how these restrictions come up in the brain. They will see how, in the part of that brain that is related to sentience (that is, awareness), somehow the overall connections allow only up to 5 plus or minus 3 "things" that can be kept all interconnected at once, and they will see how when the person attempts to think of yet another thing, one of the items loses its place in the awareness. They will duplicate the logic on neural chips.

Naturally, where they can, researchers will ultimately work on extending the 5 plus or minus 3, so our machines will be able to be "aware" – if they have awareness – of more things at a time than we can. Surely that will help the machines in figuring out things that we are unable to.

Includes handling concepts

The quantificational issues and propositional attitudes and "five plus or minus three", all these invoke pictures of a brain working with fairly arbitrary meanings. For instance some day when researchers look into the brain while the subject is using a propositional attitude, they will see, eventually, in what ways *any* proposition or concept could have been plugged into what they are watching, and they will see what would be involved no matter what proposition was plugged in. No, researchers are not going to see any of this clearly at first. But they will have ways of seeing it to some degree, and will create many hypotheses for details they can't be sure of. But they will start putting onto neural chips imperfect versions.

Biting your tongue

Consider the complex three-dimensional nature of the mouth cavity, with its jaws and teeth, and consider the motion of the jaw not only up and down but some sideways motions, and consider the complex of particular kind of three dimensional extensions and retractions and sideways and up and down motions of the tongue. Consider all of these. It must be remarkable neural circuitry indeed that keeps us from biting our tongue as rarely as we do. Considering how many chomps we take throughout the day, while at the same time unconsciously pushing food around with tongue and jaw, it is awesome. There might be some surprisingly sophisticated mathematics driven into the nerve circuitry by evolution; driven into the logic.

Perhaps there is a tendency to imagine that such sophistication would not occur in the going on's of the logics in the brain related to the other examples in this section, those examples where the brain is involved in abstract reasoning. But that may not be true. Time will tell. A long time in the future we will finally know what is going on right now in that space in our heads.

Sunset

The sun was mellow. A red evening was approaching. The desert with its panoply of alligator-skin mountains and thin shrubbery - all were awash in pinks and reds.

Presently I enjoyed the colors after the sun had set, the streaking purples, blues, and violets darkening across the floor of the earth.

In that future people will see far indeed.

I took a few steps onward, toward a single ridge that looked far below on the little dirt parking lot where my car was on the other side of the trees.

22 *Initial Notes on Logic*

One day I was talking with the god. It was in a more casual mood. The god made a request.

"Tell us about this 'logic' that you are always pointing to. I know you can't say that much because you yourself do not know. But tell us why you picked the word 'logic' instead of some other possible word. After that say a little bit about logic itself."

So I produced this little report.

At first I considered using the word "logos" from the ancient Greek philosophers. For instance Heraclitus speaks of some kind of central aspect permeating the universe and he calls it "logos". But I was afraid that this word carried too much baggage apart from what I was pointing at. Also, I wanted a plain word.

I also thought of the word "Tao", from the ancient philosophy of Tao. The book that goes with it, the *Chuang Tsu,* speaks marvelously and beautifully of how much the perceptions and sensations and awareness vary from creature to creature depending on its nature. But generally the Tao does not have a scientific image. Also it deals with much more than just the relativity of awareness.

Eventually I chose the word "logic". Generally there are two meanings of the word. At an underlying level they may have some important overlaps, but typically we think of the meanings as pretty different. One is what we think of as in the study of formal logic, or the study of logic in formal philosophy. For instance one may study Aristotelian logic, or the logic of mathematical logic. We naturally picture these as loaded with symbols. There is another meaning and it is the one closer to what I mean when I keep using in this book the word "logic".

The "logic of a situation" – people will say this. They might say "The logic of the situation was such that ..." and

231

then they will list some conditions. In these cases, here is logic going on out in the external world, and in a general way this meaning is the one that we use on our journey in this book.

"I appreciate," the god interrupted, "what you say. But it is more indefinite than I would like."

"I think," I said, "it is coming to me. Something that is more definite. But first let me just note something about 'gong on' and then finish on the choice of words."

Now there is some abstraction occurring when this book talks about logic 'going on'. At this point in our journey I fear I have to leave it at that.

I said why I did not use the word "logic" or "Tao" for this 'logic'. Let me say why I did not choose the word "information".

This word has the feeling of something that is purely an abstract creation of the human mind. My 'logic' is not a creation of the human mind. It exists out in the external world independently of humans. I know that in the definitions of the word "information" in physics it refers to something that exists out in external reality and exists there before humans existed. Still, to me, the word has that feeling.

There is another aspect of the word that made me lean away from it. To me, this that I point to, that which I use the word "logic" in respect to, has an inherent richness and inner fullness that I feel not at all in the word "information". Further, I wonder if the definitions of the word, in physics, with its measures of variability, will ever capture this inherent substantiality.

These are the reasons I leaned away from the word "information". I admit that these reasons are not scientifically precise.

Now the god has asked me to say something more definite about 'logic' – and here I glanced at the god. The god would prefer that I not just compare the word "logic" to other words. Alright, I think I can do so, at least to some extent.

When I point at something and talk about the logic going on there, if I had to choose some words for that, I would say the *meaning* goes on there. One example of logic is the *meaning* of a sequence of words. When we read a sequence of words, an idea comes into our mind. That idea is the meaning of the sequence of words. That idea (or meaning) would be logic itself.

Let us take an example, a simple sequence consisting of just one lone word, "car". When we point at an area and say "car," what we are saying is that over there, in the arrangement and dynamics of that collection of matter, a certain meaning is going on, a meaning that we might call 'car logic' or 'car meaning'. (Don't get carried away with "an object" being there, not right now while we dig further into the going on's in the brain.) There is a meaning going on in the collection of matter over there, and when we see such a situation, or in order to tell someone of such a situation, our speech processes and our corresponding thought processes behind the speech processes phrase this as something like "there is a car."

And less one get too carried away with the importance of that collection of matter, keep in mind that it is changing, so that from second to second, and much more, day to day, month to month, year to year, decade to decade, that matter that the logic is going on in, for the car, is itself different. For molecules are constantly wearing off the car, from every part of the car, external or internal. Parts, seat coverings, tires, are being removed; paints sanded off, new parts are being added, paints applied, and so on. Regardless of the details, in two or three hundred years, all the molecules that were at any time part of the collection of matter will have left the collection of matter, for most cars.

Here is another example. This time the example consists of a sequence of two words, "phase differential". We point to a specific neuron in the brain of an electric fish, and we say "phase differential". Researchers come in and confirm that the spike train coming out of that neuron encodes the following *meaning*: the difference in the phase (zero crossings) of one place on the skin of the fish versus the phase (zero crossings) of some other point on the skin of the fish. Incidentally, note that the meaning only goes on in the matter in a certain way: the spike signals must be interpreted in the right way for this meaning to be going on.

These are examples of the meanings of a sequence of words. Are there meanings that are not of a sequence of words? The logics going on in the brains of electric fish were going on long before there were any humans and long before there were any words. Meanings existed before there were any words. This is from the perspective of science. Humans create words in order to think about the meanings.

To get too much into such issues at this point would be distracting. And confusing. Let us move on with the journey.

"Not too bad," the god said, "especially for initial notes on logic."

23 Logic and Space

We do not realize what a strange connection there is of logic to space. So strange is it that maybe we should not even call it a connection but a relation, and even then, it is a strange relation.

Since the brain is so small – even the human one is maybe only roughly about a twentieth of a cubic foot – the smallness misleads us emotionally, and being emotionally mislead, we are intellectually mislead.

Because the brain is so small, it is easy for us, without thinking too much, to see it more or less as a place that is sort of like a point. But the relation between sentience and space is inherently nothing like with a point in space. And intellectually, that feels difficult.

Let us look back at our visit to the electric fish at the end of Part 2. There we looked at one particular logic, the JAR (the Jam Avoidance Response), but we looked at it carefully to see what we could learn about logics in the brain in general. From such, we concluded the following non-point-like perspective. There are logics going on in the brain. A logic hardly ever goes on at a point in the brain. It goes on in an area or region. More than that, the region can even consist of several areas not even next to each other (though there will be some nerve fiber connections between the areas). (This principle came from observing the JAR logic, where the going on of even this one logic took place in several widely separated areas of the fish's brain.)

Here we are forced to see just how far the brain is from being something like a point. Logics, and that includes sentience logic, do not go on at anything like a point.

Could sentience logic go on in a galaxy? A galaxy is a pretty large system of interacting matter, for it has tens of thousands of millions of suns, each with a complex interacting structure about it, including a swirling planetary system, swirling asteroid belts, comets, dust and rocks, Oort cloud, and so on. Each sun with the matter about it is a complex, interacting system, with a lot of logic going on. If we consider some thousands of millions of such sun-systems in a galaxy, and consider them all interacting, there is a truly vast amount of logic, even though the space for all of them is mostly empty.

Imagine ourselves on the deck of a space ship in deep interstellar space. In one direction we look at a myriad of several thousand million sun systems. Imagine that science and technology have ascertained that there is a sentience going on in that myriad, to be sure, extremely slowly according to our native perspective of time, but it is going on.

We look at that direction. Admire it. What truly vast amounts of space are before your eyes. It happens that the Earth and its solar system is one of the sun systems. From the deck of the ship, our sun and the whole solar system is a pin point in the upper left of our field of vision.

In this grand sweep, a sentience is going on. We think how strange sentience is because it is not "located" at any small point. And it certainly is not at any point in all this mostly incredibly empty interstellar space – most of the space that opens before us consists of a grain of dust every cubic mile or so. Yet the logic of the sentience goes on via gravity and this small amount of matter. You so much want to try and see one single place and say, "there, there is the sentience." But it just is not like that. Sentience, and indeed logic, has a strange relation to space. At least it appears strange from our current perspective.

As for being on a deck of a space ship in deep interstellar space, is that far fetched? If you could see the human brain from a small enough view, it would look much the same. And it is a view that is as scientifically valid as is the one from the deck of the space ship. From our native perspective the galaxy is immense in terms of its space. From another equally valid perspective, so is the brain. In neither could sentience be at a point.

(See endnote 11.)

24 All is Logic: Things

Discussion with god about everything is logic

Just how deep are some issues of things.

Everything is some logic going on in some collection of matter

On my hikes I settled down into the notion of logic. Logic – or meaning – meanings going on in dynamic constellations of matter – is going on not only in the brain – in terms of logics such as computation for finger pressures, in terms of logics that are the experience itself, in terms of logics that are sentience – but is also going on in matter outside of brains, outside of animals and plants. One afternoon, early, I ended up listing the following points to the god.

Everything is a *logic* going on in a collection of matter. This includes physical objects, but also even something like the color red.

Next point. The collection of matter in which logic is going on can itself be changing, with matter being added, or removed. Indeed, if we move out of our native perspective to the perspective where we can see atoms, one would clearly see this as a standard phenomenon in physical objects, as atoms are almost always leaving or entering or attaching or releasing themselves from what is typically an extremely amorphous surface of the object. In fact, what matter is in the physical object can be defined in only a loose, general statistical manner. In fact the matter that is in the physical object can only be defined from the logic that is going on.

As for logics that are the experience itself in a brain, for instance, what it is like for a human to see the color red, there is a duality of such logics going on in the brain, with some logic going on in the evolutionary life of the species of the creature. That evolutionary logic itself is logic going on in matter, but there the region of matter is certainly

237

larger, and the logic is of a statistical nature extending across an immense time, from hundreds to thousands to millions of years.

Much of all these issues will be made more concrete, over the coming centuries, by the gift to us of the brain, because someday we will be able to look into the brain and see all these things.

And here I listed a final point. We would certainly feel upset if what we experience was not out there at all in the external reality. For instance when we experience seeing a square, or a cloud, or red, or tasting sweet, are those real? Surprisingly, this question can be made concrete, because the experience is a fairly specific going on of logic in the brain in basically an instant of time. Is that same logic going on, "time-transformed," over the eons of the evolutionary history of the species of the creature?

"Interesting points," the god said. "They need a little elaboration."

Logics that are physical objects

Waves of logic moving through matter

That which we call a physical object is actually a logic going on in matter. In fact, which logics are perceived by a creature as having hardness is relative to how a creature is laid out in time, space, and matter. Other beings or creatures might not even pick up on (from their native perspective) things the creature would call physical objects.

As for us humans, many things that we call physical objects have some of the same aspects as a single wave moving along the water's surface. Watch such a wave. If you were able to stay focused on just one piece of water as the wave went by, you would see a cycle of motion. Before the wave arrived, the piece of water would be pretty calm and motionless. Then, as the wave started to approach, that little piece of water would start moving upwards. As more of the wave came, it would continue moving up but also start to move in the direction of the wave – let's say to the right. Eventually the piece of water would move no higher, but it would still keep moving to the right, and after that, it would start to fall but still move to the right. It would fall even lower than where it was originally, and at the same time it would start moving *backwards* to the left. And then it would start rising up to were it started. All in all, it would have traced out something like a circle! That is what happens to a piece of matter (the piece of water) as the wave passes through it. That's what would happen anywhere when the wave went through. The piece of matter would displace for a while. The wave itself keeps on going. The water that is physical substrate of the wave is constantly changing. The wave is a form moving through matter; it is a wave of *logic* moving through matter. So too are what we call physical objects. It helps to accurately see the nature of physical objects if we look at them in this picturesque, but also truthful, way.

An automobile

For instance, consider an automobile. That is certainly a physical object. It is solid, you can touch it, and as they say, "it has hardness." What could be more real?

Cars and living creatures are both made out of atoms, and recall how small these little pieces of matter are – there are about 100 thousand billion atoms in each cell of a living creature – and 65 thousand billion cells in a human body. That is how small these little pieces of matter are – these atoms. Physical objects are like a (complicated) wave moving through a sea of atoms. The car has a physical substrate of atoms – a collection of atoms which the logic that "is" a car is going on in. Like a wave passing through a physical substrate of water, so too the physical form of the car passes through a sea of atoms, always getting rid of old atoms, always getting new ones.

At a perspective close enough to where one could see the atoms of the car, we would see them wildly vibrating around, at such speeds that we would also need to change our time perspective in order to see their motion. Vast, vast numbers of atoms moving about, and in them, in a statistical manner, huge dynamic forms going on, forms that have meaning, in the statistical perspective, though we would have to back far away from individual atoms to begin to see the form or meaning. For a car, the meaningful forms are going on in only a very large scale (compared to living matter, which has meaningful forms going on even in something as small as a cell).

A car is actually "car logic" going on in a collection of matter. What is the logic? Well, in our own mind, we all have a fairly good idea what the logic is, in rough terms. If you had to explain to someone what a car was, especially in terms of the physical matter of the car, of the parts, and so on, the explanation you give would be close to "car logic" – the logic, which when it goes on in matter, is a car. For instance, some areas of the collection of matter that is a car are required to be fairly rigid: each of the four doors, the hood, the tires, the trunk, the car frame, the roof, and so on. The logic of these pieces is that they have certain characteristics of shape *and* that they pretty much rigidly maintain that shape – shape itself is one of the logics of a collection of matter. We know that the matter where these logics is going on is neither rigid nor completely free with respect to the other collections of matter – and that requirement is also part of the overall logic: for instance, a car door can swing on one side, with respect to the rest of the car, but it cannot move in other ways. For a car to be in the matter, these conditions, and many others too, are required to be going on in the collection of matter which we refer to as the *material substrate* of the car. But it is this car logic, going on in the matter, that is really the car. The matter is just a means for the car logic to be going on in.

What about the fluids in a car? Fluid logic has to do with particular kinds of spatial-time meaning that must go on in the matter for us to call it a fluid. Some of these car logics are gasoline, a variety of oils, brake fluids, transmission fluids, windshield washer fluid, and so on. Liquids do not have at all the hardness that other parts of the car have. A slight amount of pressure from your hand, and you easily start to go right through the spatial area occupied by the liquids. This is one of the meanings, or logics, of when we apply the word "fluid" to something. (There are people who extol the hard and physical – that the only things that are real are those that have hardness. But later they will add fluids to those things that are real; they just hadn't thought about that fluids are real but not hard. How much else have they not thought of? Ultimately logics are what is real.)

Matter is changing

The car is not matter. The car is a certain kind of logic going on in a collection of matter. And even then, new matter is coming into the car and old matter going out, like the single wave mentioned above moving across the water's surface. Consider the atoms that are the gasoline fluid (technically, gasoline itself is a certain kind of logic among atoms that has not only the logic of fluidity, but also the logics of energy, combustibility, certain kinds of usability, – to get the current definition one should probably go to the official requirements as listed in some petrochemical standards document). These atoms were not part of the car before one went to the gas station, and after they are burned and turned into exhaust and the exhaust is sent out the tail pipe, they again will not be part of the car.

Perhaps one wonders whether such as gasoline should really be considered a part of the car. Probably our brain has several different kinds of "part of". One wonders the same about all the other fluids – oils and the like. Are they really a part of the car? In truth though, the same issue arises for non-fluid collections of atoms in the car too. The seats of the car are constantly wearing away. Even if no one ever sat in the car, atoms by the hundreds of billions, from the seat cover, and from the seat cushioning material inside, are constantly coming loose and moving off into the air. If a person sits down in the seat it speeds up the exodus by maybe ten or a hundred fold. Even the car's paint has millions of atoms constantly coming loose and drifting away into the air. And if all the paint in an area is gone, millions of atoms of the metal underneath will be coming loose and drifting away. Tires are far worse, so bad, that every 50,000 miles or so, so much of the material substrate is gone, that the logic that is a safe tire is severely compromised, whence we are required to buy new tires. And there are other issues such as those concerning replacing carburetors, radiators, fenders, windows, transmissions, motors, frame parts, and the like, and in each case of these cases, a substantial piece of matter that was in the collection, is removed, and a new amount of matter added. And after a really big auto accident, the outflow of old matter and inflow of new, is even greater.

There is not a fixed collection of atoms – matter – that "compose" the car. The physical substrate of the car is always changing, with old matter continuously leaving and new matter periodically entering.

Thus, a car is car logic going on in matter, and the matter in which it is going on is itself always changing. Not unlike the single wave of water, the logic of the car is all the time moving out of old matter and into new.

A tree

As for another example of a hard, physical object, one may consider a tree. It also is logic going on in matter. We perceive it as a beautiful thing, with mighty roots going down into the mud, clay, and rock, and a great, physical grayish, brown trunk rising up, out of which come a multitude of branches and generally green leaves. The logic that is a tree has all kinds of these aspects and others too.

What we want to go into a little is how much the matter that is the physical substrate of the tree is changing.

Whereas some atoms in a car, for instance those deep within the frame, will be part of the car for a long time, almost all the matter in the tree is changing all the time. Maybe some atoms in the cellulose stay for quite a while, but all the rest of the tree is like a fast single wave moving along the surface of a lake. In fact, this is pretty much true of the physical body of all living creatures. I have heard that in the human physical body, none of the matter that is the substrate of our physical form stays around longer than seven years! (Nature sometimes forces different parts of our psychology to meet. It is interesting to note how the matter that makes up our body leaves. It is not the fecal matter, for that is merely left over food. It is in the urine. Some say that over a period of seven years, all the matter – atoms – that constitutes our physical substrate, all the matter that we were, passes out in urine.) (A totally exact statement in science is never totally simple: a small percentage of matter also leaves via shed skin, sweat, carbon dioxide and other gasses exhaled from our lungs, also shed nails, hair and the like, and if we have the misfortune, a loss all at once of a significant part of the physical substrate of the body. Thus is the single wave that is the human body.)

General comment on logics

Almost everything is logic going on in matter. Logic is what is real.

One day the god said, "Whenever we look at something, or even think about something, typically we are picking out key, relevant logics. And even then, any logic is composed of many further logics, with all logics and their parts continuously deformable into other logics, in so many ways. Earlier you compared some of the issues of boundary definitions of logics to volcanoes. Here is another comparison. Take a particular tree. Think of all the ways that you could divide it up into parts, into subsystems, into overlapping parts and subsystems, or into non-overlapping parts and subsystems. Any and each of these is also a logic going on in that matter. The number of logics is huge. We could also consider general categories of trees, instead of a particular tree, and look at the parts and subsystems in the category. There are so many ways that logics go on. Even so, once we understand a situation enough, there are key logics that present themselves, there are key logics that we focus on."

(I thought the god's comments valuable. After a while I wrote an additional parenthetical remark, dealing with how the brain "selects" key logics from so many potential candidates. In many places in the brain the cells are laid out in a *-topic* manner. For instance, in a number of places the cells are in a *retinotopic* map, meaning if you looked carefully at that part of the brain, you would see a surface of cells, and the positioning of the cells in each such surface would generally match the surface at the back of the eye, the retina. *Somatotopic* means that the cell layout on the surface would match the layout of the surface or skin of our body. For instance, in the electric fish there are many groups of somatotopic maps piled right next to each other. At each point in such a "surface" a cell is doing some computation for that place on the skin, and across the surface all the cells are doing the same computation but for correspondingly different places on the skin. Another somatotopic "surface" of cells in the brain will be doing a different computation, again one cell for each place on the skin. Likewise there may be areas in the human brain that maybe could be called "logotopic" maps. The cells in the map are computing some same function across a whole range of logics. Each cell is doing this all the time. So probably, in many cases, the brain is dealing with all possible logics, all at once. This is important to be aware of because we have a tendency to imagine that there is one special logic selected and focused on. Surely, in the external world, there are no doubt situations where a limited number of logics are selected. But in the brain there are many situations where the whole range of all kinds of logics is simultaneously active. In certain "logotopic" maps, only those cells that are producing the strongest response will correspond to the logics that are present in the outside world. So in these situations, the brain does not really do any selecting till later. It just works on all the possibilities with all the cells working away.)

Logics that "cause" our sensations and awarenesses

We now look at logics that are related to what we sense, and how we come to sense something.

Our senses pick up something out in the external reality. That something might be an automobile or a tree (picked up with the vision of our eyes), or the snapping of a tree that will then break and crash to the ground, or a piano chord (picked up with the cochlea behind the ear drum responding to pressure waves that are going on in the air outside the eardrum), or a piece of red (picked up with the vision of our eyes). All these are logics going on in matter.

In addition to these somethings that are logics going on in matter, there are also logics going on between that something and the sensors on our body. In the case of a tree snap, or a piano chord played, that something is waves of pressure moving through the air. In the case of red, or of light, received from the surface of an automobile or a tree, that something is light (electromagnetic waves).

Tree breaking "sound"

Suppose that a tall tree breaks with a sharp cracking sound. If we looked real close at the place where the tree snapped, and slowed the time down greatly, we would see waves of air pressure – assuming that we could see pressure waves in our native perspective – fanning out in the air from the place where the tree snapped. As stated in earlier chapters, these waves are the same kind of wave logic as occur if you drop a rock in a quiet lake. Well, actually, given the nature of how a tree snaps, with all the different pieces of wood fiber snapping in uneven patterns and sizes, the overall snap is more as if a dump truck poured a bunch of big sharp rocks all at once into the lake. If you looked very closely at the place where the tree break was happening, you would see numerous strands and jagged pieces of wood separating, due to overwhelming force. These countless strands of wood would be slowly bent, and then finally give, and with incredible energy and speed they would snap back. This is what is happening where the tree trunk is slivering into many jagged pieces. When one of these fingers of wood snapped back, it is the same logic as when a rock is dumped into the lake. Whether in the air, with the snapping tree, or in the water, with the dropping rock, the logic going on is of powerful pressure waves moving out from the source.

Generally, one aspect of this logic is a kind of description of the dynamic form going on in the matter. It is something like the idea, or meaning, that can be going on in the dynamic form. For instance, in these air pressure waves, the idea is that high and low regions of density of air molecules are fanning outward, at about 760 miles an hour, from the place where the tree trunk is breaking. This dynamic form, this logic, is going on in the matter between the tree snap and our eardrum. Another name for this logic is wave logic. Wave logic is going on in the air that is in the space between the tree snap and the eardrum.

Even before the waves get to the eardrum, the body is already transforming the logic waves via the cup shape of the ear. The ear's various folds and directions of fold are focusing and generally transforming the logic of the waves so as to be stronger and clearer when they start to reach the ear drum. (In a sense, just as the front of the eye is a lens, so too the ear is a lens. The front of the eye, a lens, focuses the light to the back of the eye, the retina. The ear, a lens, though ultrasimple compared to the eye's, focuses the waves of air pressure to the back of the ear, the eardrum. Thus across the front center of our head are four lenses, two for waves of air pressure, and two for waves of electromagnetic radiation. Behold a part of the human head.)

As the high and low pressure of the waves hit the ear drum, they start to make the ear drum wave back and forth in synchrony with the air pressure waves. On the other side of the ear drum are three tiny bones, some of the smallest in the body, the bones on one side touching the ear drum, and on the other, in contact with a small organ called the cochlea. With the ear drum going back and forth from the sound waves on one side, it starts the bones going back and forth on the other side, which in turn start corresponding waves of pressure in the liquid inside the cochlea, which, unwrapped, would be about half of an inch long. The fifteen thousand little hairs in the cochlea are of different lengths and resonate under different frequencies of pressure waves in the internal liquid. The hairs are electrosensory neurons and at the opposite side of the hairs, thousands of spike train outputs are being generated along nerve fibers and are passing outward into the big hum of the brain. At present we can only make general statements as to what happens with these electrical signals, since we have almost no ability to see the logics they become.

This joint encoding of all these spike trains emanating from the cochlea might well be just a recoding of the nature (logic, meaning) of the waves of air pressure. But the encoding is likely incredibly well-suited for the brain to do thousands of checks and computations on the coding, and the results of these are new pieces of logic, and eventually these new pieces too are analyzed in many ways and in turn result in even further new pieces of logics, and so on, until some kind of final resulting logic is constructed, capturing over a time certain aspects of the pressure waves. After that, presumably a kind of reduced version of this is passed into the logic of consciousness, and the being (the human) whose brain this is, has an experience which it calls "sound".

As an example of just one of the computations and constructions of new pieces of logic, the brain compares the air-pressure strength encoded in the signals from the right ear versus the left, and thereby computes the general direction of the sound (how much it is to the right or left). (Actually, it seems that as long as both ears are uncovered, there is also a rough computation of whether the sound is from above, from the same level, or from below, and this logic too, along with various other logics is added into the characterization of the sound, which eventually passes into consciousness).

Piano chord sequentially (the issue of time)

It is instructive to compare the rough splitting of the tree trunk, with the striking of the three notes of a *harmonious* piano chord, and further, their being struck in sequence one after another in a way in which there is a brief period of silence between each.

For the tree trunk cracking, the waves were as a load of jagged rocks dumped into the middle of the lake, producing a mess of sharp rough water radiating outward. But now, for the first piano note, it is as if a smooth rock is slipped into the lake, and a smooth wave starts out, and after a brief period of no wave (the silence between the notes), another smooth wave, of a different frequency starts and ends, (with another brief silence) followed by a third even wave. *If* the three waves were put out at the same time − if the three piano keys were struck at the same time − they would form pleasant ratios with each other, for that is the nature of a harmonious chord.

As before with the cracking tree, the air pressure waves travel from the source, through the quintillions of atoms of air to our ear, where it is somewhat focused onto the ear drum, the waves passing through the ear drum, then little bones, into the cochlea, and from thence, via spike trains over tens of thousands of nerve fibers emanating into the big hum. A primary difference from the tree cracking is that some parts of the brain are noticing the evenness of the wave of a note, and creating an additional piece of logic that will indicate that this sound (the single note) is a pleasant, smooth sound, as opposed to jagged rough sounds like the tree snapping. This additional piece of logic eventually results in the awareness logic of the pleasure of the single sound of the note.

In addition to each note being pleasant, the three notes played in order, with a silence between them, is also harmonious. Therefore, there must be logic in the brain that is constantly comparing the current sound with past sounds, indeed, with several of the past sounds, looking for examples for harmonious relations between the sounds. A special additional logic is added to various logics for the second note. This eventually results in our awareness, when we hear the second note, of the experience of harmonic pleasantness. The same happens when the third note is played. This time the added piece of logic has information to the effect that more than one note was played in the immediate past and that all the notes, even with a small amount of time between each, are harmonious with respect to each other. Eventually some version of this logic passes into consciousness, where we have the experience itself, at that one instant, of the harmony of three notes. At that instant, out in the world is the sound of only one note. But at that instant, we experience the harmony of three notes.

Start of time problem issue

Note that the *experience* we have on hearing the third note takes place at one instant in time, even though the experience derives from three notes being played over several seconds.

This led to another brief discussion with the god.
"Does what we perceive exist?" I asked.
"There are different versions of that question."
"What do you mean?"
"Well, the first version," the god continued, "is this. Does there exist something out in the world which caused our experience?

"Another version of this question: Does the experience itself exist in some way out in the external world? This question we will look at down the road."
"We will?" I mumbled.
"*You* will," the god answered. I nodded slightly, wondering how on earth could one address such an obscure issue.

"But for now you will only be looking at whether there existed anything out in the world that caused the *instantaneous*, pleasing, *harmonious experience* on hearing the third note of the chord.

"Keep in mind that out in the external world, when the third piano note is struck, there are no other sounds. The previous notes were played earlier and have been totally silent for a brief period when the third note is struck. So when that last note is struck there are no harmonies out in the world because there are no other notes at that time to be harmonious with. Nevertheless there enters into our awareness – we experience – a pleasing *experience* of a kind of definite harmony of three notes. We experience at that instant what is not out in the world at that instant."

Thus the god and I talked.
The issue seems to be whether logic out in the external real world can "reach across time." If it can, then the instantaneous harmony, at the third note, exists out in the external world, because its logic, or meaning, reaches back in time to the preceding two notes.

Whenever we get into issues like this, of what exists and what does not, it gets so difficult that it is hard to see where the ground is. Nevertheless, in this case, the question can be given a somewhat more definite form. This is one of the few places in the book where we will roam a little into mathematical formalism.

Definition of F(t)

Let F(t) be the amount of harmony of the piano notes that have been played from the time t back to say 4 seconds before t. So this F would catch the idea of the above harmony of three notes being played within 4 seconds, even if there were a brief silence between them.

Is there such a thing as F out in the world? Or is it only a construct in our mind?

Does science have any stand on this question? To be sure, science does not ask such questions explicitly about "something" such as F. Nevertheless, it casts a vote in its own way. Scientists will use a function like F in anyway and wherever they want in whatever analysis they are doing, if it assists their analysis. This would seem to say that science treats the logic F as something that exists out in the external world. After all, in our thinking we only use what exists (unless our thinking involves a proof by contradiction, and even then, it is often possible to convert such proof to one that is other than by contradiction).

Apart from science's unrestrained usage of something like F, there is another reason why F, as a logic out in the external world, exists. Logics go on across space: the matter that a logic is going on in, clearly reaches across a region of space. Why shouldn't it also be able to reach across a region of time? There is no reason in the nature of things (the fabric of the universe) why space should be allowed and time barred.

The trouble is that these are abstruse considerations. Even so, their conclusions have some importance to us. This book takes "logic" as existing out in external reality, independent of the human mind, even independent of humans (see the examples of logic going on in animals and in their brains, in Part 2). If F does not exist out in the external world, that implies that our experience of harmony at the single instant of the third note has nothing real out in the external world that caused it. This would be of some importance to us, as well as being disconcerting. However, as stated in the previous paragraph, the evidence is that F does exist out in the world – as much as one can talk of such issues. So, what we experience at a point in time does correspond to what exists out in the world at a point in time.

Before leaving this section, we include a paragraph of technical phrasing, for those interested in such. In technical areas, including science, F is called a function. Usually in science, no distinction is made between the *description* of F and the function itself, though we *do* make such a distinction here. In some areas of formal philosophy, the function itself is referred to as, "what F is naming", and in our case the function itself is a logic. If the description of F contains phrases such as "t-4" or "t-20" then we will say that the logic reaches across time. If the description contains only phrases such as "t", when referring to time, then the function or logic is said not to reach across time. We do not address issues of when the description has terms such as "t+3" and so on.

Does the function F exist out in the external world? Is there such a logic that can go on in the outside world?

If such abstruse issues of existence should seem disconcerting, we now leave them. Humans have always lived with uncertainty about some of the most fundamental issues of existence, and it seems that they always will.

Thus, what we call sound, whether that of a tree breaking, or piano chord played sequentially, when rightly looked at, is a whole continuum of logic, transforming as it moves from the origin of the wave of air pressure, going on across space, which eventually continues right into the ear, and transitions across the ear drum into the cochlea, and then into spike trains, which go out into the big hum, where it there too continues a long path through parts of the big hum, and where, from the perspective of the logic going on in the brain, it is finally realized as a kind of logic of a particular kind of meaning, that of a loud snapping noise, or a piano chord. The part of this logic in the brain outside of consciousness is suitable for the needed computations that must be performed on it, but another logic, a kind of reduced logic, goes on to the sentience logic, and that reduced logic is the very experience itself of the sound, a logic which we currently cannot even fathom and perhaps never fully will, but will be "understood" with the help of future machines of thought.

Red in vision

What about vision? Suppose we are looking at a cat in a green field of grass, the cat looks at us, and meows, and with the meow, we see its red tongue. Just like with sound, what we *experience* when we see red is a logic (in the sentience logic in the brain) that is quite different than the logic going on in the light waves that race to our eyes from the cat's tongue. Both of them are logic though. What goes on in the matter between the tongue and our eyes is waves of electromagnetic radiation (i.e., light) of a certain frequency. (Science considers electromagnetic waves to be of the physical real world. So we will speak of them as going on in matter.)

Whereas middle C on a piano, produces sound pressure waves of frequency 440 cycles a second (440 wave crests go by a point per second, or, alternately, the pressure varies from hi to lo at any point, 440 times a second), the electromagnetic waves of red, vary back and forth about 400 trillion times a second, which is to say, at any point from the cats tongue to the retina at the back of our eye, the electro-magnetic nature of light is going back and forth about that many times a second. Whereas, what we perceive as sound is transmitted on pressure waves in air, what we perceive as color is transmitted as waves of oscillating electric and magnetic fields (which is why they are called electromagnetic waves; as odd as it seems, half of each wave is an electrical field, and the next half is a magnetic field, each perpendicular to the other!). For red electromagnetic radiation, the distance from the crest of one wave to the next is about 1/3,000,000,000 of an inch. Whereas the pressure waves through air travel from the source to your eardrum at about 760 miles per *hour*, the electromagnetic waves travel form the cat's tongue to the two hundred million photo-receptors in our two eyes at a speed of about 186,000 miles per *second*. Truth surpasses fiction.

The hundred million photon-receptors (also called photo-receptors) in the retina at the back of each eye respond to certain frequencies, and one of those is that of red light. These photon-receptors are, almost all of them most of the time that we are awake, after one or two levels of intermediate neurons, constantly sending spike trains into the brain, based on the intensity and frequency of that electromagnetic radiation impinging on them. (There are three or four layers of neurons right in the eye, detecting simple visual edges or contrasts and so on. By the time the logic passes through these few layers, the number of axons is reduced from a hundred million down to only about one million leaving each eye, in a cable of such axons. The two cables, one from each eye, head toward the back of the skull, crisscrossing in the middle of the head, and it is at the back of the head where the serious visual processing is done. In the area where they crisscross, the optic chiasm, there is some dividing and rejoining of the left and right sides of visual outputs of each eye.)

Once again, we know little if anything of the details of what happens to the signals after they enter the big hum, except that somehow part of them eventually become the going on of the logic that is our experience of seeing red.

Experience logics related to evolution logics

In the last section we looked at the surprising complexity of part of what produces the experience of a sound or the experience of light. We looked at all that went on outside the skull, and at some of what went on inside. And all of it was forms going on in the affected matter, consisting of vast numbers of atoms. The forms were logics – meaning – going on in the dynamism of these vast numbers, virtually clouds, of atoms.

In this section we look more carefully at the logic in the head and its relation to the logic that goes on across the long eons of evolution outside the head.

Logics of red, tree snap sound, piano chord sound

Somewhere in sentience logic is the particular logic that is the very experience itself of what it is like for us to see red. It would be natural that this "red logic," this meaning of red, would have relevance to our evolutionary history, though we are far from knowing or understanding this meaning, and we will only do so when we understand enough of the detail logic of awareness. But the logic probably *somehow* encodes the fact that our evolutionary history had strong relations to such things such as blood, ripe fruit, and perhaps fire. Scientists suggest that being able to visually detect ripe fruit would be a survival benefit for our ancestors. As for blood, it occurred so importantly and frequently in the life of our ancestors, it could be quite likely that this too is somehow encoded in the *experience* of red. The *red* liquid, blood, was important information about an animal who was being hunted or eaten. And certainly it was important information about enemies and comrades and family and even self, if wounded. Finally, fire was important for many millennia of our ancestors, for cooking food, for protection of a campfire against animals, for awareness of the danger of forest fires and the like, and even in some cases of lava from volcanoes. When we finally fully see in the brain the logic that is the experience itself of red, we could well find that parts of *these evolutionary aspects are encoded in that logic*.

Here we are beginning to get into the idea that some of the logic going on in the head is also going on outside the head over vast periods of evolutionary time and space. In short, some day when we can look into these logics going on in the brain, we will see that the skull is not purely some encapsulation removing the brain from the world. It is only a separator between different locations of the expression of a logic.

(For a little attempt at introspection for red, see endnote 20.)

What about the experience of hearing a tree starting to break, and what about the logic that is that experience? Again, we expect the logic that is the experience itself to have nothing to do with wave logic of the air pressure waves, but rather with our evolutionary past in regard to such noises. This particular logic typically triggers our looking up and around quickly for the source of the cracking sound, but our reaction itself is not the logic that is the experience itself. Nevertheless, this sort of event must have some bearing on that logic. We must simply wait till some far future date when science has determined, literally, what this experience logic is in the brain.

What about the experience of hearing the three keys of a piano, not overlapped, but one after another, in a harmonious chord? For the experience of harmony, we can't even guess its relation to our evolution, though maybe the logic fell closer to some parts of our mind which deal purely with form and relation of form, and in general for humans with such advanced mentality, the finding of "harmonious" relations among forms in our existence indicated a nurturing environment, and so such harmony, whether in music, or in art, or in science, or in our life, or being, is a source of joy.

Evolution logics

The experience itself, of a certain sound, must be associated with the relevancy of that kind of sound in our evolutionary past. Someday, when experts are able to look into the brain, and able to understand what they are looking at, they will see that the logic that is that experience will be basically the logic of parts of our evolution. The same will be true of our other experiences, whether of the color red, or of a shape, or of all the other experiences and feelings that we can have. Much of these logics in the brain will be basically the logics of parts of the logic going on in our evolution.

Inner experience is the relevancy to our evolutionary past.

What is an evolution logic? These are logics that 1) have gone on across the long evolutionary periods of matter, space, and time; and 2) have become embedded in the logic going on in each of the individual creatures. Concerning (1), that the logic has gone on across evolutionary periods of matter, space, and time: this refers to the matter in all the ancestors of the specific individual. It also refers to the matter in the environments that were around each of those ancestors. The time and space refers to the regions of this matter. The specific mechanism by which this particular show is held together is that of the chromosomes, the detailed story of which is slowly being unraveled by experts.

Our native perspective can give us a misleading view of an evolution logic, because evolution logic is so big, in comparison to ourselves, in terms of both space and time. Our *native* perspective of an evolution logic is as the view we get close up of atoms and of their motions when what we want to see is the whole person. If we looked at individual atoms, with a time dilation where we can see them moving, our view of the whole person would be so close up and in such an enormous time frame, that the whole person would seem hopelessly unreal, and would seem like an abstract description in a text book. That is the same kind of problematic view we have from our native perspective of evolution and evolution logics. They are hopelessly vast and distant, and too slow for us to feel and perceive how real they are. Nevertheless, there are surely perspectives, from which if we could see, these aspects of evolution – these logics – would become quite real, small framed, and mentally graspable. "Real" is the keyword here. They are as real as all other logics. (For instance, from one such evolutionary perspective we might see, in what seemed like a few minutes, the billions of times over tens of thousands of years that our ancestors encountered what came to be distinguished by what we now call "red"; and at the same time the perspective might somehow show relevant logic structures out in the external world and relevant logic structures in the brain, and it might show how the logic structures in the brain came to incorporate, transformed, some of that logic out in the external world.)

This is an overall view of evolution logics. The details surely include exceptions. For instance, it is possible at some point in evolution, some kind of physical realignment of sections of the brain took place, and one experience logic became distorted and maybe even attached to some other external event. Nevertheless, the above is the general definition of the phenomenon of the relation between evolution logics and the logics going on in the brain, including those that are experience logics.

Catherine the Great. Future Brain Doctors

Catherine the Great, 1729 to 1796, from the age of 33 onward, was Czar and Autocrat of all of Russia. Supposedly tone-deaf, she never got a tune her whole life, never appreciated any music in that way (see endnote 13). When a series of notes that were pleasing to most people entered into the big hum of electrical activity in her head, what was it that happened that was different from most people? Was the ability in her brain, to experience tune pleasure, present, but some earlier piece of processing failed to add the marker for pleasure to her brain's logic for the tune? Or maybe the whole area that processed pleasure for tunes was inoperative, or maybe it worked, but an in initial part was never connected. Whatever the situation, something was different in her head, something which, some day many centuries from now, researchers will be able to look into such a brain, and say, "There! That is the problem!"

There must be other humans who have a variety of similar differences from the majority. As part of understanding sentience better, in the far future, researchers may not only be able to weaken signals in areas of the brain, but they may be able to weaken or completely stop signals in areas as small as a neuron or two, and do this for hundreds or thousands of specifically selected neurons at one time. And then they can listen to the patient self report.

"Oh, yes, I still hear the tune. Actually it sounds considerably more beautiful than it did before. Except there is some strange quality that I can't explain."

Or maybe the patient, looking thoughtful, says, "I hear the notes, and I can tell there is a tune, but ... it's interesting ... I have no pleasure at all in hearing it. It's not that it is bad. I just don't have pleasure. I can even tell it is the same tune. It's just that I have no pleasure in hearing it. Oh, and yes, the pleasure of playing the gong before, when you play it now, yes I still experience the pleasure of that."

From this last, the researchers would know that the pleasure of tunes had been disabled by whatever micro adjustments they had made in the person's brain, but not the pleasure that can be experienced by a pure sound, such as certain resonant gongs. Thusly could science unravel more of what the sentience is in terms of logic going on in the brain. And so it would go. Scientists would get better and better at understanding the logics and could build better and better devices to display in better and better ways these logics and their relation to each other.

All these logics are taking place at little collections of nerves as specifically located as are the nerves we discussed in the electric fish's Jam Avoidance Response. And so too is the enjoyment of a single moment in a symphony as well as small hierarchies of moments of beautiful sound, which in turn build further hierarchies and so on. (For an analysis of some hierarchies of musical structure, see Cooper and Meyer, *The Rhythmic Structure of Music*.)

25 All is Logic: Vinegar Bog

Other beings

The god put me into a kind of remote contact with some creatures laid out differently than us. I wondered if it was alright to do this, but the god, reading my thoughts, said that it was, given the yonder nature of the contact.

Somehow these beings perceive the following logic, strange to our native selves, but then again, what we are aware of from our native perception is quite strange to them, if they even had the cognitive mental space that allowed them to be aware of it. They have nothing like our human spatial awareness.

When they "look" in a direction, they see only what for us would be a large, general area. They inherently do not see in "points" as we do. While our visual perception is composed of thousands of such points, so that we see in great detail (though not as good as an owl for instance), this creature, this being, sees, what to us is a whole area, as just one point. The being's native perception has no more precision than that, and indeed it can't even imagine more detail, except by complex verbal analogies. Likewise we can't imagine their spatial awareness. It would be wrong to compare their vision to seeing just one point. But because of the way our brain is laid out there is simply no good way to explain to us what they experience.

Here is the most important logic they pick up. Suppose that to us humans there is a pile of papers on the floor. Consider all the words printed on all those sheets of paper – that's right, the words merely printed on those papers. Consider the length of those words and how many times words with that length occur in the pile of papers. Let us suppose that words with three letters or less occur 12,925 times in the pile of papers. Words of four letters, suppose occur 2507 times. And so on. The creature sees the area in a simultaneously two-element scale, *each* translatable into what we see as a gray scale in for instance a black and white picture, in some ways.

Generally, the more words of length 1, 2 and 3, the more pleasant the creature finds that "general direction" in which are the papers. While the more words of length 5 and 6 and 7 and longer, the less pleasant the creature finds it. If we walked through a bog of vinegar, that is sort of what it is like for the creature, when there is a lot of longer words. We could handle the vinegar bog. But it is not something one does if one doesn't have to. If there are a lot of words that are short, and also a lot of words that are long (5, 6, 7 letters), the creature perceives pleasantness *and* unpleasantness, simultaneously, in that direction. It is analogous to our sensation of eating say, sweet and sour pork, where we experience both sweet and sour simultaneously.

When the creature passes through an area of lots of short words, it might be compared, in our native perspective, to a nourishing emollient on the skin that soaks in for a while to our benefit.

This explains why the creature inherently finds it pleasurable just to look in the direction of an area with a lot of short words, though the pleasure is much more "physically palpable" if it goes into the area. (The reverse applies to areas with a lot of long words.)

For us humans, the hardness of physical objects, this particular logic going on in matter, is a central characteristic of our world and perception. However, because of the creature's substrate in the material world, it passes right through all the things that for us have hardness. As far as I can tell, from discussions with the god, who introduced me at a distance to these creatures, the physical substrate of these creatures' physical bodies has what would be for us no hard material content but is something that goes on in what might be called "energy character" fields (see endnote 14). At any rate, they have no awareness of those logics the perception of which we label "hard surfaces," the reason being just that *such logic going on in matter is irrelevant to them.*

The spatial area of their physical substrate – where there body is – is a couple of feet wide and hundreds or even thousands of miles long. Areas between different individuals frequently overlap. Because of the nature of the ground of their material being, they can only be in an area where there is neither too little nor too much energy. In terms of our sun, they comfortably inhabit any area as close as Venus and as far as about Mars. Closer or farther than this, the energy characters do not maintain the underlying logic required for the creature's substrate and required for the existence of the "world" that they evolved in and currently exist in.

While we may envy these creatures for being immune to physical objects (what we call physical objects), we must not envy them for some of the restrictions they must endure. For instance, our physical body (substrate) can easily be placed in some small capsule like a car or rocket, and moved great distances quickly; the creature has no such option. When you occupy a ropey type area a few feet wide and hundreds or thousands of miles long, you cannot just pack yourself in some container and ship it somewhere. In fact, even to this day, their scientists and their technology have not been able to produce any method to move them around faster than the natural method they have used for millions of years. Further, that method is effectual in only about ten percent of the attempts, so the creatures usually have to try about ten times before they are successful in moving their physical body to another place. On the other hand, within their physical form, they are able to "focus" their senses to any area almost without effort. Win some, lose some.

The creature has no idea of sheets of paper. It certainly has no idea of our letters and words and that they are symbolic items implanted on the surfaces of sheets of paper with the thinnest of films of black material that we call ink. The creature does not even have a native awareness of surfaces, since such awareness generally arises from "hardness", which is not in the creature's perceptual space. So it certainly has no idea of something like the surfaces of a sheet of paper.

For us there is immense logic tied up with our printed words, but none of this logic, none whatsoever, even enters into the basic level of sensory input to these creatures. Nor do the differences in letters. There is only one limited piece of logic of our printed words that affects the creature, and that is the number of letters in the word. Nor does that enter into the creature's sensation in any kind of direct way but rather it affects (the mathematical characters of) the physical substrate not only of the creature but of the creature's "world." Though the god said that information about this world would surprise our mathematicians, it would not say more.

The god informed me that there was something in the creature's long-term evolutionary environment, which, today, turns out to have logic going on in the creature's sensory experience similar to the effect of the number of letters in our printed words on paper. The creature is still capable of perceiving those ancient logics, though it does so rarely. Instead, when its sentience is focused on an area on the Earth, it can experience the sweet and vinegar bog, due to our printed words on paper.

In some ways, the creature is like us. From our native perspective there is so much we can't see and so much we don't know. We have many experiences but we used to know little as to why we had them. Only over the last few centuries has science opened a perspective onto just how immense our body is, opened a perspective onto how air is also made up of these little pieces of matter (atoms), opened a perspective onto how our experience of sound is connected with pressure waves moving through air, how our experience of vision is connected to waves of an electric and magnetic character passing through space at incredible speeds, and so on. We have no *native* awareness of any of these things. And the creatures are in the same boat. There is much they do not *natively* know.

The creatures have other similarities to us too – sort of. These beings, who have the unique sensations related to long words and short words, are most of the time unhappy. They spend almost all existence in a slightly dull sadness, from which they cannot escape. Could there be a general law that the overall amount of unhappiness for any creature has to be in a fixed percentage, and if it is not, evolution will drive it there? I don't know. But for the creature, their long periods of dull sadness are punctuated by sharp explosions of extremely intense, glorious joy, after a short time of which, they go back to their dull sadness. Overall, when you add up their short periods of extreme happiness with the long periods of slight unhappiness, the total is probably a little more sad than happy. This overall sum seems to be a characteristic of humans also. But humans do not have anywhere near the extremes of happiness of these creatures, while in terms of unhappiness, humans get far worse than the long-term slight dull sadness of these beings. Yet the overall sum for us is about the same as for the creature.

Relativity of plastic bags

For us, the creature's anciently perceived logics – I was not told what they were – are as radically different from the logic of word length on pieces of paper, as is night from day. The *sameness* with which the creatures perceive their old logics, and the new logics of word length, might be compared with us if we were to pick up two plastic bags, one being made of a standard plastic while the other is made of a new kind of material that has only recently appeared in our environment. Let us suppose that the types of bags are indistinguishable in terms of touch, sight, noise, and environmental safety. However, there could be a third kind of creature for whom the two bags are radically different objects, of radically different size due to different spatial electrical properties of the two different materials the bags are made of. The third creature's seeing the two bags as of radically different size and type compared to our seeing them as the same, would be similar to our seeing the logic of word length on a sheet of paper as radically different from the first creature's ancient logics which it perceives the same as the logic of our printed words.

Three logics: external reality, the experience, the evolution logic

Out of the categories of logics described in the previous chapter, three have been illustrated in our description of the creature. One is the logic of the number of words of different lengths in various general directions in space. This is the kind of logic that is objective for science. It is "out there," it is independent of any creature, human or otherwise, seeing it or experiencing it.

The second category is all the logics in the creature's whole evolutionary history.

The third category of logic is the logic that, when it goes on in the creature's brain, is the experience itself of what it is like for the creature to experience a mixture of short and long words. One of the ways in which the creature experiences this length of words comes from the second logic, its experiences over evolutionary time. That might be called the how, how the creature experiences the first logic. The creature's negative part of experience or feelings of this third logic has similarities (isomorphisms) to how we might experience a vinegar bog.

Science for creature versus for us

Let us look at how the creature is affected by the view given it by its study of science. Just as we have scientists, so do the creatures.

After a number of eons, the creatures' scientists finally came to a knowledge of such a thing as matter and even of the little pieces of matter that we call atoms and molecules, though their native awareness of space, so much more primitive than ours, did hinder them in ways where our scientists had an instantaneous, intuitive grasp. Nevertheless, the creature's scientists finally found the logic of (our scientific notion of) space, as for instance the sequentiality of three completely independent sorts (we call them the three dimensions of space – space which was imbued by Descartes with Cartesian coordinate spatiality – a native grasp of these structurings is so easy for us, but is not at all a part of the creature's native perspective). Eventually their scientists were even able to explain how the underlying atoms and space determined what the creature experienced. Those of the creatures who are *not* scientifically oriented are fascinated by this alternate perspective.

The creature's scientists, after many centuries of working through abstract considerations, arrived at much of the same logical, mathematical form of the external world that our scientists did. Just as we are fascinated by there being so much space between atoms, which in a confused way seemed to imply hard surfaces were not real, so too the creature was fascinated by the fact that from the perspective of science there are so many spatial directions. The way their scientists explain this to the non-scientists is something as follows. You know how when you look in one direction (they don't use the word "direction" but rather an expression for indicating what we humans call their generalized direction), and you know you could also look in the twenty or so directions around you. Well, it's as if all those directions were crammed in the one direction. Only many, many more were in the one direction. After initial amazement and confusion, the creature student learning science would grasp the idea, but in a somewhat textbook puzzle type way. However they would be fascinated by it, not unlike the nature of our fascination when science tells us that hard surfaces are overwhelmingly empty space.

As for what we call hard surfaces, it is difficult to say just how soon the creature's scientists would realize that certain statistical constellations of matter had that characterizing aspect, since their native perspective has no such thing as hardness. (Hardness logic is when a creature has a physical substrate – body – that is hindered at relatively tight spatial boundaries due to more detailed logics going on relative to that boundary, and furthermore, the hindrance can take on various aspects including harm to one's body if the speed between the boundary and one's body decreases too fast.) The understanding that the creature's scientists arrive at of such logics is akin to an understanding out of a mathematics textbook; it appears somewhat artificial and statistical. Since their scientists have no native perspective of hardness and surfaces, they have no native awareness of what we call shapes and sizes of "objects". They must *laboriously* deduce such information from mathematics and theorems. Even then, such knowledge comes piecemeal in complex theorems and statistical concepts built on top of difficult-to-grasp mathematical concepts of "space." Not even their best scientists have the broad, instantaneous, effortless grasp of shape and size that we do.

We ourselves are not without weaknesses when it comes to understanding hardness. Our native awareness gives us no knowledge that the logic we call hardness is derivable, in physics, from nothing more than these statistical clouds of a certain sort combined with the overall logic of electrostatic forces, and the like, going on in that cloud of atoms.

For the creature, the idea that matter could interfere with the movement of some other hypothetical creature (us) – this idea they treat as one of the fascinating concepts of scientists and philosophers, to be enjoyed perhaps during an intellectual digression from the harsh, real world of daily life, a digression to be enjoyed, such as when "reading" a science article in the "newspaper".

Close

This finishes our cursory visit to the world of these creatures, a world as strange to us as ours to them. The essence of everything in both worlds is logic. Each creature, they and we, has its own logics that are relevant to it. In some ways the logic of these logics interacting with the logic of the creature's evolutionary history is what enters, in a reduced form, into awareness; it is the experience. All logics are real. But only certain ones enter into the perceptual space of a creature (sentient being).

26 All is Logic: Abstract Objects

Abstract Objects

Our vinegar bog creature was content enough interacting with the world of pleasant short words and acidic long words. But something must be revealed to you dear reader. Since we were watching the creature, it became aware of us. I am not fully sure how this happened, but I suspect I have some responsibility since I should not have tried to establish too much insight into its being. As to why I have responsibility I cannot say more than that. Nevertheless, the creature, in its own way has made it clear to me that it thinks about our human world as made-up, just as we think of theirs as made up. I tried to explain to it that I view all as a form of logic. Yet the creature was not satisfied, and after some pressure from it, I agreed to look at a certain human presumption.

So looking now at ourselves as creatures that perceive forms of logic in the unique way that we do, which indeed we are and which indeed we do, consider the logic that is a car, or a tree, and let us compare that with the logic that is our bank account.

A tree, or a car, is a logic going on in matter. On the other hand, so is your bank account, stored in one or more behemoth computers scattered around your country. In this section of our journey, we want to investigate, according to the general wishes of the creature, which is more "solid", the tree and car, or the bank account? It is not any quality like hardness that we are interested in. No, not that kind of solid. That kind of solid is the one humans would automatically think of when considering a tree or car versus a bank account. That is the hardness of what humans call a "physical" object. We want to look at a different kind of solid. I mean the sense of "solid" as in a logic going on that has durability, that has a natural tendency to continue existing, a logic whose existence is not flimsy, one that cannot suddenly, accidentally, go easily out of existence.

The logic of your bank account is scattered across an enormous amount of space, quite possibly going on in some of the matter in computers in several regions of your country as well as in data archives in basement vaults, as well as in communication links ranging from telephone lines to satellite connections. Let us focus on the central logic, that which credits, debits, reports and the like, the amount in your account. This central logic is gong on in your bank's computer(s).

There are potentially all kinds of bugs in the software that could cause the computer to crash. Your bank account could cease in a million ways, from software flaws to hardware failures.

Then there is a whole other direction from which the solidity of existence of your bank account is constantly under attack. There are powerful forces of crime who love to get into the logic of bank accounts and destroy it so as to shift value away from all the accounts to their own – it's called theft. Your account may not have much, but the sum totality of all the accounts at the bank does. The companies that own the banks have to take great care that the influence of crime doesn't permeate their own organization at any of hundreds of potential places, especially at the place of software for the bank account, and at the place of the numerous people responsible for writing and modifying the complex software. If crime starts to successfully intrude at any of so many points, it can start a process of destroying the going on of the logic that is the accounts.

Thus, which existence is more solid? The logic that is your automobile or the logic that is your bank account? Is you car in danger, day and night, of suddenly flickering out of existence because of a software bug? Can a hardware failure somewhere, occurring in any of a number of places spread out over thousands of miles, cause the logic that is your car to totally end – i.e. you car would just disappear?

The logic that is your car, or is a tree, is based on the laws of nature. Is there a possibility that at any second the laws of nature, the very laws of physics, will change for the material that is the substrate for your car, and the logic that is your car will cease to go on in that substrate, and what was your car will suddenly meld with the breeze? Hardly.

There is no large company required to have elaborate policies and procedures in place to keep criminals from interfering with the laws of physics less your automobile suddenly convert and become one with the air. There is no maintainer of physics who can slip up in writing code one day, whereupon suddenly a little later the material that is the physical substrate of the logic that is your car partially but repeatedly oscillates between the usual laws that make hard metal stay metal and some slight modification of the laws wherein now the metal becomes like highly electrostaticized liquid. Thus as you are driving down the main drag of your town, the supporting frame of your automobile, which carries all else, and keeps the thousands of plastic parts that are the substrate of the logic in their relative positions, that metal frame suddenly becomes like an oily electrostatic substance, which being highly charged the same in all its parts, flies apart from itself.

Thus do you and your car, moving down Main Street, go flying about. What usually looks like the underneath and lower part of the automobile now looks like some oil suddenly flying apart in all directions, partly a squirt, partly an exuding, partly an explosion. And with that, all the other pieces of the car, mainly the semi-liquid underlying frame, go off in their own directions, including you. Not to say that many of the plastic and non-metal parts do not have some cohesion to each other independent of metal and the underlying frame, so that as the oil, or what looks like oil, flies out in all directions from under the car, the various rest of the pieces of the car, somewhat holding in place, with some groups of pieces more than others, some holding together for as long as maybe forty seconds, go swerving, spinning down the street.

No, we do not have to worry about the logic of matter changing. It is solid. It is durable. And so is your car, your car being the logic based on the material substrate and on the laws of physics of that material substrate. Those laws are *solid*. They are not going to change. No one has to worry about keeping the laws going. No one could change the laws of physics even if they tried.

Of course someone could blow up your car, thus ending the logic that is the automobile. Or the vehicle could slowly rot away over many decades in a junk yard, also ending the logic going on in the material substrate simply because the material substrate would no longer be there, having all drifted away. However, all these are strong, active assertions against the car. But how different from the bank account, whose existence has to constantly be fought in behalf of, less it go out of existence. Thus, in one case, the logic's existence is so solid that only strong, active assertions will cause it to go out of existence, but in the other case, only constant active assertions will make sure the logic stays in existence. Which logic has the more solid existence?

Here is a major difference between a car logic – where when it goes on in a substrate of matter we say there is a car – and the bank account logic going on in matter usually in widely dispersed and shifting parts of the country – where when it goes on in matter we say our bank account exists. The car logic interacts with our various neurosensory cells to produce in us experiences of touch pressure, vision, and the like. However, as for the logic that is the bank account, our senses do not pick up at all where that logic is going on, they do not pick up whether it is going on in some computer in London, or Atlanta City. Nor can we do anything like move the matter that is the basis of our body to where the logic of the bank account is going on, and then have that electrostatic clash at the surfaces that occurred when we make contact with the car, or the tree. Nor is the information shown by the electromagnetic radiation (light) tied up with the logic that is the bank account in the same way as it is with the logic that is the car (To be sure, there is a kind of vision related to the bank account, but it is not as "intrinsic" as it is for the car; what we see for the bank account can shift wildly, from computer screen to computer screen, or as to what we see on bank statements – as the bank may change the format of the statement. Such electromagnetic radiation for the bank account originates far indeed from where the logic for the account is going on. It is not like the car logic in this regard.)

As for our touching a tree, clouds of trillions of atoms from the tree would intermingle with clouds of trillions of atoms from our body. Such intermingling starts up electrostatic force interactions, built on larger statistical electrostatic force structures of the clouds, and that would cause certain kinds of relative displacement with respect to our body, of the atoms in the touch receptors of our skin, and that would set of a long reaction that would eventually result in our experiencing touch. So from this perspective, the solidness of the tree, and our act of "touching" it and the subsequent awareness in our sentience, an awareness that we call "touch", comes from a statistical process of huge interacting clouds of atoms.

There is this characteristic of most of what we refer to as "physical" objects. There is a mathematical, statistical surface boundary, dividing (three-dimensional) space into an inside and an outside, and the boundary is such, due ultimately to electrostatic forces of matter, that it impedes our movement if we attempt to bring our physical boundary through it. I say "statistical" because nothing of this sort is exact, neither the spatial aspects, nor the matter that is "part of" the object, nor that the same matter remains "part of" the object. So we must speak of "statistical."

Thus, physical objects as well as abstract objects such as bank accounts, are equally logic, and both are logic going on in matter. While the difference in the solidness of their existence is relevant to us, as is also the difference in the solidness of our touch sensory interaction with them (our physical form coming up against where their logic is going on), this is not enough in the grand scheme of things (the metaphysics and ontology of the universe) to claim that one exists more than the other. They both have equal existence (even though the continuation of the existence of the bank account, in the media that we are in, is flimsy compared to that of the car).

They are both logic, and logic is what exists.

In these cases, what we call abstract is primarily what the "eye" of our thought, but not of our native perspective, can "see." What we call abstract is a logic that the "eye" of our thought can pick up on, but it is not a logic that our sensory neurons, for instance of vision and touch, can pick up on.

The Human Physical Body

If you access your bank account from your home computer, then at that point in time, one terminus of the physical substrate of the bank account logic is your keyboard and monitor. At least the situation could be seen in this way. The physical substrate, unlike that for a tree or car, can vary wildly as to the space where it is going on. (From a scientific perspective, the space where the bank account logic goes on can vary faster than that for a tree or a car – merely faster.) Another way in which an abstract object like a bank account varies from something like a car or a tree is the size of the definitional boundary: if you could look close enough, you would see that the surface of a tree trunk or leaf or root was a fog of atoms, a fog in which it wasn't clear whether the atoms were or were not part of the tree. Some abstract objects may have a much bigger area of fog or indeterminacy (is the keyboard and monitor of your computer part of the physical substrate for your bank account – well sort of and sort of not – is the bank computer that is the current primary one for your account – is that computer a part of the physical substrate of you bank account – well certainly more than is the keyboard and monitor of your home computer). But objects such as trees and cars also have these boundary regions of fogginess or indeterminacy, and so do objects such as bank accounts; only the bank account's is much bigger.

Issues of indeterminacy or definitional boundary can strike our thought processes as amusing – in this case, "amusing" being the experience due to different kinds of ideas coming together that had previously been classified by the brain as having no possible chance of connection. For instance, look at an apple. The apple is not you. Look at say your elbow. Clearly that is you – it is part of you. You take a bite out of the apple. As for that bite in your mouth, is it you or not, is it part of you or not? Well, no. At least, pretty much no. You chew it up enough to easily swallow it – unless some of the processes in the brain get a little off and you swallow some of it before it is sufficiently chewed up. So you swallow it. As it goes down your esophagus, is it you, is it part of you? Well, probably not. But it is getting there. And it is more part of you than when it was in your mouth, and certainly more than before it was put it in your mouth. In your stomach, it turns into acidic slurp, mixed in with all kinds of other stuff. What about then? The slurp moves into the intestines, where most of the desirable stuff of the apple is pulled through the intestine wall into the blood stream. Now it is really part of you. Or is it? All of this is the definitional boundary issue, that is, the area of fog or indeterminacy of what is in the physical substrate of a logic going on and what is not – the logic in this case being your physical body.

Is anything in your mouth, esophagus, stomach, or intestine you? After all, the whole thing together is just one long tube running through you. Your whole body is like a donut with a quite longish hole in the donut – that hole being the tube of mouth, esophagus, stomach, and intestine; and if you hold something in a donut hole, that something is certainly not a part of the donut. So nothing in that long tube through your body is in you. All of that is on the outside of you in the same way that the area in a donut hole is still outside the donut.

What about the air that you breathe into your mouth or nose and into the sinus areas, and throat tube; but now, rather than moving down the part of the tube that goes toward the stomach, the air moves off into a tube that branches to the lungs. There the air goes through the lung walls into the blood. Close up enough, the lung walls are filled with openings to the blood tubes, and so in this perspective the blood tubes are simply the same open outside space as the mouth, esophageal, stomach, intestine, trachea, and lungs.

Is everything "in" the body actually on the outside? Is it all just one big vast open porous drafty labyrinth, with everything open to the outside, and so everything being outside?

What about the piece of apple that helps build the bone in your little toe, or the other piece of the apple that helps build the muscle in your arm that moves your forefinger. It is still apple. Your body has just moved it to that place so that now the piece of apple can play its part in the logic going on that is your body. Your whole body is made of nothing but pieces of apple, asparagus, chicken, bread, air, lake water, river water – all of them arranged to play their role so that the logic that is your physical body keeps going on.

All these issues, these perspectives, all these are part of the definitional boundary issue, part of that issue of what is the size of the area of fog or indeterminacy of the physical substrate of a logic. We have no right to look down on the physical substrate of many of the things (logics) that we call abstract objects. As for the physical human body and as to what is part of it and what isn't, it is a matter of perspectives. Seemingly the only thing that is relevant is the overall logic that is the body that is going on.

One or Two Objects?

Why do we instantly perceive a car door as being *a part of* the car but a stone just next to the car as being *a different physical object than* the car? Logic in our brain, outside our awareness, immediately gives to our awareness this information. To see that this brain logic outside of awareness is not trivial, try to explain to someone else or to yourself, why you perceive the car door as part of the car and hence as one object with the car, whereas the stone next to the car as a different physical object than the car and hence the car and stone are perceived as two objects. Why don't you count the door and the car as clearly two physical objects?

Which kinds of logics are picked up by a creature is determined by the creature's evolutionary history. That includes the logics that give the relations between other logics, as for instance the above logics of the *part-of* logic, or the *these-are-two-separate-things* logic, or *this-is-a-physical-object* logic, and so on. These are special, important logics because they apply to a wide range of other logics. Someday when experts gaze well enough into the human brain, they will read exactly these fascinating logics, and furthermore, once these logics in the brain are understood, the experts will understand where and how it is that we perceive a car door (logic) as being a "part-of" the car (logic), whereas we perceive the rock (logic) on the ground next to the car (logic) as being "two-separate-physical-objects". The experts will understand not only how the logics in the brain work in this specific case but how they work in general.

In addition to the above, there must be a wide range of brain logics that give the relation between other logics. These logics are also outside awareness though they go on inside our head. Someday experts will not only see and understand these relational logics in the brain but will understand what principles they correspond to out in the world. One overarching principle we can divine right now. It is a principle applicable to all creatures no matter how they are laid out in the universe. It is how much different logics are (statistically) *independent* of each other – for this typically will determine how much the creature views the two logics as two separate, different "things". This is why the car door and the car are seen as one object while the car and a stone right next to it are seen as two objects. The logic of the door is tied up quite a bit with the car, whereas the logic of the stone is relatively, quite independent of the car. This relational logic between other logics, this relational logic that tells how independent other logics are from each other, this relational logic is an overarching principle for any kind of creature.

Whether a logic gives rise to a creature perceiving one versus two physical objects comes out of the creature's evolutionary history where it was, and whether that logic can be decomposed into two logics that have an independence of each other that was important in the creature's evolutionary history. Stated briefly, whether a creature sees one versus two objects depends on the creature's evolutionary history.

These issues (logics) are so much a part of our being and are so much outside our awareness that we take them as a given of the universe. But they are only a given relative to how *homo sapiens* have been structured by evolution, which is to say, by the coming into being of all our ancestors.

27 *All is Logic: Close*

Whenever we perceive something, that something is a logic going on out in the world. Whenever we perceive something, whether it is something like a physical object or is something like an abstract bank account, what we perceive is a logic going on out in the world. Whether it is a tree, a car, a table, or the body of another person, or our own body, or the long finger nails of a check-out clerk, or the kindness of another person yesterday at 3 p.m., or our bank account, each of these is a logic going on outside in the real world, a logic, a characteristic, a meaning going on in a dynamic of a constellation of matter out in the world, a logic going on even in a shifting constellation of matter.

What we perceive are logics.

Whenever we perceive something, the logic that is the something may be perceived at first with only one of our senses as for instance when we first see a car or a tree, we pick up the object with our eyes (photons, retina, photon sensory nerve receptors as either cones or rods, 100 million nerve signals from each eye reduced to 1 million channels of electrical spike trains going out over nerve fibers into the brain, and so on). But most of the time, the thing is perceived either right away or after a while with a whole range of our senses. More than that, there is a good deal of processing of all those electrical spike trains from all the senses, even before they result in the experience in our consciousness of the perception.

Whenever we perceive something that is right now out in the world, in truth we are not perceiving *it*. There is also a relevant logic going on in our long evolutionary history that we are perceiving. We are perceiving some kind of "put-together" – like a vision through another vision – of the two logics, one is the logic that is the thing out in the world right now, the other is the evolutionary logic that went on over a long time in the environment of our ancestors. Both logics in the external world come to be mirrored – one through the other – in a single put-together logic inside the brain – eventually the experience-itself logic.

"The logic could even go on in a way radically different from how science and mathematics might typically take logics associated with our body and being to go on," the god said, and then continued.

"The model furthermore allows the mathematical structure to contain logics that are not going on."

"Model?" I asked.

"That is something that will appear much later in the journey."

"Alright," I shook my head.

All these are logics. How many logics are there, how many can there be? What is their limit? Perhaps whatever is consistent is a logic.

Everything except matter is logic. Some might wonder whether the sole function of matter is as a means for logic.

"The vinegar bog creature?" the god said.

"Yes?"

"That is something I should not have mentioned. That is not something you would have found out on your own."

"But it illustrates the idea I would have eventually found out."

28 Logic to Physical

So far on our journey we have looked at a wide range of logics going on in matter. We might write this as

$$P \leftarrow L$$

Figure 20 Logic to Physical

indicating logics, L, going on in an area of matter in the physical universe, P. A logic goes on in a place, in an area.

In line with mathematical terminology, we sometimes speak of "$\leftarrow$" as a *mapping*. Logics are being mapped to the physical universe.

An amoeba is a certain logic going on in a cloud of atoms, perhaps a cloud of about 100 trillion atoms. If you build a play house out of blocks, that is certain logic, or relations or meaning going on in the set of those blocks. If you build a toy car out of a children's construction set, the logic going on in those pieces is the logic of a toy car. All these are logic going on in matter.

If about 65 trillion cells are arranged in a certain way, then a human body will be going on in those cells. A human body is a certain logic going on in matter. It is a certain meaning going on in the dynamics of matter. One might also say that it is a meaning going on in a cloud of about 7,000 trillion, trillion atoms, or one might say it is a meaning going on in a sea of molecules. But in all cases it is a logic, or meaning, going on in matter.

A tree is a certain logic going on in matter. When that logic is going on in matter, we say, "there is a tree." Our native perspective detects many of the characteristics and consequences of such a logic going on in matter, and our thought processes classify that as "there is a tree" (to be more precise, our thought processes put together "stuff" coming in from our native perspective in such a way that we use sentences such as "there is a tree").

279

A house is a certain logic going on in matter. When that logic is going on in matter, we say, "there is a house." When the logic that is a toy block house is going on in toy blocks so arranged, we say, "there is a toy block house." When a suitable logic is going on in a cloud of about 100 trillion atoms, and when we have suitable extension to our native perspective, and when we know enough, we say, "there is an amoeba."

All these are logics going on in matter. P ← L is a conceptual picture of this sort of thing. It is a very large perspective. Very large pictures, and they can occur in mathematics and science, often shift one's thinking just a little bit. P ← L. For instance, there is a logic in L, and the logic is being mapped to some place in the physical universe P. The amoeba logic is going on in a certain place, so conceptually, amoeba logic is being mapped to a certain place in the physical universe P. We have arranged some blocks so that the logic that makes for a toy block house is going on in that place, and more abstractly, from the P ← L, a toy block house is going on in a certain place; that logic that makes for a toy block house is going on in a certain place, conceptually it is being mapped to a certain place.

The phrase and idea of "mapped to" comes from mathematics, but the idea is clear enough from the above. Well, maybe it isn't clear enough. Or maybe it is. Depends on how you feel and how you set up your frame of reference. So maybe we will just say that it is clear enough in some ways.

Let us go back to the examples from the last several chapters on "All is Logic."

Any physical object, a tree, a car, are logics going on in matter. They are logics in L going on in a certain place in the physical universe P. P ← L. Ditto for a human body, a house, a leech, a *Euglena*, an electric fish. They are logics, in L, going on in certain places in the physical universe P.

There are the logics going on in the matter in the brain, going on in the oceans of electric signals zipping round in the gray slime. For instance, we looked at the JAR logic going on in the brain of the electric fish. Going on where? It was going on in a region that consisted of several pretty separated areas in the fish's brain. The "Devices to See Logic" took some of the logic going on in these waves of signals and created visual versions of them that our native perspective could pick up.

Other logics too, that whose going on we refer to as sound of a tree falling, or a piano chord playing, these too are logics, a variety of logics, going on at the source and going on between the source and our ear drum, and in a partial sense, afterwards in the electric signals in our brain. All these are logics going on in matter, in different places, but all going on in matter; all are P ← L, logics going on in matter, logics going on in matter at places.

Even a bank account is a logic going on in matter, P ← L. The area and matter where it is going on may shift far faster than for a physical object, but whether bank account or physical object, the underlying matter and area is always shifting; only dynamics of the speed of the shift are different for the different logics.

A few brief words about the P ← L concept and symbolism. The concept and symbol of ← comes from mathematics and is widely used there. Symbols are innocuous. Symbols are not innocuous. They are both. They can shift your mental perspective. The P ← L can shift one's mental focus to just the idea of logics, the logics going on all about one, going on in the vast clouds of matter, or vast sea of matter, or vast collection of matter, or however you think of it at different times. Mainly the symbols P ← L shift one's focus toward the character of the abstract dynamic going on in matter, and after it does that, it is hard not to use words like "meaning" for what it is that is going on in these clouds of matter. In this book, we introduced a more technical term, "logic". Then we use the word "meaning" to help give an idea of logic. Logic is meaning. A logic is a meaning. None of these terms can be defined fully well, yet we have a feeling for them. At times, by focusing on P ← L, we, and mathematicians too, can tend to see logics as something that starts to have substance on its own because it is something to study, to think about, to analyze: from a mathematical scientific perspective, they become things which can be studied, thought about, and analyzed as things in their own right. (That combined with the fact that our being, our sentience, is one of these things, should make them quite real to all of us).

But this moves away from the goal of this chapter, which is simply to introduce P ← L. Just as at a party, someone introduces you to someone else for the first time, you say a few words around and you now have an idea of this person. So too, this chapter has introduced P ← L, we've said a few initial words, and we have an idea of P ← L.

The

$$P \leftarrow L$$

fits the methodology and symbology of physics and mathematics, where one can postulate almost any abstraction, in this case a "logic" in L, and then analyze it, work on it, and think about the properties and characteristic of the going on of that logic in some area of the physical universe P.

In addition to issues of mathematics and symbology, P ← L matches what really goes on. All the universe is just clouds of matter. So important is the logic (the meaning) going on in the dynamics in parts of the great clouds of matter.

29 Definition of Evolution. Omega

Discussion with god

The god wanted me to write a chapter on evolution. I was talking about some ideas.

"It's 'through,' not 'from'. Some thinkers believe everything comes from evolution. But it's not *from*. Everything comes *through* evolution."

The god nodded.

"Furthermore, since this journey has revealed the central role of logic, let's give a new definition of evolution. I'm not sure that the old one was ever defined anyway. Not solidly. A new definition that is simple, to the point, and includes all the old possible definitions."

"That could be good for the first section."

"First section?"

"What about regular evolution?" the god asked.

What about it, I thought.

"Some statement about it."

"Oh, and what would that be, about cats and kittens?"

The god nodded.

"Cats? Cats!" But then in my mind I started to see a picture, and it was not from the god. Maybe one could have a valley of cats. Suddenly I imagined cats running wild all over the place.

"Something like a valley of cats." The god seemed thoughtful. Or otherwise it was accessing information. I couldn't tell.

"Alright," I said, realizing I would go along with the need for another section. Either I would go along, or I would make up my mind that way anyway.

"Here is what this chapter will consist of," I continued after giving some thought as to what would be a wise layout. "First there will be a statement that everything comes *through* evolution not *from*, and this will include a general, simple definition of evolution, that implies all other definitions. Then there will be a section illustrating the usual evolution, using a valley of cats and kittens in the wild. Then there will be a theoretical section on whether one could prove, in the abstract, that an environment's logic passes, transformed, into creatures." For some reason, as I said this, an image of a dinosaur popped into my thoughts, but just as quickly jumped out. "Then close the chapter with a note on the general, simple definition."

Definition of Evolution. Logic. Omega

We define evolution as the means by which logic enters into matter. It is the means by which more and more logic comes to be going on in matter. Some scientists say that everything comes from evolution. So they want to make it be an omega. But the idea of logic includes ever ascending rungs of logic, which in the earlier religious conceptions, were viewed as ever advancing toward an omega at least of logic. Evolution is merely the mechanism of moving logic into matter. In fact it seems one could prove theorems to this effect, once one formulated possible mathematical idealizations of the universe and evolution and a space of logic. Everything comes not *from* evolution but *through* evolution.

Evolution is the conduit through which logic moves into matter.

It should be noted that no matter how deep a logic appears to our understanding, generally it is as capable of passing into matter as is the logic that appears simple to our understanding. All logic, no matter how advanced, is already there to influence the physical. We might see an apparently happenstance characteristic in a form or in a creature, but in reality it may be a logic, or the start of a logic, which we will not be aware of nor understand for 10,000 years. Since all matter is randomly vibrating and so are all the logics built thereon, the random is always ready to start picking up, through the process of evolution, on all possible logic, no matter how off into infinity it is. Pure randomness in the physical reaches out to the infinite in logic.

One specific kind of evolution is that which Darwin presented, the process whereby the living things on Earth came to have their present form, that process entailing cyclical multiplicative individual instances of logics fanning out with stages called reproduction, birth, living, dying, all hung together with those little blueprints (DNA) in each individual instance of a logic, along with excessive reproduction as it interacts with the environment, the excessive reproduction being wildly culled away to leave a generalized imprint of the environment's logic. The next section illustrates regular Darwinian evolution, using house cats running around wild in a valley.

Valley of the Cats and Kittens

Imagine an open field with cats, or even a whole valley. We will assume there are no humans in the field or valley, or at least that none of the cats belong to any humans. The cats roam around free. We assume that the number of cats stays pretty much the same. That is not too unusual a situation.

It may vary on the type of cat, but let us assume, conservatively, that a female has, say, five litters over her lifetime. Further assume that there are five kittens in a litter. Some litters might be four, some might be six kittens. We will assume five. These are reasonable numbers.

Although the males move around, *computationally, we will get the same numbers* if we assume that a male stays with one female its whole life. So we can assume that we have two cats, a male and a female, having five litters times fives kittens in a litter, equals 25 kittens over the lifetime of the mother and father.

Here is the clincher. If the number of cats stays about the same in the valley over time, that means that of the 25 kittens only two will live into adulthood where they can start having their own kittens. This is on average. Some mothers and fathers will have more than two kittens live to adulthood, some will have less. On average though, only two will make it. The other 23 will not!

Thus 92% of the cats will die in kitten-hood. Some years it will be less. Some more. But if the number of cats remains overall about the same in the valley, then only 8% of cats will make it to adulthood. Only 8%. This is an important fact in the dynamic of the theory of evolution.

Which 8%? Those that are the best 8% at surviving and themselves multiplying, in the overall environment. The environment may throw various challenges at the kittens. Bad weather. Bad location for the nest. Maybe the parents were already weak in some way. Other parents having more offspring that make it to kitten-bearing years. Bacterial and viral infections. Other animals that want to attack the kittens. Other animals that want to eat the kittens. Food not easily gotten. And so on. The best 8% at successfully surviving these challenges of the environment will be the only ones to go on and themselves have litters. All the rest die. This is on the basis that the number of cats in the valley stays the same overall over time, and as stated, this is not an unreasonable assumption.

You didn't know evolution was so cruel, did you?

Where does the chromosome and DNA stuff enter all this?

Blueprints (DNA, chromosomes, genes)

In terms of this little vignette of cats running around this valley, DNA, chromosomes, genes, all mean the same thing. They are referring to a little blueprint that is in each cell of the creature. The blueprint is for how to build the whole creature, even though the same blueprint is in each of the trillions of cells of the creature (as stated toward the beginning of the journey, the blueprint is in a rather odd form: logically it is one big long sequence of letters from a four-letter alphabet – in humans this blueprint is a little over three thousand million letters long – that is lot in each cell –

but as also stated at the beginning of the journey, there are about 100 thousand, thousand million atoms in each cell – so there are plenty of atoms, and plenty left over even after what was used for the three thousand million letters).

Blueprints from parents, characteristics, environment's logic determines blueprints

One of the keys to the evolutionary process (on Earth) is that our blueprint is made from one-half the blueprint of each of our parents. The characteristics expressed by our blueprint are related in many ways to the characteristics expressed by the blueprints of our parents. These ways range from the simple to the totally complex to the no relation (there may be a combination of letters in the parent's blueprint that didn't get used at all because of the way in which they occurred in the blueprint, but the same combination of letters did get used in the child's blueprint because there they occurred in a different way). Some blueprints tend to give rise to creatures (cats) that are in the 8% of those who will make it to adulthood and have kittens of their own. These are the kinds of blueprints that, as the generations pass by, more and more cats will be built upon. What blueprints are in this 8% is completely determined by the logic going on in the environment. It is in this way that a generalized imprint of the logic going on in the environment is made on the surviving 8% of the blueprints; in this sense, a certain amount of the environment's logic, transformed, moves, over generations, into the creature.

Evolution sort of scary

Even though this little story is about cats, we can't help think about our own human blueprints and our own situation.

These are disturbing issues out in reality. Nevertheless it has to be kept in mind that the real situation is incredibly complex. The truth rides on a statistical fuzz of considerable magnitude. The logics going on in the environment are huge, and there are important variations between different areas of the environment. Hidden in every step of the above analysis are a multitude of logical complications. But the overall mathematical averages cannot be avoided.

Cat has no awareness of what it is part of

Incidentally, it should be pointed out that the cat has no awareness of the evolutionary "forces" and logics that determine and compose its very being. For instance, the cat's wonderful tail, which it can wave in so many disjointed ways, combined with tail behavior emanating form structures in the brain, along with the unique sensations and awarenesses that a cat must have regarding this appendage, derive from the environment of countless thousands of years. But the cat has no awareness at all of what is going in this larger picture. And we probably do not either, of our own environment, except for those small areas which we are able to figure out.

Abstraction of environment. Logic passing transformed into creature

Almost "magically" some hikers appeared, and conversation started.

"I wonder if you could prove that the logic in the environment passes into the creature, in some way.

Transformed."

"Transformed? Oh, probably. Time and all that. And probably more too."

"But could one prove it?"

We all talked back and forth. After a while, one of the hikers said, "OK, we look at this very abstractly, as a mathematical space. Everything. Including the environment.

"In fact, we imagine the environment to be a mountain. A mathematically clean-cut mountain in multi-dimensional space.

"And the creature is also mathematically clean-cut, as to what states its DNA could move to. And we have a lot of creatures, and those who don't keep climbing the mountain – that will be treated as ceasing to exist."

"Maybe if you wanted to investigate in a simple way environment characteristics that existed for a long time, the mountain should be replaced by a surface. A straight surface that slopes upward forever. With some logic embedded in it, whatever that means, but that would be the logic of the environment.

"Of course, eventually, according to the usual evolutionary perspective, those creatures that were better at climbing this surface, in multidimensional mathematical space, would over time spread their DNA to the whole population."

"Yes, because the others would have been 'outcompeted,'" another hike added.

"Well, here's the thing. In some sense, the continuing logic of that incline, over enough time, moves into the future generations of the creatures. Transformed though."

"That's what would have to be characterized. The nature of that transformation of logic from the environment, in this case, highly idealized and abstracted, to inside the creature, who also has a limited space of DNA variations."

"Well, then what would you have? Seems to me, it would be a very limited, highly theoretical result."

"But after one got it, maybe the mathematics would be such that it could easily extended to complicated situations.

"Who knows, maybe one could prove that any of a wide range of abstract environments could be mapped into these mathematically conceptual inclined surfaces. That mapping might give a lot of information about the transformation of the logic as it passes into the creature over many generations."

"What is an environment? It is all the logic that is going on in which the creature is physically immersed. In a sense, reflections of some of that logic are passing into the creature, and since a reflection of logic is itself logic, though transformed, the logic of the environment starts going on internal to the creature, but in a transformed way."

"Perhaps some of it isomorphic," I chimed in.

Longer logic in environment, more creature imbibes

The logics in the environment that affect how many successful offspring a creature has are the ones that the creature imbibes over time.

The longer such piece of logic has been in the evolutionary environment of a creature, the more will that creature imbibe it. There are always a wide number of variations in the creatures in a species. Some of the variations will get along better with the logic and others not so well. Of those not so well, they will be hindered, by definition, in having successful offspring, and of their offspring the ones hindered will have fewer offspring, and so on, until those hindered by the logic will be fewer and fewer. And the reverse will be true of those variations that get along better with the logic, so that over time they will be more and more. In this way, the species imbibes the logic. The logic moves into the creature.

The complexity and intricacies, of all kinds, including statistical, are great and many. Nevertheless, overall, over many generations, the law applies.

Since the longer such a piece of logic is in a creature's environment, the more the creatures will imbibe it, evolutionary psychologists sometimes concentrate on a logic that has been going on for a long time. For instance, in humans, the female has a much higher investment in each offspring than the male (This is because the female can have at most one offspring every nine months, and in most of our history, females could only have children for about two or so decades of their life.) The results of this logic, this meaning, this inequality of energy investment between females and males, valid over the incredibly long time of over the whole evolution of humans and of the ancestors of humans, make it a good subject of analysis and investigation in evolution.

In rare cases a logic is imbibed not because it has been around almost forever. An extremely powerful change in the environment can enter its logic into creatures very quickly. Such a change happened when a great meteor crashed into the Yucatan coast of Mexico, 65 million years ago, devastating the Earth's whole environment, and the resultant impact on life was the removal of those logics that characterized creatures such as dinosaurs, together with a flowering of those kinds of logic that eventually characterized the life we label as mammals and humans.

Dinosaurs and humans

Later on I went home from the desert, then to bed, when I had a dream.

I was in a restaurant, rather up-scale, with linen table cloths, napkins, many goblets and shiny silverware, with windows along the whole wall to my side looking out onto the night. When I looked out the window, there was nothing. No city lights, no automobile lights. I couldn't even see any stars, or moon, or trees or grass or the ground.

I looked back inside. The din of all the talking seemed gruff. It was then that I noticed the odd shape of the eating utensils. One looked like an immense ice pick. Another appeared to be a silver rake of a sort.

Then I started to notice that the din of conversation was guttural. But after a while it seemed completely natural. Even my own voice.

Slowly I looked up from the fine linen, away from the windows. All the patrons were dinosaurs, scaly, growly, of all kinds of sizes. None was too big, but big enough. Me too. I looked at my claws. They were large and kludgie.

Though we were dinosaurs, the conversations were witty and subdued.

Then I realized what had happened. The meteor that hit the Yucatan Peninsula 65 million years ago and wiped out the dinosaurs had never occurred. The dinosaurs had never been wiped out. We restaurant goers were the logic that entered slowly into us, through the ages. We had become the somewhat suave, sophisticated dinosaurs that we are now.

Having this thought made me put my hand, or I should say, claw, up to the side of my head. But there I felt a button. I pushed it. Suddenly the logic of all the bodies in the restaurant changed to that of ... for a second I couldn't place myself ... what were these smooth-skinned, uniformly sized creatures called ... I used to know. I looked at the table. My claws had become like those of these other creatures about me. And then my mind changed too. I knew right away what these were called – hands and humans. The logic that I now saw was the logic of the crashing meteor interacting with the logic of that ancient environment, and eventually entering us, in the transformed shapes that I now saw ... and was ... humans.

Wittiness, and the like, were the same. That is because those logics (pre-transformed) in the environment were the same, whether or not the meteor crashed into the Earth.

Again I whipped up my … hand … was it a hand … yes … good … to the side of my head. But there was the button again. Should I push it? I started to shake so much that I accidentally repeatedly pushed the button. Everything in the restaurant was going back and forth between humans and dinosaurs. My whole body was sore, especially my stomach. Then I had a horrifying awareness. There was button on the other side of my head too. I let out a long yell.

Thankful and sweaty, I was again lying in my own bed. Starting to get up, I noticed a pain in my stomach. Pizza was too greasy. I flicked on the light, for an odd second, wondering if there were any dinosaurs in the room. None. Good. Good.

As I went to the kitchen I turned on every light along the way, and had a nice glass of milk. One has to be careful with reality.

Is it fanciful to go back and forth between dinosaurs and humans? Is that because we can't see all the logic that is the environment? We can't see all the logic that we are. Both are big.

A Logic as It Occurs across the Whole Universe

Evolution is the means by which logic comes to be in matter. Consider the evolution of logics – all logics – across the whole physical universe. That physical universe is pretty far ranging. You can see that if you have ever looked at one of those picture books of the billions of stars in the heavens, where almost every page looks as if a person grabbed a handful of diamond dust and tossed it up to the night sky. Each spec of diamond dust is a sun with lots of matter swirling around it, matter in the form of asteroids, comets, rocks, dirt, planetoids, planets. There are about 500 thousand million solar systems for each person on earth. Here is the physical universe in our perception.

Consider one particular logic in all the places that it occurs in this whole physical universe. How well is that one logic surviving? Is the logic slowly changing in certain places in the universe, due to the environment? Are occurrences of it replacing occurrences of another kind of logic, or vice versa?

30 Deep Physics, Newton, Some Religion

All page references in this chapter, unless otherwise noted, are to the book

Newton, Isaac, *The Principia: mathematical principles of natural philosophy / Isaac Newton; a new translation by I. Bernard Cohen and Anne Whitman, assisted by Julia Budenz; preceded by a guide to Newton's Principia by I. Bernard Cohen*, Berkeley: University of California Press, 1999.

Idea of force was break with past. Newton.

Modern physicists take the conceptualization of something called a force as so fundamental that they don't realize what an intellectual breakthrough it was at the time. In Newton's day, the standard metaphysical predisposition was that there is matter, and there is motion of matter, and implicitly, nothing else. Hence it rationally followed that the only way matter could affect or influence other matter was to come into physical contact with it. There was no gravity or such a thing as force acting from a distance. This is the strict metaphysical position of what was called in those days, "mechanical philosopher." At first this was Newton's belief too. (See 1st paragraph of Section 3.4, page 56. Also, for a clear statement of this previous metaphysics, which goes back at least to Leucippus and Democritus, versus our modern one, see Guthrie's *History of Greek Philosophy*, vol 2, page 498.)

From Newton's own mind and apparently from whiffs of partial ideas floating around in his intellectual environment (see paragraph between pages 56 and 57), Newton early on, but not instantly, came to accept that, conceptually speaking, something of a much more "abstract" nature should be allowed in addition to the notions of matter and motion, an abstraction which today physics calls "force".

To be sure, in time, he went far beyond this early conceptual leap, with a multitude of derivative concepts and specific formulas and formulations of gravity and equations of motion and force, and a whole tool box of intellectual techniques for using these equations and the idea of force. But none of these would have come about without the initial shift in perspective due to this new concept that played the central role in understanding and analyzing the world. Without this shift, he would never have opened a new system of the universe.

The key was shifting away from the concrete. Physical objects and their motions – that is concrete. The idea of force, that is not concrete. You don't see it. You see the results – namely motion – but you don't see force itself. And you don't feel force itself, only the effects of force, the movement in one direction or another of material in your skin and body.

(You can see and touch physical objects. Technically speaking, your feeling and seeing of force is indirect; you only feel and see the *results* of the force; you only feel the displacement of the molecules in your body due to force; you do not feel the force itself; you only see the motion of objects *due to* force, as for instance a rock dropping toward the earth; you do not see the force itself. Something like this would have been the perspective from Newton's mental landscape. Why do I say "perspective"? If you feel the results of a force, isn't that a way of feeling the force itself. If you see the results of a force, isn't that a way of seeing the force itself. What else are measuring devices? What else are sensory neurons?)

Logic, force, derivation, fallacy about ontological primacy

For any logic that goes on in the physical world, its going on is traceable down to the forces between atoms.

These logics start to appear in the dynamic form of vast collections of atoms getting caught up in the dynamics of expressing logic. However, just because logics can be explained in terms of the forces between atoms doesn't necessarily mean that matter is primary and logic derivative. Perhaps it is like the notion of proof. Just because we prove A from the statements B, C, D, does not mean that A has in any sense a derivative existence, or ontologically, a secondary existence, compared to B, C, D. Indeed, it could *also* be the case that B is provable from the statements A, C. Derivation does not show ontological primacy. Proof is a human technique by which humans see in spite of their limited mental faculties. But whatever one sees exists, as long as one validly sees it (as long as the proof is valid).

Comparison of force and logic

In several ways, one can compare Newton's idea of force with that of logic as it is developed on our journey. Here we look at only two such comparisons, humans' ontology and that of spatiality.

The shift that Newton did in his thinking, to introduce the idea of force, is mathematically trivial compared to what followed. But in terms of ontology (what exists in the fabric of the universe) it is everything; everything else is built on it. Here is a case of great success by extending one's ontology. Might the idea of logics, as developed on our journey, have a similar role?

With logics, at least you can point to an area or region of areas where the logic is going on. But with gravity, you cannot, unless you want to point to the whole space of the universe, for gravity pulls between any two atoms anywhere in the whole universe. If this seems comforting for the ontological basis of logics, nevertheless as earlier sections of the book have indicated, it is only the *going on* of the logic that has a region. The logic itself is not so clearly connected to a location in space. Indeed, the same logic, or same sentience, could be going on in a small possibly quantum computer, or in the sentience part of the brain, or if it is a simple sentience perhaps in a segment of a galaxy.

The going on of logic is odd. You cannot indicate any one point where it is going on. It goes on, inherently, in a region. That is quite opposite the usual laws of physics, which "go on" at points. Thus, both the notion of logic and the notion of force have issues with spatiality, with logic having the greater.

(There may be a way in which the laws of physics are beginning to indicate a "weakening" of the "absolute" nature of the concept of distance in physical space. Ingeniously verified by experiment, a characteristic called *quantum entanglement* has two particles, with arbitrarily great distances between them, instantaneously related or connected. When one measures something called "spin", of one of the particles, the other particle, no matter how far away, even if billions of light years away in another cluster of galaxies from our own, will be *instantaneously* affected by that measurement. This appears to contradict the usual principle of physics that no "connection" between particles can occur faster than the speed of light. Indeed, this seems to indicate no distance at all – not from our native perspective of distance.

One idea for explaining this seeming paradox is to view the two particles as projections of a single particle that is in a higher dimensional mathematical space; and there are other ideas too for explaining the paradox (see for instance, the discussion in Martin Gardner's article, "The Guided Wave Theory of Louis de Broglie and David Bohm.") But all the ideas so far yield perspectives seemingly contrary to our native ideas about the absoluteness of distance.

All of this is in terms of just one pair of particles. But in reality there must be huge numbers of such particles. I do not know if furthermore there are characteristics of more than pairs, that is, it may also be triples, quadruples, and so on. Could this be seen as regions throughout regular space, each region identified as a single point in the higher dimensional space. In that case, the spatial issues of physics move closer to those of logics. Quantum entanglement, along with the regional nature of the going on of logics, suggest a perspective of distance different from that of our native perspective with our native layout in the fabric of the universe. This different perspective of distance would provide an alternative perspective of many things that we base on distance.)

Newton's *Principia*

In his great work, *The Principia*, which appeared in three editions, 1687, 1713, and 1726 (page 11), Newton describe this new system of the universe. But only more recently are we on solid foundation as to understanding some of his motivations in formulating this system. I wonder if perhaps these helped him not only in formulating but in withstanding some of the intellectual insecurities around the notion of "force" and gravity. As for Newton's work, not only was it of genius, so was it of concealment of motivation, which now, after 300 years, we understand much better. This information is in the new full re-translation of *The Principia*, in 1999, with extensive commentary and analysis, the first full re-translation in English in almost 300 years. It has a wealth of scholarship of the highest level. Books as these will never make much money, but a civilization that does not produce such will die.

The final paragraph in the third and last edition of *The Principia*, 1726, is the famous general scholium about what is gravity. Interestingly, the several paragraphs before that are the only place in *The Principia* where Newton talks about God.

Because this 1999 *Principia* with commentary is almost a thousand pages long, I have included some pointers that might be of special interest. The following are my phrasings, interpretations, and commentary:

How Newton came up with his ideas. Pages 14-21.

Some problems with the previous English translation of *The Principia*. Pages 25-33.

Newton's scientific ideas came out of background of several sets of ideas including those of his religious beliefs. Last half of page 58 to page 63. My guess is that these background ideas also helped him weather the insecurity and controversy about his idea of a force, gravity, acting across vast regions of empty space.

Hiding of motivations. Page 60.

Newton's concern about the strangeness of gravity – "action at a distance." His lack of explanation for it. Discussion of the legendary general scholium and his famous "Hypotheses non fingo" comment on gravity. Pages 274-280.

The translation of his general scholium and his comments on religion and theology. Pages 939-944.

A pure concept is rarely as simple as presented but instead gets its fuller meaning from its multiple incarnations in the world. To get a nice summary of the various incarnations of the concept of force, as it appears in Newton's usage, see pages 54-56.

Nice summary of Newton's and Leibnitz's place in history. Pages 369-370.

It took almost a century (page 215) of mathematical derivation from Newton's fundamental equations to predict, carefully, the orbits of Jupiter and Saturn. In other words, mathematics, I would point out, can be a slow oracle, a demanding provider, or a reticent handmaiden, depending on one's point of view.

A detailed statement of all definitions, laws, corollaries, lemmas, propositions in *The Principia*. Pages 945-971.

31 *Astounding Sentience Logic*

Introduction

What kind of sublime logic must sentience logic be? We feel amazed that something like the brain, something that should be nothing more than a dead machine, nevertheless, is us. It must be a strange logic indeed because we can't seem to imagine any process or logic in the brain that would truly be us. This chapter looks at possible paths to the answer.

(In many places in this chapter we go back and forth between two different ways of using the word "logic". One is the way the word is used in almost all of this book. This is where "logic" is going on in animals and in brains, where logic exists out in external reality independent of us. The other way is where "logic" is as conceived by formal logicians, such as Goedel. The logic of formal logicians is an idealization of important parts of the real logic out in the external world. These formal logics "contain" formal expressions that basically become the subject matter of investigation. Even so, one would think that once the essence of sentience logic in the brain is understood, either by humans, or by our machines, then the understanding could immediately be phrased in the logics of formal logicians. It is on this basis, in this chapter, that we look at the logic out in the real world as having essential aspects that are in those logics of formal logicians.)

Goedel and Chaitin

Is there some profound, earth-shaking new theorem about pure logic, a theorem waiting to be discovered by a future Goedel or Chaitin, a theorem which would literally revolutionize our understanding of the universe, a theorem that says how sentience can be in logic? In 1931, Goedel shook the worlds of mathematics and philosophy by proving, as a purely mathematical theorem, that no system of logic that had a certain minimum strength of expressibility could be used to prove that the system was consistent (i.e. that there was not inconsistency or paradox lurking in the system). Could there be another astounding theorem, about logic, that is at present as unimaginable as was Goedel's theorem before 1930? (For reference and comment on Goedel and Chaitin, see endnote 15.)

Cartoons or Movies

One reason that the logic of sentience must be rather strange is that it would seem that it cannot be of this or that form nor of many other forms. If it cannot be of any of a number of forms one has to wonder, well, what form could it take? In this section, we argue why it cannot be too close to any logic which is similar to a movie or sequence of logics "one after another."

Oh. I see someone walking this way. "Good afternoon."
"Good afternoon"
"Would you be, I mean are you..."
"I'm just hiking but I felt like arguing. Oddly, about sentience and the way cartoons or movies are just one frame after another, shown real fast. Though I have no reason why I should particularly want to argue this." The hiker shook their head as if they themselves were embarrassed at having said such a strange thing out of the blue.

"Nor is it that the logic of sentience is like a cartoon."
"Why not?"
"Well, suppose you draw a little person on sheets of paper, with the little person moving a little from sheet to sheet."
"OK."
"Then when you flip through them, you will see ...."
"I know, I know, you will see the little person moving. It's a cartoon – same technique used in movie cartoons."
"Right. Well, each single sheet is dead. I mean if I tossed a single sheet in front of you and you picked it up,

you are not picking up anything that has feeling, you are not picking up a sentient being."

"Obviously."

"Now suppose I take all the sheets and flip through them so that we see the little figure moving about. There's not an aware being in front of you. Is there?"

"No. Of course not. It's just pictures on sheets of paper."

I scratched my chin.

"Well, it's the same with logic in general. Suppose I have some logic in a computer, or for that matter, in a neural net. And for the sake of where I'm going with this argument, suppose that this piece of logic is not sentient - no being is aware here. Think of it as a single sheet in the cartoon," he waved a single sheet of the cartoon at me." I nodded.

"OK. Now suppose that I have a whole series of pieces of logic in the computer"

"Or neural chip."

"Now let's add another piece of logic, which just so I can refer to it, I'll call Lmovie".

"Lmovie steps through the other pieces of logic."

"What do you mean 'steps through'?"

"Well, somehow executes one after another"

"OK"

"Well, if that's all that Lmovie did, you still wouldn't have any being there. Just stepping through them. No being would suddenly be in that computer or neural chip. Not if each piece of logic was dead, and if all Lmovie did was just march through one piece of logic after another."

"Well I'm not completely sure what you mean, but I guess so."

"Hmmm, suppose you had 100 chips, and the logic on each was dead - there was no being that appeared on the chip when the logic was executed."

"OK."

"Suppose we took another chip, and wired it to the hundred others so that it executed one chip after another, in sequence."

He looked at me.

"Well, let me rephrase a little. Suppose I have a 100 cars around me in a circle. Suppose I start up each car and then turn it off. Then I go to the next one, and start it up, and turn it off, and so on around the circle. And then when I turned off the last car, I went and got a little baby and brought them into the center of the cars."

My eyebrows raised.

"When that little baby was brought into the center of the circle, you would know that a being, with awareness, had appeared. When the cars were being turned on and off one after another, there was nothing like a being in the assemblage of cars."

"Obviously. A car is dead. Having a hundred cars go on and off in order is no closer to having the being of a baby than one car is."

"Exactly. And it's the same with the neural chips. If there was no baby there with a single chip, there's not going to be with a whole sequence of chips that you simply turn on and off after each other. "

"So you're saying, if you don't have a being in any single chip, then turning on an off a sequence of a hundred different such chips is not going to cause a being to be there."

Cartoon Summary

So here the moral is, sentience logic cannot be some simple sequencing of logics. The meaning of "simple sequence" is left somewhat vague, but it seems clear in this argument that anything too much like just a sequence of logics cannot make a being, cannot make or be associated with sentience.

Puppet and Other Forms/Constructors of Logic

When I was a child I thought that if only I could make a cartoon, with flipping sheets of paper, then the conundrum of sentience would be resolved. Finally, one day I made a little cartoon of stick figures. The only thing that instantly became clear was that this was not the answer. There was no sentience there whatsoever.

After that I became convinced that the conundrum of sentience would be solved as soon as I saw a puppet – the kind with strings attached to the body and by pulling the strings you move around the different parts of its body. As you can see, a child can be quite naive, and this was before grade school. (Sometimes I wonder if, on these kinds of issues, we are just as naive as adults as was our child-self, and that is why we need machines to do our understanding for us.)

One day my uncle brought over a puppet and moved it with the strings. It was instantly clear that the puppet and strings, all of it, was deader than a door nail. This wasn't the answer either.

This can be put in the form of a statement about logics. A puppet logic is basically where some logic X, is controlling or "pulling the strings", of another logic Y. This gives no answer as to what is the amazing aspect of sentience logic. It only pushes the problem back to string-puller, logic X – we would still have to figure out why logic X is a sentience logic. (This is the "ghost in the machine" fallacy.)

There are other standard forms of logic, and it likewise seems that none of these offer any answer either. Depending on the formalism, these other forms can be built from constructors such as AND, OR, NOT, IF, DO-WHILE, etc. Below, we will look at the general statement.

Summary statement of cartoons and puppets

A sentience logic is not just a sequence of logics or a problem pushed back to another problem, nor is it any kind of simple combination of logics. As Parmenides (in the period around 480 BC) stressed: you cannot get something out of nothing – only nothing comes out of nothing. As for sentience logic, there is some essence of it, some conundrum in there, that brings it all together. Some day, when it is the first time the conundrum is understood, some human, or likely machine, will gasp, "Wow!" Then a second and a half later, in hushed, reverential tones, the person or machine will add, "that's it!"

A More Formal Approach. Constructors

One can approach the astonishing character of a sentience logic in a formal, mathematical way. Some parts of the approach need to be modified, however at the end we will point out which. But the results are so interesting. One single mathematical formula leads to the different ways that sentience might inhere in the fabric of the universe.

The logics used by formal logicians are built out of "constructors" such as *and*, and *or*, and *not*, and, depending on the formalism, perhaps constructors for actions, such as *if ... then ... else* , and *while ... do*, or the constructors could be a machine that joins component machines, and indeed the constructor could be anything which the logician imagines, or for that matter, which program language designer imagines, or a neurobiologist designer of formalized specifications of brain processes imagines. For instance, perhaps a cartoon movie logic might have a constructor called sequence-play, and the idea is that

Sequence-play(L1, L2, . . . , L1000)

would be something like flipping through the logics L1, L2, . . . , L1000. As the argument stated earlier suggests, such a constructor does not help explain sentience. Indeed it is hard to see how any constructor could.

In fact let us look at the more general case of a formal logic. Consider

Constructor (L1, L2, . . ., Ln)

No matter what constructor the formal logician has come up with, it is hard to see how this could be a sentience logic unless one or more of the L1, L2, ... Ln were already themselves sentience logics. But it turns out that this implies that nothing in this formal logic could be sentience logic (the mathematical proof would go something like this: pick an expression for sentience logics that is minimal length; as just pointed out, that expression cannot start with any constructor; therefore it must be a primitive in the formal language; but that defeats the whole purpose of which was to show the structure of a sentience logic).

There is a seeming impasse here. Our journey, this book, is from the perspective of science, and so we assume there is sentience logic going on in the brain. (Even if we step outside the perspectives of science and we assume there is only a correlation between our sentience and some logics in the brain, the problems discussed in this chapter will apply to that logic that is tightly correlated with sentience, and the same astonishing character will arise. Further, as regards this correlation, there are many places in the brain where signals introduced by extermal electrodes will cause no experience, that is, they do not affect consciousness, that is, in an appropriate display, many other logics can be activated but they cause nothing to enter consciousness. The seeming impasse is that there seems to no logic that can be a sentience logic. Yet clearly there it is in the brain.) (We also assume that one day when sentience logic is understood, either by ourselves or by our machines, it will immediately lead to being able to cast the key characteristic of sentience logics as a formal logic.)

There are several ways out of this impasse. One is that there is some quite amazing constructor that creates sentience logics out of logics that are not sentience. And amazing it must be, since at present we have no convincing ideas as to what it could be. Perhaps that constructor will come from some future theorem as amazing as Goedel's. For just as Goedel's theorem shook the world of mathematics and philosophy, by showing that formal logical systems of sufficient power could never prove their own consistency, so too will some future theorem show that something amazing – sentience – can arise out of non-sentience logics when put together (with the right constructor).

Another way out of the impasse is that the constructor will not be so grand as another theorem like Goedel's, but instead will come out of further knowledge from neurobiology. For this approach, see below.

A third approach seems strange, but who knows. It is that certain constructors add a little bit of sentience to the sentience already in the components (the L1, L2, ..., Ln, above). This is in line with a form of *panpsychism*, that sentience inheres throughout "everything." Panpsychism will be discussed below.

(For one consideration about this perspective of constructors, see endnote 16.)

(The idea that a huge number of constructors, each adding almost nothing, can add up to a lot, is illustrated in your personal computer. When you log onto that little machine, *everything* that it does, *all* the interactions that you have with it, *all* the colors and letters and shapes it displays, *all* the sounds it generates, *all of that* comes solely from an unbelievably huge number of simple constructors, "machine commands", applied to an unbelievably huge number of 0's and 1's. But talking in this way is potentially slippery and tricky. Every computer programmer knows that the "real" meaning takes place at a much "higher" level than the machine commands and the 0 and 1 bits.)

(One can replay an argument in different ways, in order to solidify understanding and to address vague questions. Suppose that at some future date, we or our machines or some being or some beings' machines understand the sentience logic as it goes on in our brains. Is there any "component" of that logic that is itself already a sentience logic. If so, look at that logic. And ask the same question. Is there any component of it that is itself sentient? If so, then focus on that logic. And keep doing that until we arrive at a logic going on in the brain that is sentient, but no component of it is sentient. Look at *how* the sentience logic is put together from the non-sentient components. That 'how' would be the magical constructor sought above, that would be what turns non-sentience components into a sentience logic.)

Neurobiology

Where might the answers come from, what might they be?

Could some understanding of this astounding thing in logic come from neurobiology, as maybe a new scientific awareness about what can go on in the brain (and consequently also an advance in formal logic). For instance, at times in the past there was some concern whether the working of the cerebral cortex would ever be understood in terms of the working of individual cells. At the least there was a concern that possibly even after understanding a great deal about the nerve cell, still not much about the workings of the cerebral cortex could be understood.

"It has long been realized, of course, that knowledge of the cellular properties of neurons is essential for any detailed study of the brain. Nevertheless, there were no clear indications of how an understanding of membrane properties, signaling, or connections could help to explain intricate psychological phenomena such as depth perception and pattern recognition." (Nicholls et al, *From Neuron to Brain, Third (not Fourth) Edition*, 1992, page 2.)

Nevertheless, eventually, it *was* seen how the workings of individual cells could explain these other aspects of the brain.

"Such a pessimistic view has now lost much of its force." (Nicholls et al, *From Neuron to Brain, Third (not Fourth) Edition*, 1992, page 2.)

The answer, or resolution, lay in the *concrete* knowledge gained from cells applied to the connectivity of groups, or neural nets, of cells.

"Although new functions that are not to be found in a single cell do emerge from networks, the modeling of neural circuits becomes a sterile exercise unless it incorporates the known properties of nerve cells." (Nicholls et al, *From Neuron to Brain, Third (not Fourth) Edition*, 1992, page 2.)

So too in the future, possibly, some concrete knowledge that we are not now aware of, will contribute, in context, to bringing into our understanding more sophisticated aspects of sentience then we are now able to imagine.

We have the gift of the brain, in external physical form, in which is written so much of what evolution has discovered in a journey of billions of years.

Are Discoveries Such as Monkey Mirror Cells Part of the Answer

Let us mention a path that sometimes seems to be part of an answer but if it is, it is unfortunately not a simple answer.

We have brilliant knowledge of amazing things in brains, one example of such is the monkey mirror cells. Many neurons in the premotor cortex fire when the creature performs a single task. The mirror neurons also fire if the monkey sees another monkey doing that task (For references, search internet for e.g. "mirror neuron").

Does this wonderful knowledge advance our knowledge of sentience? An un-sentient machine could do the same as the monkey mirror cells. The neural chips of the machine would just need to carry out this awesome categorization and feature detection logic. But if that is all the neural chips did, the machine would be dead, there would be no sentience there, it would just be a big dead feature detector and categorizer.

Now that we know there are such things as monkey mirror cells does this bring us closer to understanding sentience. Not directly. But presumably these cells have paths reaching into areas where the sentience logic is going on. On the other hand, so do many other cells.

Perhaps our current knowledge of what is sentience logic would be like having a few small points on a sea. Even if sentience logic were only directly involved with a hundredth of the brain, it would still have a thousand million cells - it would be a thousand million dollar house, just by itself. That is an awful lot. Maybe the problem of trying to imagine what is sentience might lie in the sheer size of the logic.

Panpsychism

What if the answer lies in none of the above? What if there is simply no *astounding* aspect of the logic that is sentience logic

Yet, as pointed out earlier, whatever your beliefs, at the very least the sentience logic in the brain has deep, extraordinary connections to sentience.

What if we went ahead and said that *all* logic has or is correlated to some degree with sentience. But where there is greater depth of logic, there is a greater amount of sentience. Perhaps the amount of sentience is only what that logic can experience in the mechanical sense of experiencing.

For instance, the logic going on in the human brain is so vast, so deep (even what it is like to experience red may be quite deep in terms of logic), that this is why, we, a sentience, have what seems to be a miraculous range of experiences - miraculous to us but not to some advanced, super machine or computer that can figure us out.

I take panpsychism, generally, to be the belief, that sentience resides everywhere throughout the universe. The specific version of panpsychism suggested in this section is that sentience resides throughout the universe, in degrees, with greater degrees at those regions where deeper logic is going on. We say "region" because logic typically goes on in a region of, not at a point in, space (see the section "Structure of Logics" at the end of Part 2).

(For one "problem" with panpsychism, see endnote 17.)

(For references on panpsychism, search the internet. One good site is
http://plato.stanford.edu/entries/panpsychism/#Bib)

Part 4   Onwards
Some More

32 All Souls are Waiting Right Now

Introduction

Another part of the book had been finished and I looked forward to giving the report to the god, but in the usual little desert clearing, I saw no one. For a second, the thought went through my mind that maybe I could complete the journey on my own.

"Ah," came a voice.

I glanced down. There was the god's head, by itself, lying sideways on the ground, looking at me. A little distance away was one of its arms, and further in another direction was the other, and around them were scattered more parts of the body.

"Ah, I'm not feeling too well."

"Yes, I can see," my eyes opened wide. "What?" But I received a not surprising no answer.

"Ah, there is a lot to do. *You* have a lot to do."

"*I*? A lot has been done. A lot of the journey has been covered already," I pointed out.

"I want more on sentience," the god said. "I *need* more on sentience."

"*You*? *You* need help," I said looking around at the various pieces of its body.

"I need *your* help."

"You need somebody's help." Then I added, "I've done enough. And I've been told enough!"

"Only need some of your help. Most of it is done. Just a little more."

Perhaps I acted too quickly. Was it wrong? I turned on my heel and walked out of the little clearing. I admit that I was a little surprised the god didn't say anything more. This was the most human the god ever was.

To get far away, early next morning I went a whole day's drive to the Northwest, far out into nowhere, to a park in a gorge amidst mountains with cliffs that dwarfed humans, rising toward bright whiffs of clouds against blue. But still the desert heat was there. After pitching my tent, I drove to town to eat and in the evening hiked around the park. There were virtually no people. Possibly there was the caretaker and maybe there was another party way on the other side, over a flat across from the cliffs. At 10 p.m. I took a flashlight and hiked down into a deserted valley next to the cliffs

Loneliness is a great way to commune with the earth itself in certain of its manifestations, and in back of those is the universe. Being so upset with events and the god, I probably got the date wrong or the park wrong, but there was supposed to be some rock-finding club there, or so I thought. Whatever the reason, I saw not one soul there, except at the start and end.

At the end, the late afternoon sun was streaming down. I was talking with the woman caretaker, who was raking some of the gravel areas. "There are organizations that can freeze you immediately when you die and, *if* all works out well, will unfreeze you at some far distant future date, when, with the advanced state of future science, you can be brought back to life. Some people think that it's a big 'if – if all works out well'," I said.

It was later while watching the wind stir around little eddies of dirt and twigs, that the idea came – the theorem – as if from out of the desert.

Every sentience, after its physical form dies, is in a state of waiting.

Here starts the technical material of the chapter.

This result of this chapter is in the form of a theorem. In plain language, what that means is that we have some assumptions and on the basis of those assumptions we prove a result. But beyond it being a theorem, the key idea is that we can assign *any meaning* to the terms in those assumptions, and the result will still be provable. In the kind of theorem we are proving, this is important because the "true" meaning of some of the terms may not be known for hundreds or even thousands of years, or maybe never. But that meaning is still there.

Now someone might ask, how can we prove something from assumptions without knowing the exact meaning of the terms in those assumptions. The answer is that the proof depends not on the meaning of the terms but purely on the logical form of assumptions independent of the meanings attached to the terms.

Still, it is necessary to state a caveat. In spite of depending on the logical form of its statements, the theorem is also about the real world, and hence it will not fit perfectly into this ideal mode. One can assign plenty of different meanings to the terms, but one cannot apply just any meaning whatsoever. Moreover, we do not go too much into what must be subtle issues that restrict the allowable meanings. Even so, the theorem presents a thought provoking result about the "death" of the physical form that is associated with a sentience.

To explain the theorem we must first describe a process called *cryogenics* or *cryogenicization*. That is a process of deeply freezing a person at death - preferably just before the moment of death – so that some day, when technology and science have advanced enough, they will be able to bring the frozen body back to life. When we say "frozen," we mean *really* frozen: hundreds of degrees below zero; it is close to absolute zero. The atoms, and even less the molecules in relation to their size, in the body hardly move at all. Truly, everything is almost absolutely still. So the body, including the brain, and all the connections in the brain, stay basically exactly as they were at the time of freezing, even if the body should stay frozen for thousands of years before science figures out how to successfully resuscitate it. We will not go into the technical issues of whether the freezing can be accomplished well enough today so that the body will ever be able to be successfully brought back. Nor will we go into the possibility of alternatives to reducing the temperature to close to zero, as for instance possibly spreading some substance throughout the body that accomplishes the same thing as freezing: to stop the motion of all the matter indefinitely. All the theorem requires is that someday there exists the technology to successfully freeze, or essentially to stop all molecular and atomic motion in the body; and then at some distant time further beyond that, when science has progressed even more, science be able to successfully thaw the body and bring it back to life. This is all that the theorem assumes.

By the way, though it is *not necessary* for the theorem, we still might ask why would anyone want to be thus frozen at death. Well, today, if one were frozen, and the process worked, then at some time far in the future when science had figured out all the things needed to successfully bring the body back to life, you could be brought back to life in a world that might be able to keep you alive forever, and in a much healthier world. Or if today you have a terminal illness, you could be frozen and stay that way until science had figured out a cure for your illness, even if that took centuries or millennia. Then your body could be brought back to life and be cured. These would be some of the uses of cryogenesis.

All Souls Theorem

Consider a physical body that has a sentience associated with it (for instance, you yourself are a sentience, and that sentience is associated with your human physical body). Then suppose that the physical body is cryogenicized. (Your body is deep frozen in the cryogenesis process). Suppose that after a period P of time, the physical body is successfully "brought back to life," that is, the sentience is once again present. During time period P we will call that particular relation of the sentience to the fabric of the universe, *waiting*. Just for this chapter, the term *waiting* will be used is this sense, i.e., with this meaning.

Let us state this formally. Consider the physical body that a sentience is associated with. Supposed that body is frozen as part of cryogenesis, and especially important, suppose that later the body will be successfully unfrozen and brought back to life. By definition, we will call the relation of the sentience to the fabric of the universe during the period that the body is frozen, waiting.

Theorem. Suppose a sentience is associated with a physical body and then that body dies. Then forever after, that sentience is waiting.

That's right. Even though no cryogenics is involved, the sentience that was associated with the now dead body is *waiting, it is waiting just as much* as is the sentience with the body that was cryogenicized and will come back is waiting during the period that the body is cryogenicized. Forgive the awkward grammar, but the theorem is saying that after death, a sentience is waiting no less than if it had been part of a body that was successfully frozen *and will be successfully brought back.* You say that this is not obvious. And I say, that is right, it is not obvious. That is why it is a theorem. That is why it has to be proved, and that is what we will now do.

The idea of the proof is to look at "right now" in three different situations, and we will argue that the relation of the sentience to the fabric of the universe must be the same in each of the three different situations at the "right now". To make this more concrete, we will assume that the sentience is our dear Aunt G. In the first situation, Aunt G was successfully cryogenicized ten years ago, and right now is now ten years later, and further, in this situation, it is given that Aunt G's body will be successfully brought back to life a 190 years from now.

In the second situation, Aunt G was successfully cryogenicized ten years ago, and right now is now ten years later, and further, although no one knows it now, in this situation, it is given that 189 years from today, a terrible mishap occurs. A thousand pound slab falls on the profoundly frozen physical body, crunching it as if it were powdered snow. So much for Aunt G.

In the third situation, Aunt G died a natural death ten years ago, and right now is now ten years later.

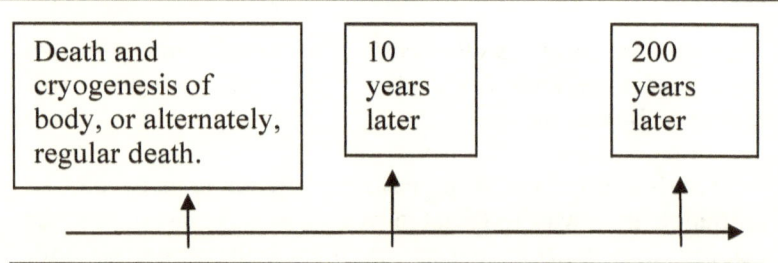

Death and cryogenesis of body, or alternately, regular death.

10 years later

200 years later

Situation 1. Aunt G frozen and 200 years later successfully brought back to life.

What is the nature right now of Aunt G's being, or sentience, or soul, or whatever you want to call it, in the fabric of the universe ("right now" is 10 years after she is frozen)?

Situation 2. Aunt G frozen and 199 years later there is a mishap: Frozen body accidentally destroyed.

What is the nature right now of Aunt G's being, or sentience, or soul, or whatever you want to call it, in the fabric of the universe ("right now" is 10 years after she is frozen, and it will not be for another 189 years that her body is accidentally destroyed)?

Situation 3. Aunt G dies a regular death. No freezing.

What is the nature right now of Aunt G's being, or sentience, or soul, or whatever you want to call it, in the fabric of the universe ("right now" is 10 years after she died)?

Figure 21 All Souls Proof

Situations 1 and 2.

Look at Aunt G's situation, 10 years after cryogenesis, where she is later brought back successfully at the 200 year mark, versus when they botch everything up just before the 200 year mark. At the 10 year mark, the two situations are still the same. So Aunt G's being in the fabric of the universe is the same, after 10 years, in the two situations.

Situations 2 and 3.

Now look at Aunt G's state of being, 10 years after a regular death ("regular death" meaning no cryogenesis). Compare that to Aunt G's 10 years after cryogenesis, where 189 years further into the future the frozen body will be totally destroyed, without ever having been brought back to life. In both situations, at this 10 year mark, the matter in the body is meaningless, is dead, as far as sentience being there. In the one case where she died a regular death, the matter 10 years later is clearly meaningless, because there is nothing left of it in a relevant organized state except maybe some bones. In the other case where the resuscitation is botched far off in the future, the matter 10 years later is at virtually an absolute standstill. Even if you looked close up at the atoms, there would be virtually almost no motion. That body is deader than a door knob (at least the atoms in the door knob are moving a little due to not being so cold). After another 189 years the form of the physical body is totally destroyed. In one case, the body is quickly decomposed. In another case it becomes like still, dead twigs and dirt, and then staying absolutely still for another 189 years, is then decomposed or destroyed. I would think of the body as nothing more than a dry, dead, crackly shell. There is nothing there. There is no logic going on in that matter. No being is there any more than in a hill of rocks, at least not as far as sentience in the fabric of the universe. So Aunt G's being in the fabric of the universe is the same after 10 years in the two situations.

Since 1 and 2 are the same (at the 10 year mark) and since 2 and 3 are also the same (at the 10 year mark), it follows therefore that *1 and 3* are the same (at the 10 year mark). In other words, the state of Aunt G's sentience or being in the fabric of the universe, 10 years after a regular death, is the same as 10 years after if she had been cryogenicized where a 190 years into the future she will be successfully brought back.

As for the numbers 10 and 200. They are arbitrary. The same proof would work for any two numbers whatsoever (provided of course that the second was larger than the first). So in place of 10 years after her death, the proof could be for *any* number of years after her death. What that means is this. Consider the situation where Aunt G has died a regular death. Look at *any* point in time after her death. At that point in time, her soul is waiting to come back. In other words, at that point in time the state of her soul or sentience or being or whatever you want to call it, or however you want to phrase it, in the fabric of the universe, is the same as if she had been cryogenicized and will be successfully brought back in the future. That is the theorem.

End of Explanation of proof

That then is the theorem. It seemed natural to call it the *All Souls* theorem since it says that any soul (i.e. sentience) that ever existed, and whose body died, is waiting right now. Or stated in a more casual manner, all souls are waiting right now.

Discussion and Close

The above then is the fascinating theorem. Just as a textbook sometimes presents a result and then follows it with a series of exercises for those who optionally want to think about further issues, so too we now present a set of open-ended discussion points.

The sameness of situations 1 and 2 at the 10 year point is obvious. It is obvious that after 10 years the state of Aunt G's being in the fabric of the universe is the same whether or not the resuscitation attempt is later on botched 200 years off into the future (situations 1 and 2). Still, it would be interesting to see what kinds of assumptions about time, space, matter, logics, and state of being (sentience in the fabric of the universe) lead to (imply) this sameness.

(If one were to think about possible assumptions, in a form that physics and mathematics uses, then the *waiting* would probably become *waiting(t)*, meaning that at each point t in time, *waiting(t)* is the relation of Aunt G's sentience to the fabric of the universe at that particular point t in time. Admittedly, it may be hundreds or thousands of years, or never, until we know fully what *waiting(t)* means, but as the theorem shows, we can still prove some theorems about it.)

The sameness of situations 2 and 3 at the 10 year point is not as stone solid – this is the situation of the plain, regular death versus the situation of the cryogenesis with the botched resuscitation a further 189 years into the future. Nevertheless, the line or reasoning presented in the proof above seems more solid than any alternative; any alternatives would seem to be abstruse. That is why it would be all the more interesting to see how something called axiomatizations would apply to this sameness or lack thereof. In exploring such formalizations and axiomatizations, I suspect one will be plunged into issues of in what way a sentience has a whereness in our regular time, space, and matter.

Regardless of what the truth turns out to be, it must be admitted that all these issues are of the utmost strangeness.

I had written the above and was hiking amidst stones and some trees too, for by now the way was getting to a higher altitude. Suddenly I heard two hikers come across the rocks behind. They had that usual slightly befuddled look, though they said how beautiful the area was. I handed each of them a copy of what I had written and explained the proof. But the second hiker spoke up right away.

"I disagree with the second part of the proof, that the plain, regular death is the same as a failed cryogenicization."

"Well, there are more reasons for it than I wrote, and here are the reasons why failed cryogenicization is the same effect as a regular death.

"Looking at the cryogenicized body is like looking at nothing more than a piece of paper with print on it. It is like the page that you are holding in your hand right now," I pointed to the papers the hiker was holding. "That page. There is nothing to it in terms of it itself being alive. It is dead. Even though it has letters and words written on it. It is the same with the cryogenicized body. So to speak, there are letters and words printed there, but the body itself, the page itself, is utterly nothing." I felt I had made my point. But now the first hiker intruded but went in a different direction.

"The cryogenicized body. Ah, but it is not like a sheet of paper with letters and words printed on it! The body is not stopped. Everything in that frozen body, the atoms and electrons, is still moving, only *extremely* slowly. Maybe 1 second is spread out to 10,000 years. That's how slow the molecules in the body are moving. So the body really is still in its living state, only it has been slowed down by an incredible amount. So situation 2 and 3 are not the same at the 10 year mark."

This upset me. "Your argument *sounds* good. But I don't think it is valid in some larger context. Still, I have to admit that I can't say clearly what I mean by 'some larger context.' If something has slowed down that much, then for all intents and purposes, in the real world, it has stopped. It has ceased. It is as dead as leaves and dirt and twigs swirling around in the wind."

I was feeling somewhat unhappy that I did not quickly came up with a good counter argument. To my surprise, a third hiker jumped in – third hiker? – where did this one come from?

"Oh, he's real alright," the first two hikers said in unison, but immediately they looked even more confused that they had said such a thing.

"What if," the third hiker said, "one of the axioms were this. Any collection of matter that has no motion whatsoever has nothing more related to sentience in the fabric of the universe than does the collection of matter that once constituted the physical body of a person that has since died a regular death. As for the issue of the atoms still moving slowly in the cryogenicized body, though very, very slowly, why, I say, that nothing in the real world is totally still. So once things get *really* slow, it's the same as if they ceased motion. And anyway, who is to say that the really slow motions of the atoms in a cryogenicized body are anything like a slowed down version of the *motions in the living body*? In fact, I bet they are not, not at all. Therefore the logic of sentience is no longer going on in that extremely slowed down moving matter – because different motion is going on – not just slowed down motion – but different motion."

This was good, I thought.

Now an earlier hiker spoke again, referring to what was already said, but going to make a larger point.

"Someone said that the difference between a temperature of absolute zero and a temperature a little bit higher *is* significant. In the argument of the second hiker, this small difference was made responsible for whether Aunt G's sentience was or was not going on in the frozen body. But to me, this sort of reasoning seems to violate a general principle.

"The principle is this. A hair's breadth of any kind of physical difference cannot be associated with a major difference in conceptual issues around sentience and the fabric of the universe. If such a hair's breadth in a particular analysis results in a major consequence, then something, somewhere else, has to be adjusted somehow in the analysis. If one still puts forward such an analysis, one must additionally, explicitly address the issue of why a hair's breadth of physical difference should make a major difference in the state of the sentience and the fabric of the universe."

Even though the god was in terrible shape, it somehow sent me the following mental communication, with more open-ended points for discussion. So I will just copy them here.

The P ← L framework will be developed more on our journey, but the basic idea was already introduced in the chapter "Logic to Physical." This was logic L being mapped to the physical universe P; in this framework, logic is thought of as being mapped to that place where it is going on in the physical universe.

Remember Aunt G was the person cryogenicized in the proof of the theorem.

The time during which Aunt G's sentience is mapped to the physical universe consists of two different time periods. One is up to the time that she is cryogenicized. The other is the time that starts after her physical body is resuscitated or brought back to life, or however you want to say it. Apart from that, in the P ← L framework, her sentience would seem not to be mapped to the physical universe, and would that not make the meaning of *wait* problematic? Maybe and maybe not.

And if sentience is not mapped to the physical universe during the time that she is cryonicized, that too is interesting. Look at the region of the physical universe where Aunt G's sentience is mapped to. That region consists of two *separated* periods of time (they are separated by the period during which she is cryogenicized, and this could be arbitrarily long).

And a separate point. In the next chapter is described a concept called "neural net people" (see Detailed Table of Contents). The concept can be used to recast many results and theorems in a more concrete form. That might apply also to this theorem.

Well, that was the god's "thought-gram." I guess all these open-ended discussion points could be pursued if one wanted to dig into issues regarding the theorem. But then, my goodness, it is like a list of exercises in a text book. Let's move on to the next chapter!

By the way, just to state it again, the theorem works – is true – for the full true meaning of *waiting*, even if it takes centuries or millennia before we know fully what that meaning is, or even if we never know. Further, one can see that the three-situation proof would work for many other

meanings of *waiting* – so the theorem would be true there
also.

33 *Teletransportation and Evolution of the I*

In this chapter, we look at how teletransportation yields information about sentience.

In a way, this is a continuation of the previous chapter on "All Souls." That was about sentience "jumping" across time, this chapter about "jumping" across space. I remember back in 1969, when I was already thinking about teletransportation and its odd implications for sentience, and went to talk to a philosopher – I seem to recall his name may have been Bonner – at the IBM Research Center, at Yorktown Heights, where I got a summer job in mathematics. That was a long time ago. It was not the same as my father's service station. The philosopher said that I had a new context for issues that had been around for a long time. It's funny how life moves along but some of our drives remain the same. Though the ideas move on.

Explanation of Teletransportation

'Teletransportation' just refers to taking a person apart, atom by atom, and reassembling the atoms somewhere else, fairly quickly. For some time it is an idea found in science fiction, and many people feel that someday it may be reality. It is a fast effortless way for a person to move around. One simply walks into a teletransporter, and the machine "beams" you to your destination, fast and easy for you. It was the method used in a famous television science fiction series, to beam people up and down from their spaceship to the current planet they where near, or to beam people to and from nearby spaceships.

The desert sun was pleasing and so was a delicate breeze.

The problem – Atoms Replaced by Other Atoms – Is It Still Me

Basically, the teletransporter, through some kind of advanced technology and science, rapidly takes all the atoms that you are made out of, and sends them quickly through space to the destination, sending them all to the right place so your body appears in the new location. Let us suppose that you are about to ride the teletransporter. What if during the teletransportation process, one carbon atom that was in your body is replaced by a different carbon atom that was not in your body. Changing one atom among the 6,500 trillion, trillion in your body would certainly not be a problem as to whether the same sentience came out at the other end. But what if you replaced 10% of the carbon atoms with other carbon atoms? No problem. What if you replaced all the carbon atoms with other carbon atoms, but everything was of course put in the right place at the destination, so "you" walk out of the machine.

However, now you wonder before using the teletransporter. If all the carbon atoms are replaced by new carbon atoms, will that really be me coming out at the destination, *or will it just be someone who is an exact copy of me*? Still, you might think that since all the oxygen, hydrogen, sodium, sulfur, phosphorous, and all the atoms other than of type carbon are the very same atom that was in your body when you entered the teletransporter device, that it is probably alright. You might think, "How can I possibly be partly me and not partly me. And anyway, it's exactly the same atoms, except for the carbon type atoms, and in that case each carbon atom is being replaced by another carbon atom. So, OK, it must be all right to ride the teletransporter. I mean, *I* go in and *I* come out."

But what if all the atoms are replaced by other atoms, though atoms of the same kind. So a hydrogen atom is replaced by a new hydrogen atom, a sodium atom by a new sodium atom and so on. Now you are a little more concerned about riding such a teletransporter. But after some thought, you sigh and say, "yes, sure, it is OK, I would ride the teletransporter." Perhaps you think, I have decided to go this far, why not all the way. And again you think, it's the same kind of atoms, and they are all put in the right place. Perhaps you even think about an important point: how the material in our body is constantly being replaced by new material, so that after seven years our body is made of a completely different set of atoms than it was seven years earlier; sure I've changed in seven years, but I'm not at all a different being than I was seven years ago; my being is that being, it's just that seven years has intervened. These thoughts make you comfortable about riding the teletransporter even when all atoms are replaced by others of the same kind.

I always enjoy the pleasant sunlight on the cactus and its soothing warmth on my skin. As the sun moved toward the horizon, a layer of flat cloud took on the hue of rose-colored flame reflecting across the desert closing at day's end.

Problem of Two People

Now we arrive at our teletransportation problem. What happens if there is one entrance teletransporter device but two exit teletransporter devices? Just as before, where you had concluded you were so safe, the device uses new atoms at the exit station, but now it makes two of you, each with new atoms.

"Oh no," you say, "I wouldn't ride that; which person would I be?!" But then we would argue, "why not?" You rode it when it used all new atoms to make one of you at the destination. You believed that the person coming out of the destination would be you and not just an exact copy. Therefore, you must believe that both of these bodies are you, and not just exact copies. Now there are two of you.

Now, even more alarmed, you cry out again, "oh no, I'm not going to ride that!"

Someone repeatedly points out that you were willing to ride the machine when only "one of you" came out at the destination. So you should be willing to ride it now.

Exhausted by hearing this line of reason too many times, you lapse into distraction. Such a reaction is not uncommon in the philosophy of mind – which is what we have been calling the figuring out of sentience. Aristotle, about 345 BC already noted it – "To attain any assured knowledge of the soul is one of the most difficult things in the world" (Aristotle, *On the Soul (De Anima)*, near start of book 1, for instance in J. A. Smith's translation in McKeon's *Introduction to Aristotle*.)

After a while you make a full confession of error. "I guess I was wrong. Anyway you sort of lead me on. When you replace the atoms with all new ones, well, that is *not* the same person that comes out at the destination. It is just an exact copy." But I can go back to some of the arguments that it is the same person that walks out of the teletransporter, at which point, you and I can only conclude how really odd the whole situation is, and how odd are the issues of sentience. (Of course, one might go all the way back and desperately claim that sentience itself is illusory. Yet what a convoluted nihilist position this would be. It is a false way to avoid profound, real aspects of the fabric of the universe.)

Consequent Cognitive Collapse

Consider the two human bodies (people) that come out of the one-entrance two-exit device. Imagine if the two meet. Or imagine if all they do, is know of each other; of course they each know that there are two exit machines and hence that there is "another of them." Each of the two has exactly the same past and memories. And yet now they are two "different" people. Each has incredible knowledge of what the other is thinking and feeling. In fact, they have more than incredible knowledge of the other. It is a little bit too weird.

If the person wasn't strong (either one of them, since after all they have the same character, past, experiences, abilities, and all), or maybe if the person was too logical, the person - well both - would go insane eventually. They couldn't handle it. The situation might sound innocent as we sit and read about it in this book (like for instance reading history can make things seems innocent but if you are there it is a different story), but the actual experience could be deadly hell. The situation might eventually be so weird that they could simply not come to grips with the existence of this "other person." And that would be the problem. The most extreme identity problems, supported by the most solid logical reasoning, problems that would only deepen and overwhelm their whole being as time went on and they thought more and more deeply about it. "Am I the other person? Are they me? Are they me you? Maybe the other one is the real one, and I never existed till the teletransportation" - this would be the most deadly thought to gnaw forever at the whole essence of who or what they were, and even worse, would profoundly impact their belief that they were even real in the sense that the other is real, and would likely lead to a severe psychological problems and suicide. We will call this situation *cognitive collapse*, which is where the most profound basis of the mind (in the psychological sense) collapses into hell and self-destruction.

Hiker. Memories, Reality, Beliefs, as Logics

Suddenly there appeared a group of hikers, who as usual seemed to come out of nowhere. Nor did they seem at all befuddled – as had been the case in the past with some of the hikers – as to why they were there, but on the contrary, they had great intensity on their faces. I gave them copies of the above writing, which they quickly read over, but seemed to know already. Quickly we sat on some logs in a little copse of trees, a little auditorium so to speak, and one hiker never even sat down but immediately started to speak.

"Our beliefs, and the beliefs of the society we are in, influence us in the way we feel and interact with the world around us. In short these beliefs determine, in part, what is reality. We also have hard-wired into our brains aspects of what is reality and what isn't. For instance, things that recently happened to us and enter into memory, are one way or another captured in certain logics in the brain. A hard-wired working principle of the brain is that the things at these logics are recent reality in our environment. And this working principle is hard-wired into many dynamics of the brain. On the other hand we also have things like dreams, and daydreams. These are placed at certain logics, and it is hard-wired into the working of our brain that the things at these logics are "not real." These distinctions are maintained solely by the hardwiring in the brain. There are also certain principles that are likely directly or indirectly hard-wired into our brain, such as, in the past there was (reality consisted of) only one of us, or, in the future there will be (reality will consist of only one of us). To even explicitly enunciate these principles seems strange because it causes us to become aware of the contrary to the ideas hard-wired in our head, which in turn causes the feeling we describe as 'strange.'

"These principles are hard-wired in the brain because they were absolutely important in the survival game of evolution: in mentally advanced creatures such as our selves, if these principles were weak or messed up, the mental processes of the creature would poorly handle its resources. It would not have so good awareness of the difference between memories and dreams."

Hiker. Time

Even before the speaker got seated, another jumped to the center of our gently blowing, green trees, and started to speak.

"The previous speaker spoke of our sensation, so to speak, of reality, and how that is surely to be found in logics in our head. I want to make a brief comment about time.

"There may be some difficulty in understanding issues about time that are nevertheless important. Our experience of reality is absolutely an experience across time, from our past, through our present, into the future. Yet, from another perspective, those logics are all encoded in a transformed way into logics in the brain that are total in the single instant of time, the current instant, with nothing outside that instant. There is an element of the piano chord, played one note at a time with a brief silence between each. The experience at hearing the current note is of the framework of all the notes. Even though only one note is played at a time, we *hear*, we experience, not only the note but the harmony of the note with the previous notes. We experience harmony even though the notes are across time and there is no harmony at that *instant* out in reality. The framework takes place across time. The experience of the whole framework is at an instant in time.

"Admittedly the language of the preceding paragraph is strained because the tenses of our verbs do not handle these ideas. But the situation is something of this sort.

"Again, maybe some axiomatizations can yield interesting theorems pertaining to these matters. Mathematics has some amazing techniques for wrapping up whole processes into a single point."

"Axiomatizations?" said a hiker sitting and listening.

Hiker. Two Have One and Same Experience. Axioms.

Now a third hiker jumped up and took the previous' place. This hiker spoke of an experience, a sensation, that the person had before going into the teletransporter, and how the two that came out had that one and same experience.

"Consider the experience of biting into a greasy bit of chicken yesterday. Let's suppose that is what the person who went into the teletransporter experienced. But as for the two bodies that came out of the teletransporter, *I say that here was one experience, but they both had it.* Maybe this could even be proved from certain axiomatizations. For the brain, at this instant in time, is, in *some* sense, everything; not in all senses, but in some sense. Not only would the two have had the same experience, but, at as deep a level as one wants to go, all of their being was the same being, up to the teletransportation. Even though we may not be able to understand the full depth of what is going in the brain, or in the fabric of the universe, maybe it is still possible to prove, from certain axiomatizations, that the two coming out of the teletransporter had in every sense possible the same being, up to the teletransportation experience."

"I think," I tried to interrupt, "it would be good to discuss a little bit the idea of axiomatizations." But no one seemed to hear. I wondered, maybe some of the hikers will discuss this by the end of the chapter.

The standing hiker continued addressing our little group.

"Suppose that one of the people coming out of teletransporter had the same atoms as the person who went into the device, and that the other person who exited was made of new atoms, but of course of the same kinds.

"Of the one who had the same atoms, could we not argue that they have nothing better in the fabric of the universe than their duplicate has. Not in any way whatsoever, even if we could understand everything. Again, consider some earlier experience, had by the person who later went into the teletransporter, as for instance that of a greasy bite of chicken.

"The logic going on in the brain of the one with the original atoms, that logic contains the memory logic of the experience, as well as all other logics (such logic of what is reality, what things in the brain were experiences of reality, and all kinds of other dynamics, in the brain, of related logics that we are not aware of) in the brain that have *any* relation to that memory – all of those logics are no better or more solid in the body made out of the original atoms than in the other body. Even more, the going-ons and where the goings-on are taking place, are no better in the body with the original atoms. None of the logics going on in the brain of the body with the original atoms are in anyway superior to the corresponding ones in the brain of the other person.

"That is why it would seem that the two bodies had the same experience, no matter what reasonable meaning (or axiomatizaton) that you assign to the term 'same experience.'"

Hiker. Evolution of Logics of I

Now another hiker spoke, this time about how the logics that are the basis of I were derived, like everything else, through evolution. Then the speaker interestingly related this to teletransportation and duplication of the person. It turned into a remarkable conversation.

"I want to continue a theme of the previous hiker's, but take it in a much different direction, that of the evolutionary aspects of the logics in the brain that are related to there being a functional unity in the physical body that stretches through time. Maybe I should phrase this as the evolutionary aspects of the logics in the brain that are part of the basis of the I.

"As for the person going into the teletransporter – and here I want to speak of *us* going into the teletransporter as that will make it easier to describe – so, let me say, as for *us* going into the teletransporter, and as for us being 'duplicated', it is hard for creatures as us to conceive of this and its aftermath. Maybe it is not possible. Our evolution has not made us able to think in this way. In all our evolution, the future in our brain has been only one of us. This, and its consequences, have been embedded at a deep level of our mental processing, our I logics, and so also of our being, likely, even hard-wired in some ways into our brain.

"The ideas of the theory of evolution say that possibly much, perhaps all, of the thought-processes related to I-ness, for our being as it is on Earth, derive from resource allocation. Most, or all the logics, that are part of or related to I, come from optimizing resources, though these are advanced, abstract resources indeed. But this is because there are a hundred thousand million neurons in the brain and these are involved with some pretty deep logics. Resources are about establishing situations that are beneficial to 'ourselves', or alternately, working against situations that are the reverse of beneficial to 'ourselves.' The reason I put the word 'ourselves' in quotes is because here I mean something more than our usual meaning of the word. Here I mean that physical form, what we call our body, that extends through time. Consider this physical form extended through time and consider the thrust toward optimizing resources that takes place at each point in time. At any point in time the being associated with the body optimizes for the future of that physical form, it (we) optimize on the basis of the resource situation bequeathed to us by the past of that physical form. The mental processes and forms and concepts that are the basis of I-ness are the ones that evolved as the most successful at resource allocation, from an evolutionary perspective, for this kind of time-space continuation of body that we have. The logics – the mental processing – that we have, includes that in the future we do not duplicate into two, for it simply has never been that way for us on Earth, and those are the conditions under which those logics and processing developed in our particular evolutionary past.

"If however, we had always had in the evolutionary history of humans something like teletransportation which could take one body and duplicate it into two, then the basis of I-ness would have evolved differently to be optimal in this other kind of environment.

"But with the evolutionary environment that we have had, we ask with implacable insistence, well, if I were teletransported and became two, then which one would I be? The insistence is implacable because the reverse conceptualizations (logics) have been hard-wired into our brain: there will only be one of us, or none, in the future; and further, towers of *derivative* logics (principles), that we cannot understand because they are a basis of our I and because we cannot introspect to that depth, are also hard-wired into our brains. If we had an alternate evolutionary history, where one body could split into two, then *that* principle would have come to be hard-wired into our brains, and towers of derivative principles to *that* would have also come to be hard-wired.

"That is all I have to say on the subject. So I will sit down."

At this point another hiker swiftly leaped up, continuing on the same topic.

"I think that mathematically it is necessary to look at all of this as follows. We need to look at it as a fully spatiological thing. Forget about time. Think of a human as nothing but a cylinder in a space that we imagine in front of us. Maybe it's a little bit wavy, like a snake. But a cylinder. At one end of the cylinder is when the person – the physical body – is born. At the opposite end of the cylinder is when the body dies. Think of it as a cylinder laid out from left to right, with birth at the left, and death at the right.

"Now there is something special about the communication within that cylinder. It can only go from left to right, never in the reverse direction. One slice of the cylinder can only speak to slices to the right. The slices to the right know about the slices to the left, they get communication from them, but they can never say anything to them. Communication is only one-way in this cylinder. That puts some unusual restraints on the cylinder, to be sure. But there it is. That is just the reality of what such cylinders have to deal with. That is just the environment of their evolution. They have no other choice. This is a severe restriction. And it clearly influences the kind of logics in the cylinder that have grown up out of such an evolutionary environment.

"Each person is a cylinder, and you can picture all these cylinders, and picture the usual laws of evolution between cylinders. The work at optimizing resources can only take place at a slice and be passed to the slices on the right, never can the results be passed 'backwards' to the left. And the foundational logics of the particular kind of I that we humans posses are due to evolution in that environment.

"Now what about an evolutionary environment where now and then a person could become two persons – just like in teletransportation, where one person goes into the entrance part of the device and two people come out the exit locations. Well, instead of just a cylinder we have something like a "Y". At the bottom of the "Y" the person is born. Moving upwards, where the "Y" divides into two arms, that is where the person became two physical bodies. At the top of the Y is where the two people that the person has become die. Well, if we want to keep the view the same way as earlier, we should lay the "Y" down to the right, so the person is born on the left, splits into two bodies in the middle, and the two die on the right.

"As before, a time slice through any part of the 'Y' can only communicate with slices to the right. And the same is true of the results of resource optimization.

"There is a wrinkle to the communication issue. As usual, the communication from one slice on the left to another on the right is massive, for it corresponds to the communication that can take place between the billions of parts of the brain through all the billions of electrical signals. However, the communication between the two arms of the Y is millions of times less, for this is done with the same efficacy as two physical bodies talking to each other. Such communication is done not with billions of electrical signals but with hundreds of spoken words. This then is the nature of communication between the slices of the 'Y'. This characteristic of communication is their evolutionary environment. It determines the logics that are going on in the brain, including that are the foundation of their I. As you can see, they are going to have a different kind of I than we have.

"Well, that is all I want to say. Thank you. I will sit down now."

But another hiker jumped up and started to speak. This was taking longer than expected.

"I was thinking," he said with a twinkle in his eye. "We use the word 'I' to refer to ourselves. The letter 'I' is basically like a cylinder, going from the bottom to the top. The people who can become two, as in the teletransporter, should refer to themselves as Y, just like we refer to ourselves as I. So where we would say, 'I think I have to go', they would say, 'Y think Y have to go.' 'I felt I had to say I couldn't,' would become, 'Y felt Y had to say Y couldn't.'" With a grin he could now hardly contain, he looked around at the other hikers. "Instead of saying, we need to get going, they would say, 'ywe need to get going.' The 'yw' is pronounced like a slightly longer 'w'." "Ha, ha, sit down already, ha, ha," another hiker hollered. The speaker sat down, but yet another hiker jumped up and started to address our little group.

"To be serious again. I just wanted to say, you noticed this approach lead to some new insight. These, uh, Y people," and he briefly smiled toward the person who had just sat down, "we *initially* assumed that after they become two people, that these two Y bodies would behave and feel toward each other as two bodies with our kind of I. But given the nature of their evolution, the two I's in the two exiting bodies would have a great deal of deep identity with each other. It would be similar to the amount of identity we have with our different selves as we picture our selves at different times. For instance, it might be similar to the amount of identity, in feelings, and backed up by logics, that our 8 p.m. self has with our 4 p.m. self. That's how strongly the two resultant people of a Y person would feel about each other. The two would feel a lot toward each other the way we feel about our different selves.

"Look, let's back up. A long way! Let's step far outside. Let's step outside of even sentience, speaking as if you looked at everything as beyond dead or alive, or as purely mechanical systems, or as purely mathematical systems. From this view, there are a number of logics that create the functionality of identity, across time. They are physically in the brain, and like everything physical, these logics arose through evolution. Mathematically, a time slice through our 4 p.m. body results in one system with logics, and a slice through 8 p.m. in a *separate* system. But each system, the 4 p.m. one and the 8 p.m. one, have I logics that treat as if the two systems were one.

"These logics that treat as if the two systems were one, we could say that these logics are the foundations of the structure of our I. The Y creatures would have I logics that have a good deal of the same identity between the two duplicate exiting people. If our evolutionary history had always been like that for Y people, the two duplicates that came out of the machine would treat each other with a great depth of identity. And the reason we struggle so much with this thought experiment is that we plunk ourselves down in the teletransporter, ourselves who have this I logic from our own particular evolution, and then the teletransporter does something to us that is at odds with our whole evolutionary history, and what comes out of the machine are two bodies that are in line with a radically different evolutionary history but in our thought experiment still have I's that are derived from our evolutionary history. Cognitive collapse could follow. (Notice however, that even if the cognitive collapse where extreme, even to the point were I logics would cease, even if sentience itself eventually ceased, that nevertheless, sentience would still have taken place for at least a little time after the duplication, like in some extreme particle accelerator experiment.)

"I want to speak more generally. We may get confused because of our particular evolutionary history, which has generated logics that are the foundation of our particular kind of I. We may get confused and think that these aspects are a *requirement* of sentience. But they aren't. They are merely characteristics of the kind of I that we are. Certainly, removing what is accidental and not intrinsic to sentience will put us in a better position to understand sentience.

"(All our strong feelings about death must also derive from our evolutionary history. Those feelings are not intrinsics of sentience.)

"On the other hand, do not think the feelings from these logics of I since they are abstract are not real, which is hard to do when thinking about the teletransporter experiment. When you eat some fine food or have some other great pleasure, that feeling is real, oh so real. If a huge brick where to fall on your hand, that would not be a feeling that you could avoid. Few things could be more real. So too it is with the feelings derived from the logics of I. What you *feel* about death is oh so real. Do not imagine otherwise. In fact all these feelings, from those from foods to hurtful bricks and death, are logics, going on in the brain, going on in the signals going around in the brain.

"The same is true of the feeling that you are the same person as yesterday. It is true that this feeling derives from our particular evolutionary history. It is true that this feeling includes a component feeling (component logic) of being unquestionably absolutely real. These feelings themselves are real, as real as for instance the terrible pain you would feel if that huge brick fell on your hand, or of the pleasure of the experience itself of eating a scrumptious desert. I admit that these considerations drive one to deeper wondering as to what is going on in the fabric of the universe. As for these logics, it is not them but solely their going on that is in the brain. Perhaps it is only this one going on of them that is in the brain. We do not know.

"I think that these issues are beyond our capability to understand. Problems of whether a mathematical system can understand itself – that sort of thing. A system cannot understand at the depth required to understand itself. These are problems that can only be resolved someday by our machines, which will be able to dig deeper than we can.

"On the other hand, perhaps we can still get insights, in spite of our mental limitations, by looking from a very general perspective at evolution of the logics that are the basis of I. Earlier in this get-together of hikers, the Y I evolution brought us insights. Perhaps one might try to think about what kinds of evolutionary environment would produce people with different kinds of I's. This would lead to insights into sentience today, rather than having to wait 7 or so centuries till we, or our machines, fully understand it."

Summary
(1) Taking slices of us in time. These cylinder slices, mathematically, logically, and physically might as well be totally separated from each other. Logically there is no reason at all that a slice to the right should have the same being as an 'earlier' one to the left. This says that the being we are right now is different than the one associated with our body (the time slice of our body) yesterday, and will be a different one from the one associated with our body tomorrow. It is only that at a deep level of logics physically going on in the brain, we have an absolutist and unshakeable belief, hard-wired, that we are the same as that person yesterday and will be so tomorrow (If not for this belief, the allocation or resources that derives from all the

concern we have for the person tomorrow would not take place and other brains, that did believe that the today and tomorrow person were the same, would out compete us – after all, if you don't have to worry at all about tomorrow, you aren't going to put much work into helping the person tomorrow, whereas someone who has this belief will work very hard to make sure that the person tomorrow isn't in too bad a shape.). In other words, this utterly powerful belief comes about solely because of evolution. In reality, we could be completely different from the person tomorrow. It is only evolution that has unshakably programmed in this belief, programmed it in so that it is absolutely powerful.

But maybe this line of reasoning isn't so. For if sentience is deeply and purely in the logics – those "things" that appear so abstract to us – then, that the logics extend across the different even separated time slices, makes no difference. It is the same logic, so it is the same being.

(2) As for the 2 copies of a Y person, in a Y evolutionary environment, they have the same positive and concerned feeling and dedication that we have toward ourselves (toward our self of 4 p.m., toward our self of 8 p.m. and so on). The two even have the same logics of I, but enveloping both bodies, save for anything violating the laws of physics removed.

(3) Whatever will turn out to be true for sentience in the fabric of the universe, the above will still be true, and hence the truth in the fabric of the universe will not contradict it.

(4) The approach above is an interesting experimental method. In thought experiments we grow a sentience in evolutionary environments different than ours. Though we humans may never understand in depth the logics going on in our brain related to our I, these thought-experiment grown sentiences can help us see much farther than we would otherwise (though many centuries from now, we, or our machines, will understand all this directly, without the need of these thought experiments).

(5) In this section, we have repeatedly referenced the ideas of evolution and evolutionary environments. The kind of evolution we are talking about is built around a blueprint carried in each body, that blueprint being the DNA. When

one looks at the reasoning used to analyze this kind of evolution, it is all built around this DNA, this blueprint. Arising from these blueprints and their connection to bodies and the competition of bodies with each other for resources, come all the physical characteristics that we are, and that includes the logics in the brain, which in turn includes the logics relating to the I. The Y person, the one body that is born and the two bodies that it becomes, all these three bodies are of *one* blueprint, one DNA. In terms of the processes of evolution, which derive from the blueprint, this three-body Y is a single entity. And that is why the multiple bodies of a Y person have much the same kind of positive affect and identification toward and with themselves as we do toward and with our "one self", or toward and with our different time slices of ourselves.

The DNA should not be identified in some metaphysical sense with I-ness. What can be said with certainty and clarity is this. The role that DNA plays in our evolution gives rise to there being one single I-logic (one single I-ness) that envelops all the time slices of our physical body. The role that DNA would play in the evolution of Y people would give rise to one single I-logic (one single I-ness) that envelops all the time slices of the three physical bodies "of a Y person". (This could be phrased without the term "physical" and it would mean the same thing. But adding "physical" seems to make it clearer.)

(Evolutionary psychology is another area that uses, in a central way, ideas of evolution. See endnote 18.)

Hiker. Neural Net People in Thought Experiments

Now another person jumped up to speak.

"The issues I now discuss are nowhere near as grand as with the previous. I will discuss a method to tighten up certain thought experiments. This turns out to be also related to the issue of the discrete versus the continuous in mathematics and its relation to sentience, for there are strangenesses when you think on these things more carefully.

"Some of the thought experiments carried out with teletransportation can be simplified and strengthened by using 'neural net people' instead of standard biological humans.

"Suppose that our technology is advanced enough that at least it can focus in on a tiny area of the human brain and after a certain amount of observation and analysis it can figure out the logic going on in that area. Then it duplicates the logic in a neural chip. (A neural chip is like a common silicon microprocessor chip, except for this difference. The typical silicon chip has one active element that does one computation at a time. The neural chip is like a collection of neurons in that it has a whole collection of active elements, all of them capable of doing computations all at the same time, and passing the results between them.)

"The technology to copy logics in a tiny area of the brain to neural chips does not have to be able to understand at any deep level the logic it is copying into the neural chip. It only need copy the essential going-on's of the signals, based on the brain structure and nature of that tiny area, to the neural chip. We presume that by doing one very small area of the brain at a time, it copies the whole brain to neural chips and suitably connects the chips together, so that a copy of the brain, with all the logic in it, is produced. It is a frequent belief of the scientific perspective that such a collection of chips, connected to an appropriate body, would have a sentience being "in" it. (The question of whether the neural chip copy has the "same" being going on it as the human it was copied from will be irrelevant in the following discussion.) We will call such sentient beings, *neural net people*.

"Neural net people can be a huge advantage in certain thought experiments. For instance in the teletransportation thought experiments, we did not go into the issue, but there is a disquieting, questionable aspect about taking a person apart into all their atoms, moving the atoms to another place, and then getting all the atoms together and working in the right way. It is likely that someday, awfully far in the future, technology will be able to do it. But the questionable feelings we have about the possibility of such a device weaken thought experiments using such devices. But! A neural net person has no such problems. Teletransportation becomes a conceptually simple, solid, unquestionable process of taking all the chips apart, moving them over to another area, and putting them back together the way they were (and if they have something like "states", then setting the states to how they were when the chips were taken apart.) If one wants, one could have lots of large chips or far more much smaller chips. If one wants two people exiting the machine, just create duplicate chips. The deductions from the thought experiments with neural net people would not have those lurking feelings of questionableness, but instead would be solid. Keep in mind that neural net people have genuine, full-fledged sentiences. So thought experiments with neural net people give genuine results about sentiences.

"In the teletransportation thought experiments, we could use a neural net person (a neural net being) instead. Thus, the teletransporter takes apart the neural net person's brain into the individual neural chips. It then moves those chips to the destination, and at the destination, reconnects the chips, in the same way that they were originally connected. Perhaps some of the chips can be replaced by duplicate chips. Perhaps we can duplicate each neural chip and connect up two copies of the original neural net person. And so on. Neural net people could even be used in other thought experiments, such as in the chapter on 'All Souls.'

"Two points can be made. There is a philosophical position that perhaps it is inherently impossible to understand the human brain (See endnote 19 on Frazier's interview with Martin Gardner). The creation of neural net people from standard biological humans does not violate this philosophical position, even assuming that the position is true. The creation does not involve an understanding of the signals outside of a tiny area; it only involves a sufficiently detailed duplication of the brain functioning in that tiny area. We leave un-discussed the issue of how to obtain what are the essentials of the signals that are going on there. It is also possible that, since these are involved in thought experiments, all we need is that there *exist* these essentials of the signals, and that these have been put into a neural chip. After that, we can come back to the world of the concrete, physical, with the teletransporter disassembling the chips and reconnecting them somewhere else, and so on.

"(The second point has to do with mathematics. There is Discrete Mathematics, and there is Real Analysis. Discrete mathematics has to do with discrete quantities and measurements, such as 5 apples, or 16,224 possible ways of arranging certain playing cards on a table, or quantized energy states of some system. Real Analysis has to do with 'continuous' quantities and measurements, such as 3.4521 inches, or a length in inches equal to the square root of 2. With continuous measurements there are virtually an infinite number of values getting ever smaller. (Using the terms 'Discrete Mathematics' and 'Real Analysis' may not be the only way of pointing to this kind of distinction, but it works.)

"The neural net people are the discrete versions of brains and logics and sentience. The brains in standard biological systems are squashed, three-dimensional, tangled jungles of immense size, and everything about them is continuous – is capable of having an infinite number of values of ever greater precision. The logics and their going-on's and everything about them involve Real Analysis.

"Just as discrete sets of fractions, when allowed in ever increasing numbers of them, ever closer together, come increasingly close to any real number, so too neural net people, as the neural net captures ever smaller details of the going-on's of the signals, comes arbitrarily close to the logics in the standard human brain – the logics in the immense, squished jungle. Indeed, at a certain smallness, the neural chip might as well be nothing more than an individual atom. This is similar to having increasingly accurate fractions approximating a real number, especially an irrational number.

"There may be some ways in which a neural net person is inherently different than a standard biological human. For instance, when enough differences between the neural net version and the human accumulate across enough chips, there may be some differences of behavior, experience, and so on. Note however, that this does not deny there being a sentient being "in" the neural net brain.

"There is one more area of difference. The standard human brain has an uncountable number of logics going on in a very weak way that would never get copied into the neural chip. As noted in the previous paragraph, when these logics combine, they can sometimes produce noticeable differences in behavior, experience, and the like. Also, one might think that these 'low level' incipient logics could burble into more substantial logics over long periods of evolution. Such could never take place in the neural net people, since they simply do not have this vast panoply of incipient logics burbling under the surface. Yet even this argument seems not as strong when one considers that the brain derives from the human genetic code, and that this code of about 3,200 million letters is a discrete affair; (there are a *finite* number of possible 3,200 million letter sequences, thus it is discrete, not infinite, not continuous). Hence the brains that can exist have only a finite number, though large, of specific physical foundations."

Hiker. Reverse Reasoning

Another hiker quickly jumped up, though the body crouched a little and voice cracked.

"Ah, you know how the brain has all these logics going on in it. The brain and the nervous system, and the logic going on in the body too, outside the nervous system. Call that a computer, a brain, whatever.

"Separate it from the body, and have all the signals from all the senses, all the sensory neurons, on, and in, the body, sent to this separated brain or computer or whatever you want to call it. Maybe one could call it a logic focal device, LFD. I'll just call it a brain. But it could be a computer too. Like I said, I'll just call it a brain."

The other hikers sitting in the little audience about seemed to be slightly restive.

"Ah, OK, so this brain gets all the signals from all the sensory neurons in the whole body. It also sends out all the signals that activate all muscles in the body (the signals touching directly on the muscles). Let's remove the brain to another place – this whole thing is a thought experiment and we are assuming everything stays alive and there is no discomfort.

"Maybe the brain is removed to another room, or some other part of the country or of the earth, but all the signals would be radioed back and forth between the brain and the body with its sensory neurons and muscles.

"The sentience logic, along with lots of other logic, is going on in the brain, wherever it is on earth. But the being is 'in' the body, no matter how far apart the body and brain are on earth. The being is 'in' the body because that is where all the perceptions are occurring and where all willed actions are being effected."

Someone in the audience interrupted.

"This has slight interest. Is there a point you're getting too?"

"Ah, yes. Ah, ah, suppose the brain is far from the body. First, suppose the sentience logic in the brain 'knows' that its brain is far from its body. Or second, suppose that the brain is in the skull of the body but again the sentience logic in the brain 'thinks' that the brain is far from the body. There would be no difference between the two situations. And the same applies to the two situations if the brain was in the skull."

The eyes of the little audience were intently on the speaker, but little grimaces and furrowed brows appeared. The speaker continued.

"So it is *irrelevant* whether the brain is in the skull or far away.

"Ah," and here the speaker gasped in a little breath, "ah, let me just jump to one of the main points. How do we know that two people – say two people talking to each other – aren't the same person. How do we know they aren't one and the same being talking to itself in different modes."

The furrowed brows of the audience turned to big eyes while the grimaces lengthened.

"Ah, ah, I mean, you can't just say that two physical bodies are obviously two different beings. You must eventually come up with an explanation of why, and then you must open that explanation to questions. That is the way of science."

A person in the audience grumbled.

"Ah, I mean I'm saying, for instance, what if *one* brain far away interacted with radio waves with *two* bodies. In that case, the two bodies would definitely be one person.

"But now, what if the brain were 'not unified' in its processing, but consisted of two separate systems of logic going on, with only one interacting with one body, the other with the other body, but with no communication between the two systems in the brain, except via communication between the two bodies. In this case we would say each body had its own being."

Someone in the audience cleared their throat.

"Ah – let me slightly change the perspective here – in everyday regular life – ah let's just consider one person for a second – the logics in the brain – and this includes the sentience logic – can be divided into coming from three sources: the logics going on right now in the external world about the body; the logics that have gone on in the external world about the body since birth; the evolutionary logics that have gone on in the external world around all the ancestors of the body. I suppose one could consider these as three types of 'vision,' what we take in with our senses right now; what we take in since birth; and maybe the most foundational, what has been taken in from the logics going on in the environments of all our ancestors, this last being 'radioed' to our physical body – our current physical body – our physical body via the physical entities of genes. And I use the term 'radioed' as I used it earlier, when speaking about the brain communicating with the physical body, no matter where the brain really was in the thought experiment."

A round of coughs and throat-clearings went up from the audience.

"By the way, in asexual creatures – and now I am looking temporarily at non-humans too – the ancestors are a measly single string stretching back in time, but in sexual creatures it is an expanding web spreading ever wider back in time. It is like two very different kinds of antenna. The variety of logics absorbed by the string antenna is paltry compared to the variety absorbed by the expanding web antenna. That's an advantage of sex."

The audience was fascinated and distracted by this digression, but then the speaker went back to the original topic.

"To repeat, all the logics, including the sentience logic, come from these three sources: the external world now; the external world back to birth; the external world of all our ancestors. These three sources come to be reflected in the brain by different processes. For two people, the logics in their brain that are a reflection of their common ancestors' environment is the same, and *to that degree*, the two are literally the same being. For two people, the logics in their brain that come from the genetic code blue prints, which were foundationally shaped by the logics going on in the evolutionary environment of the common ancestors, would be the same."

A grumble went around the audience, interspersed with a few hmmm's.

"Ah, so to some degree we may be the same being. But I talk about a specific situation here. Let us not dwell on it but move to the more general."

"Specific!?" groused a hiker sitting on a log by himself. "If that is specific, what is general?"

"Ah, ah, well, I..., generally, such considerations as all these may help us understand – why two people are two different beings. (Or possibly, partly the same being.) We can't just assume such things as obvious – we must put forward an explanation, and then that explanation must be opened to questioning, as is the way of science.

"All these considerations are important lines of argument that might lead to insights about the relation of space and sentience. Or the *apparent* relation of space to sentience."

Some hikers frowned.

"Or the identity of sentience,"

A mumble went up from the group.

"or the apparent identity of a degree of sentience of different individuals."

A louder mumble went up.

"Those are my comments," the hiker added quickly, then sat down just as quickly.

Sighs and more mumbling came out of the audience. One hiker said, "You have three sources, and that's right. But you left out one other. Physical intrusions into the brain." The speaker responded, "OK, that too." Then a different hiker said, "Just because identical logics are going on in two different places, that doesn't mean it is the same being. So too with the third part your talking about. It doesn't mean to even that degree we're the same being." A further hiker closed the discussion with, "Nor does it mean the two bodies are *not* the same being. So too with the third part your talking about. We just don't know either way. We simply don't know these things yet."

Hiker. Response to previous hiker.

A hiker responded to the one who had been speaking.

"You talked about the brain being a computer, and about how that computer could be anywhere in relation to the physical body. Typically it is inside the head, but it could be anywhere, near or far. That is interesting. And important.

"When you talked about this computer, or brain, or, or what did you call it, an LFD? – when you talked about this earlier in your presentation, you always had the brain – or computer – in the same time as the body, but communicating across space. I'm sure there are important directions to explore with this. But I want to look in a different direction.

"When you talked about a computer being the brain and moving it to different locations, with remote communication with the body, you said it made no difference whether or not that computer was far or near or in the head.

"Why talk just about across space? Why not across time? Can a computer – brain – be in a different period of time and communicate with the physical body?

"For that indeed is where the computer is going on that is our perceptions right now – and our being too. Our perception (and being) is the past, with the present poured into it – the past is evolution perceiving its environment – is a computer which is running our physical body now.

"Well, clearly there are issues to be worked out." The hiker's hands waved, "this is just partial."

The hiker fell silent for a second, but then brought up two more points about the "time" issue, if that is what to call it.

"Sometime when you are before the mirror, look at that image. Keep in mind that that thing, that physical form, that body that you see, really is nothing but a huge number of rocks bouncing around, if you could see close enough, and whether or not you can see close enough, that's what it is anyway right now – at this very instant. We are the logic going on in all the motions of those rocks. Logic, logic being meaning, meaning going on in the motions of all those rocks. As if that weren't special enough, keep in mind that what our eyes see, as we look at our physical body, and what the process in the brain behind those eyes eventually moves into consciousness, show nothing of that one and same logic, that same meaning stretching back millions upon millions of years, going on in all that other matter too. You see only a terminus in the current point in time. Such is our being! Such is the nature of space, such is the nature of time, such is the nature of logic, and of the fabric of the universe."

"When looked at in terms of space, and time too – and I am not talking about the special issues of the theory of relativity, I'm just talking about plain old four dimensions, three of space and one of time, and mathematically, that is just a product of four dimensions, and in that sense the dimensions can be seen as the same. Well then, when you look at this, one sees the DNA has literally the same function as physical nerve fibers – the axons and dendrites of neurons – as they carry meaning from one place to another. The nerve fibers essentially carry meaning from one spatial location to another. The DNA carries meaning from one time location to another. But they both are (natural) physical devices that carry meaning from one place (in either time or space) to another."

Hiker. Axiomatizations. "Definition of"

"I want to present a further technique for studying sentience," said another hiker, who now stood up. The hiker expressed the following ideas.

The method of axiomatizaton could be a powerful tool in studying sentience. The value is that if you prove something from a set of axioms, then that is true no matter what the terms in the axioms mean – provided the axioms are true. In our thinking about sentience, we often do not have a precise idea of the meaning of terms, though we have vague intuitions. If something is provable from the axioms, then it is true in the fabric of the universe, no matter what the terms may mean in the difficult to access ontology of this fabric (provided the axioms are true in the fabric of the universe). In fact, the theorem would be true even if humans didn't know what the terms really meant for another thousand years.

All mathematical approaches have a number of angles. One angle of axiomatization is the one just stated, that stresses that the proof works no matter what the terms mean. A different angle of this approach is that one may think of sets of axioms as defining those interpretations that you want to consider.

Some possible terms used in axioms about sentience might be *being, sentience logic, brain, body, evolutionary environment of a body, experience itself.* And as thinkers explore the issues, they will come up with further terms. Some terms we might be thinking of as fully real in the fabric of the universe – maybe we feel we understand the term – maybe we feel not too certain about what the term means – maybe as for other terms we might see them as capturing some formalist meaning – while other terms might turn out to have a very definite meaning in the fabric of the universe though it will not be understood even by our machines for many centuries. This is the beauty of the method of axiomatizaton. It makes no difference to the theorems what the interpretations of the terms are, or whether we even know for sure how to interpret them.

In this book we cannot go into a full description of this method, yet a few things can be said. The idea of axioms traces back to Euclid in about 300 BC with the mathematics of geometry. Euclid listed a set of axioms, which contained terms (words, or symbols) such as *point, line, angle, intersect*, and so on. From these axioms, he proceeded to prove a vast array of theorems. What was quite nice about the overall method is that all the theorems traced back to the small set of axioms: as long as the axioms were true, then you could be confident that all the theorems followed. This is a fine characteristic of this approach.

All this was already in ancient times. In modern times, mathematicians concentrated on the perspective that the terms in the axioms could mean anything; as long as the axioms were true in some world or model, then all the theorems would also be, and this would be so no matter what kind of interpretations you could come up with for the terms in that world or model (technically speaking, mathematical logic uses the term *structure* in place of our terms *interpretation* and *model*, and it uses the terms *interpretation* and *model* in a somewhat different way than here, but the same ideas are being talked about).

Modern mathematics has determined the deep connections between looking (1) at axioms, (2) at proofs of theorems as pure manipulations on meaningless symbols and (3) at all possible interpretations of those symbols.

Modern mathematicians have also become exquisitely competent at determining equivalent but ultra-minimalist variations of a set of axioms, and sometimes such a set of axioms can give powerful insight into the larger set of axioms. And mathematicians developed exceedingly powerful tools for analyzing the connections and relations between different but similar axiom systems. And all of this is independent of what interpretation of the terms is used. This is why the axiomatizaton approach would seem to be so beneficial in the area of sentience. The benefit of a theorem being true no matter what the terms really meant in the fabric of the universe would appear right away. The benefit of the attention and work of mathematicians would appear over time as some axiom systems started to establish themselves, for this would draw mathematicians to apply their skills.

Whole ranges of questions would be reduced to orderly schemes of great generality and power of understanding or quasi-understanding. Instead of re-arguing a question in multiple guises, one might grasp it in great generality as it appears underneath a range or perspectives.

With our limited vision, perhaps this approach is required.

Hiker. Axiomatizations. "Writing of"

Another hiker stood and expressed the following.

Unless you are interested in formal languages for their own sake, you should not get too involved in some of the technical issues regarding something called predicate languages and the like. The right level of detail in writing an axiom system is probably illustrated quite well by Euclid's axioms of geometry. It is not that you want to write axioms of geometry. But that sort of way of phrasing the axioms is a good level to aim at. True, now and then you may have to learn about some new issue. But that should not occur often.

Thinkers about sentience, logic, and the universe should not get pulled into terminology and issues that specialists in areas such as mathematical logic are involved with. Now there is a somewhat vague line between what thinkers should or should not be aware of. Specialists, being enthusiasts, will naturally want to pull you into all kinds of issues, and there are plenty of such concepts and notions and terminology and intellectual methods in areas like mathematical logic.

Perhaps, over time, the goal may be to set up an interface of language and concepts between the mathematicians and the non-mathematician thinkers who want to explore axiomatizations of sentience. There might be a tendency over time for this interface to become more complicated than is beneficial. The only answer may be eternal vigilance.

Thinkers who explore various axiom systems would think of *interpretations* of the terms in their axioms as, well, ... *interpretations* of the terms. Ah, but a specialist in mathematical logic, upon hearing the word *interpretation* immediately thinks of a specific, technical concept that is a relation between two formal languages, with the intent that one language can talk about what the other language is talking about. Speaking somewhat loosely, this is what an *interpretation* is in mathematical logic. Moving this meaning of interpretation into the interface language would ruin the interface. Other examples are that the term *structure* in mathematical logic is close to *interpretation* for the thinker; the term *term* in mathematical logic has only a jagged, distracting connection to the same word as used by the non-specialist thinker.

The interface should be mathematically precise language and concepts *that are natural and conducive to the non-mathematician thinker*. The goal of axiomatizations in this case is to help thinkers. It is not to create a system of intricacy that ends up drawing energy unto itself and away from the true goals of the thinker, most of whom are not specialists in these particular formalist, mathematical areas.

One might explore axiomatizing issues in this chapter, or issues elsewhere in this book. Theorems might be proved about relations between different axioms, or about consequences of the axioms. People could not argue with the theorems – as long as the proof was correct. They could argue with some axioms, but they could not argue with the theorems themselves.

34 Standard I, Non-standard I

Standard and Non-standard I

The "standard I" assertion is that our sentience reaches across and envelopes all the time slices of our physical body. In the standard I, our 4 p.m. being and our 8 p.m. being and our 4 p.m. being tomorrow are all the same being. The "non-standard I" assertion is that there is a separate sentience for each major time-slice of our body, but all the sentiences share the belief system that they are the same, even though they aren't. They all share what in reality is an illusion that they are the same person.

Presumably, the non-standard I would be more exotic. Still and all, it is the unexpected that appears from the plain old standard-I.

Story-like Immediacies

In straightforward cryogenesis (as for instance in situation 1 of the chapter on "All Souls"), Aunt G dies, is cryogenicized, and an arbitrary length of time later is successfully brought back to life. In terms of the $P \leftarrow L$ model, her sentience is mapped to the physical world up to her death, and then the same sentience is mapped further to the physical world after she is successfully brought back to life. In between, from the cryogenesis up to the successful resuscitation, it is not mapped into the physical world P.

Thus, the same sentience goes on in two different space-time areas completely separated by an arbitrarily large break in time. Now one can play with, and explore, different axiomatizations that imply the standard I, different axiomatizations that affect the before and after cryogenicization state, and axiomatizations of that odd state during cryogenicization. But the issues and consequences are thought-provoking all the same. Let us touch on one of them.

Some speak of "the mind leaving the body", as for instance, when a person dies, whether or not the person is cryogenicized. Some might speak of the "mind returning to the body" when a person is successfully brought back from being cryogenicized. These phrases would naturally suggest themselves to a person living through such an experience with a cryogenicized relative, especially if the relative had been close. In this book, we have not used such phrases or terminology nor given any definition for such. What are such phrases?

Truth sometimes offers challenging indirections where we initially create story-like immediacies. These phrases about "the mind leaving or returning" are story-like immediacies. But the truth – the reality – is more conceptually challenging. It is indirect, and involves what appears to us as abstract issues such as mapping sentience into a region in space time, even mapping, via P $\leftarrow$ L, one sentience into a region consisting of several separated areas, as in the cryogenesis of a person. No wonder the reality challenges our understanding.

Mind continuing beyond the body

The last section dealt with our usage of words and phrases. This section deals with the fabric of the universe that the words attempt to get at.

From the perspective of the standard I, the physical body and brain is conceptually cut up into slices in time each separate from the others. Yet the sentience logic going on in the brain across the time slices is of one single being. At present we do not understand how, but in the trillions of electrical spike trains interacting in the brain, fully is our sentience logic, a logic which nevertheless pulls in all the time slices. The logic being reflected is just a matter of all these interacting signals. The going on of these signals could be slowly deformed into a vast range of other logics, from a logic that had no more sentience than a rug, to one that had no more awareness than an ant, to just all kinds of other logics, including likely a logic that made for a truly different being during every major time slice but all sharing the belief that they were the same person.

Perhaps a person's changing beliefs can affect such changes of sentience; perhaps only hard-wire brain changes can do so. But clearly one way or another, these deformations or changes in the system of signals can radically transform the nature of the sentience, to that of a rug (i.e. no sentience), or of an insect, or to one of separate beings occurring every major time slice and so it is all just a matter of shifting these electric signals, for then the changed electric signals change the nature of the sentience being "reflected."

If a single being can envelope the disjoint time slices of the body, provided the signals are right (the right sentience logic is going on), then there seems to be no reason why the appropriate signaling or changes in the system of signaling should not give rise to a sentience logic that would be that of a single being that would envelope time and physical form larger than the current body. For a single being would again be taking place across different physical systems.

Some religions posit that changes in belief systems can do this. Some spiritual systems of belief posit that this is the case independent of one's beliefs. Some belief systems posit that this is not the case, and that it is independent of one's belief systems.

Perhaps ten or so centuries forward, we, or our machines, will understand all of this.

End

At the beginning of this chapter we said that the non-standard I seems more exotic, but that all and all, the unexpected appears from the plain old usual standard-I.

The reason that the standard I, the one we presumably experience in every moment of our being, has deeper consequences is that it makes a stronger statement about logic and sentience: the logic inherently reaches across time; it, the one being, goes on in different physical systems across different time slices. By comparison, the non-standard I, though it seems exotic to us, is simple and mechanical: it is what you get when, ala the mathematics of the Cartesian coordinate system, we cut time up into slices, along with the physical systems in time, and we do nothing more.

35 *Brain Mirror Theory*

The god seemed to be giving me my next assignment. It was not unreasonable, and maybe I shouldn't have seen it as an assignment. I looked down at the head as it spoke.

"So far you tried to build carefully on what is known, or on what went before, or on what we can point at in the external world. Now make a leap.

"Be free here. For once, let go of the usual requirement of science that everything must be carefully analyzed and built up. You haven't let go in the past on this journey," and then the god quickly added, "and you must not in the future. But right now, loosen up, and come up with a theory that just might be right. Come up with an explanation that doesn't rest on posits of mathematical formalisms. Come up with a theory that is graspable in our native being, that can be understood and experienced in our native awareness."

Why not another thing to do, I thought, as I started roaming later that day through the rocks of the desert, and the next day too. The following seemed natural.

The Theory

The brain is a mirror. When we go through all kinds of consternation about the brain being a dead machine and how can we be in such a machine, well, the answer is to go ahead and accept it: the brain really is dead. But it is a mirror. It is an exceedingly sophisticated, complex mirror, but a mirror all the same. It is what the brain reflects that is important. It reflects the logic that includes us. It reflects us.

This theory presupposes that there is another "realm" which the mirror – the brain – is reflecting.

Now it is true that both the mirror and what it reflects are very complex. At present we humans don't understand the details. Perhaps we never will. The brain is an exceedingly complex inanimate device, mirror, in our head. What it reflects is exceedingly deep. On this journey we have traced out some of that complexity. This is the theory.

Advantages of Theory

Here are some of the explanatory advantages of this theory.

Consider a person standing next to an operating table. This is in a hospital in our present day. The patient on the table has a powerful local anesthetic but is otherwise conscious. The person standing so close to everything, this viewer of everything, looks back and forth between the patient and the doctor, a neurosurgeon who has a precise electrode device that can apply an appropriate electrical stimulus to different parts of the brain of the patient, and upon the most delicate application of the electrode to those different places, the patient reports feeling a cool breeze on their arm. Or when applied to a different place in the brain, the patient reports hearing violin or piano music, but that they could not make out a tune. When the electrode is applied to yet a third place, the patient reports a remarkably vivid memory formed years ago, so vivid it almost seems to replace the presence of the operating room.

The viewer, looking back and forth at the patient and at the electrode being applied to different places, is in awe. There is such incongruence between what the patient is experiencing and the moist areas of the gray brain tissue that the electrode is being applied to!

The brain mirror theory gives an answer. The brain is merely, literally just a mirror. All that the doctor is doing is tweaking one little point of this complex mirror, and that affects the aspects of the patient's I that are being reflected from the realm beyond. Tweaking the mirror tweaks the mode reflected for that being – for the patient's being.

If the viewer wanted to know a little bit more about what the theory meant in terms of the journey in this book, we might say this. The moist brain tissue has oceans of electrical signals going on, oceans that are like a mirror reflecting logics that are out in the environment right now, also logics out in the environment back to birth, and most foundationally, logics out in the environment back through all the ancestors of the creature. These oceans of signals are merely a mirror for all these logics. That brain, those oceans of signals, are nothing but a mirror, and it and they are different than what they reflect. Yes, what they reflect is indeed quite difficult to understand because of the hugeness of the brain, because of the hugeness of the environment, and because of, and this is something we underestimate, the logical depth of that which the brain is reflecting, and because, although all of this is there, none of it is visible to our native awareness, at least not in this form. This is how the theory would go according to the journey so far in this book.

For the viewer standing next to the operating table, the theory has a great advantage: no longer need one struggle with a messed up sentence like "a person is in that brain;" no longer need one suffer the mental confusion that goes with such a sentence.

There are other advantages too. The theory raises logic and sentience up to something that exists independently of time and space and matter. It also offers a special explanation of "birth and death." Birth is the creation of a new mirror. But it is merely a mirror. Likewise death is the shattering of a mirror. Yet what a mirror reflects does not cease because the mirror ceases. We will talk about this at greater length below.

Proofs

Can a "proof" be given for this theory? The theory, not being solidly based, neither can any proof be so. A proof is an explanation of why. In mathematics, which is much more absolutely based, the proofs are also absolutely based. Even so, an explanation for the why of the theory can be adduced, but it will not be as solidly based. Still, sometimes, when the mood takes us, such explanations or "proofs" can be more convincing than mathematics.

Here are three "proofs" of the brain mirror theory.

Proof 1. Quit playing a back and forth game. Take a stand.

The first proof starts out by noting the feeling that many people have, that the body is a dead machine, and so is its brain. But our proof does not then proceed in the usual cycling, which is to become consternated at how the brain is us, which is alive, not dead, and how can this be; whereupon the thinking goes back and forth between the body is a dead machine, but we are alive, but the body is dead, but we are alive, but the body is dead, ad infinitum. Our proof is based on not going into this oscillation. Instead, it looks at the body, and brain, and says, yes, they really are dead, so quit trying to fight it. Quit going round and round in circles. Accept it and move on.

(We could use different words. The brain is an inanimate machine, yet we are alive, yet the brain is an inanimate machine, yet we are alive, yet and ad infinitum. Our proof is based on accepting that at a profound level the brain is an inanimate machine. Accept it, accept it all the way, and move on.)

Given this first step that the brain really is as dead as a door nail, then where does one go. One answer is to see the brain as a mirror. Of course the mirror is dead. But it is reflecting a being, which is alive.

This is the first proof.

Proof 2. A feeling proof.

This is a "feeling" type proof. It moves in a different way than the rest of the book.

You wake up in the middle of the night with the most powerful certainty that none of "this, " the world of everyday, is real. The "real" reality lies underneath all of this illusion.

Well, what can I say? That's it. That's the proof. Like I said, it is a "feeling" proof. This whole world, and everything about it, and science too, all these are artificial surface constructions of what is truly real. If we somehow pulled this into the science perspective, then we might say that this deeper reality is what the superficial mirror – the brain – is reflecting.

The god looked at the mountains, then turned and looked at two glasses of the purest water. The glass and the water inside it were clearer than diamond. They made the sand and log behind the glass clearer than looking directly at them.

"Here," I put one glass next to the gods head lying on the ground. Somehow it was drinking it.

"Thanks."

I drank. Suddenly some of the thoughts of the god were in my mind. No, there was no magic about the water and nothing special was going to happen to me because I drank it. Nor was it prelude of any future discussion. It was totally of the present point. Rather the water was a distant symbol of … I couldn't quite catch it. It had the most unadulterated taste of water I ever had. It didn't even have the interfering taste of water.

Proof 3. Timeless equivalence.

"Hmmm. Here's a *non*-feeling proof.

"If logic is timeless, if logic exists outside of time and space, if logic exists outside the mathematical space of physical time and physical space, then in terms of function, the brain is like a mirror, and scientifically, function is almost everything. We simply define "reflection" as what we have so far called a logic "going on". In that way, logics going on in parts of the brain are being reflected by that part of the brain. Such a formulation is equivalent to the brain mirror theory, provided logics exist independently of whether or not they are going on anywhere in the physical universe."

Evolution of Mirrors

We might consider the evolution of these mirrors.

Over the vast eons of Earth history some of these brains – these mirrors – for some species – have reflected more and more logic. For other species their mirrors have reflected more or less the same amount of logic. Forces in the long-term environments have somehow determined how the mirrors have evolved for different species. But it is in the mammals that the mirrors have evolved to ever more incredulous size, reflecting ever more logic. Out of the mammals, the mirrors of groups such as monkeys, apes, gorillas, and chimpanzees, raced ahead in even further feats of size. And finally what is us, *homo sapiens*, broke out of this pack and fared far ahead. Behold the brain, behold the mirror, the dead, machine-like mirror, in our dead, machine-like body.

Is Logic Timeless

This brain mirror theory presupposes another realm, of what in this book we have been calling logic. The theory presupposes that this other realm *exists outside and independent of the brain, of any brain, of matter, of time, of space*.

"That is what worries me," I said to the god. "That is quite a leap from what I was previously saying earlier in the book."

In spite of being in poor form, the god tried to comfort me, pointing out that the god wanted me to come up with a hypothetical, but bold, theory. Eventually I wondered what could it *mean* to say that logic is timeless, or that it exists as a separate realm.

If logic is timeless, would that by itself imply that it exists outside of time and space. And as for it being timeless, consider the logic of the interference waves of two electric fish near each other. Doesn't the nature of that interference, the logic of the interference, didn't it always exist, doesn't it inhere in the fabric of the universe independent of time, and hence is timeless. Well, who knows how to think these issues? This is getting pretty abstruse. And yet it is related to our being. Is our true being timeless?

(The god says be bold. Alright. Many things "can be seen that way" – there is a certain sense of this phrase. We can see logic as being timeless; logic can be seen in that way. But if a tree is not at a certain place, we cannot see a tree in that way. Perhaps anything that is pure logic that can be seen in that way – as long as it is consistent – really is that way. Maybe one could prove this in certain axiomatizations. Since being, or sentience, is in pure logic, then … Maybe there is no place to be bold here. Prehistoric humans walked about in a world that seemed absolutistly primitive and void of our modernity. Yet back then, the ability in matter was there for freeways, television, radio, movies, magazines, modern medicine, warm houses. Quantum mechanics, the theory of relativity, quantum entanglement, electrodynamics, radio waves were there too, along with thermonuclear reactions in the sun and stars, and neutrinos flying through space and most of matter. All of that was already there. It was dense all around them. But it was not in their awareness. They were simply unable to bring it about in their mind or in matter for them. It would take a long time for mirrors, through the weeding out process of evolution, to advance to those capabilities.)

Death is Shattering of Mirror

What is real? Our *experience* of death is real. All experiences, are real, as an experience. If I'm in a theater and I'm frightened by a volcano erupting in the movie, my experience of that fear, is real.

Likewise, around death we have an experience of extreme negativity. That feeling, that experience, is real as an experience. We really do feel that.

That feeling comes from evolution. All creatures, not just us humans, have a tremendous proclivity to keep living, a tremendous energy to keep going. Generally creatures fight tooth and claw to stay alive. Those creatures that did not have so strong a drive had fewer surviving offspring, on average, because they just didn't fight as hard to stay alive. And over enough generations they disappeared. Therefore the feelings about death are a mere artifact of our evolutionary history. The experience of the feeling is plenty real.

Since we can't say scientifically with certainty what sentience is, we can't say scientifically with certainty whether death of the body (and its brain) is the end of sentience.

I think we try to say that I-ness ends with the death of the body because in some muddled way we "locate" I-ness in the brain. But once the brain is seen as merely a mirror – as is the case in this theory – then that muddled "location" thinking disappears. When a mirror shatters, the mirror ceases, but what it reflected continues. Only the mirror no longer reflects it.

"I wanted you to loosen up a bit, and this brain mirror theory does that," the god's head spoke.

"That's what you suggested. To let go a bit."

"That does it," was the last thing I heard the god say for a while.

36 Pillars of Judgment

I came into a thicket of trees, an area which looked suspiciously like where meetings would take place. This one seemed larger and nicer than others. Which didn't help much because I was a nervous wreck.

A breeze blew now and then.

Then I noticed. Odd, they didn't even appear in the usual way – walking and greeting. Five hooded figures sat there. I couldn't see their face, just all kinds of folds of cloth, thick coarse brown flannel cloth, almost like medieval monks.

Oh this is weird! And on top of everything else!

One started to say, "We are here..."

But I just knew from this voice – something is wrong. Got to get out of here! Now!

I turned and darted.

But suddenly I'm back in front of them, I'm struggling, with my arms and body, but I can't get out of there. The five continue to sit there,

"We are here ... to decide"

"To determine" another said

Then the center one added, in a voice that must have had the last drop of human warmth, or being warmth, in the universe, drained from it.

"To judge."

Simultaneously the figures stood up and pulled their hoods back.

The sky turned black as the blackest night in a totally deserted desert. The stars came out in incredible multitudes, and our milky way galaxy stretched more beautiful and terrifying than I had ever seen it, purple and reds and oranges, all of it diamond suns. No, there were many milky ways. All over. In every direction.

The figures shot up in height to the heavens. They were all robe. Many planets big as they extended upward in space. Uneven pillars rising infinitely upwards. I felt as if my stomach fell out of body. I desperately wanted to get out.

Now I stood on a plane of smooth marble that went to the horizons, the five pillars rising up to the front and sides.

Marble so smooth, black, glistening and terrifying. The marble became three-dimensional with the most beautiful colors ever seen, radiating from different depths inside.

Against this, I was so small against this. Then I found I could move, so I started running. But the marble plane was cosmic and it made no difference. I could run for quite a while and I was still on the marble surface with the five columns about me.

One of the cosmic robed columns began to move, as if in preparation to speak. The other columns started to respond as if they were opening the session. One probably unaware – I don't know how I knew it was unaware – maybe it wasn't but that is how I felt – did something as if casting a vote about not being valid.

With all lapping fingers of orange flame, a huge fire ball appeared, racing from outer space down on me. I ran as fast as I could. The thing was as big as a thousand foot ball fields. My God, what chance did I have!

The thing went crashing toward the marble – rather it went right into the marble – no marble buckled up or exploded or such. It should have totally missed me or totally caught me, because it was so immense. A gaping crater sat in the marble, the size of a thousand football fields. I felt the heat and flames. I think part of my clothes were burned. Yet the crater was pretty far from me. I stopped. There was smoke coming out of my clothes from the side that had been in the direction of the fire ball.

I screamed. My thumb and forefinger of my left hand were burnt off, and the rest of my hand was charcoal-like.

The great column seemed to indicate it had done that accidentally. That was supposed to be for the end of the trial if things didn't work out. Yet it meant nothing to the robed figure that it had made the mistake.

"We are here to judge," said the center figure. The figure started to talk, in this way and that. I neither knew or followed much of what it said. I looked at the crater. Later my attention returned to its speaking.

"We are here to judge," and they all mentioned a specific name. What is it talking about? I am not the person so named. Maybe they mean someone else. Maybe they got the name wrong. Maybe it doesn't make any difference anyway.

"We will determine and ask you questions. Whether you have found the right answers."

My clothes had quit smoking, and my hand was unfortunately char-coaly looking, excepting of course the missing finger and thumb.

Suddenly thinly etched lines, about four or so, appeared in the marble, as if dividing everything up into vast areas, for the judges, for discussion. The crater became farther away and not so distracting. And there were little lines around my area. Sort of like ... a witness stand with the defendant in it!

Surveying this sad situation I looked up across the marble land. Here and there, scattered over quite a distance were five or six little mounds of charcoal, or at least that's what they looked like. I thought, that's what happens to people who don't make it after they're brought here.

Suddenly I pictured that this must happen all over the universe, now and then, people – beings – like myself. And they're brought here. To determine ... did they get the explanations valid. I looked at one of the piles of charcoal. So that was my fate if they judged that way.

I fell down and fainted. This I was aware of as I picked myself up off the glimmering marble plane.

37 Judgment: Does What We See Exist

Far away on the black marble plane, the five irregular pillars rose up around me, rose for the size of several planets upward toward the heavens. This session of judgment started with their asking me several brief questions, whence they focused in on one particular issue, and asked, "does what you humans see exist?"

This chapter is the answer that I wrote for them over several days.

I handed it to the central pillar. Rather I should say that I extended my arm forward with the report, whereupon the document ascended upward so high toward the top of the middle pillar, against the starry, diamond-studded heavens

The middle pillar intoned out loud what I wrote. Here it is.

Specific Statement

Does "red" exist "out there"? If you are looking at something that has some red – if you can do so right now you might look around and see if there is something red or some such color – push your hand outward toward the red, and we are asking, does that red exist in some sense out there. Otherwise, the red you see is nothing but a construct in your mind, it has no meaning outside the feeling in your mind, it has no real existence at all out there.

Now at first, one might respond, well, of course the red exists out there, it is a certain wave length, and *that wave length is out there* ; in the scientific perspective the wave length of the light is objective and is out there, as opposed to being in our brain.

However, the logic of the "red" we see is not the logic of a wave length. The logic that is in our brain when we see red, the "red experiential logic", the logic that in some sense enters or appears or is invoked by our sentience logic, the logic which is the experience itself, will be totally different from the kind of logic that is going on for a wave or wave motion or wave frequency or light of a certain frequency. Certainly that wave logic is going on out in the external world. But *that* logic is not the red experiential logic.

What is red experiential logic? What characterizes it? We can make some guesses, but we have almost no darn handle at all on any of it. It's there, but we just can't get at it or see it. But in a couple of centuries.

At present, scientists suggest that the ability to see red helped our ancestors spot ripe fruit (our evolutionary history comes into the experience itself, even though that history where red played this role ended many tens of thousands of years ago, we still see it; it still determines much of what it is like, the inner experience, for a human when they see the color red). Of our distant ancestors, those who had the capability to see the color, versus those who had to get to the fruit and taste it before they could tell if it was ripe, had a distinct survival advantage, and so eventually this capability became a part of almost all humans. This is the theory scientists put forward. We will accept it. For the purposes of discussion here, we will assume that this capability contributed to a structure of logic which may very well be connected to our awareness when we see red, and which is what it is like for a human to see red, the inner experience itself.

It would not be surprising that when we finally understand the structure of red experiential logic – red in the consciousness – it will *somehow be connected or have relations with the general categories of the events in our evolutionary past associated with red*. This is not to say that we have any consciousness of this; it is only to say that there are connections with the structure of the logic that is our experience of red.

Red is not only the color of ripe fruit. It is the color of blood. In fact, there is hardly any other substance, or situation that occurs, where the color is so close to that special color we call red, than blood. If you look at fresh blood, spread out, as opposed to for instance, in a test tube, it is surprising at how intense the color is (Indeed, the possibly special way this sentence stands out for many readers may itself be indicative of the important role blood plays in our psyches and the role it plays in the structure of experiential red logic). Why should we be so attuned to that special color? It hardly takes much imagination to guess. Keep in mind these aspects of our consciousness, like the physical parts of our body, have been formed mainly during past stretches of evolutionary history stretching back tens of thousands of years, and before that, further hundreds of thousands of years. Except for the last thousand or so years, which by evolutionary standards is not much, for most of our history the sight of blood was much more common than today. The color of such a substance on our person, or a substance of that color coming out of our person, must be coupled with an intense reaction, from immediate concern, to horror, to panic, to stoic bravery, depending on the situation. It would hardly be surprising if, when we finally do see the structure of the logic that is what it is like for a human to see the color red, that in someway that we cannot fathom at present, these aspects of blood should be in the meaning of red. And it is not only the color of blood on our person that causes reaction. That color on kith and kin, and comrades and enemies in battle, must likewise be vital, as is the sight of the color on animals when we are hunting. This color must be a central aspect of the life of our psyche, judged from our evolutionary history, which determines much of what the inner most experience is like.

Finally, red has some proximity to the color of fire, and with the powerful role that fire played in the whole psyche of our ancient ancestors, from the importance of warmth (there were no invisible sources of furnace combustion or electric heat), to the light provided by fire (it was the only human-made light there was), to the establishing of security from dangerous animals, who are afraid to get too close to fire, and to possibly the camaraderie around the camp fire at night, so different from today when we have so many choices of what to do day and night. All these from our history, which might be called, "our human history with 'red'", may well be somehow in the meaning, in the experience itself, in what it is like for a human to see red. As to what this logic in the brain is, at present we can't begin to fathom. But wait five or so centuries.

By the way, when we play that game of how do we know if another person sees the red we see, and how do we know they are not seeing a different color, the answer is that the logical structure in their brain that corresponds to their experience in their consciousness when they see red will be basically identical between them and us. And yes, some aspects of the experience itself, the logic, will most likely in someway relate to evolutionary factors, such as fire, ripeness of fruit, and blood.

(For a little attempted introspection on the experience of red, see endnote 20.)

Does our perception exist in external reality

We are ready to look at whether what it is like to see red for a human exists in someway in the external world, or is purely an illusion in our mind, or to what degree it is both. For the sake of argument we are going to assume that the inner experience itself of seeing red, the logic in the brain that is this, has some involvement with ripeness of fruit, of blood, and of fire.

In our modern world, there is not the amount of blood, in relation to ourselves, to others, to animals, that is likely connected to the logic that is our experience of "red" - unless we are in a war-torn area. Thus, barring war, what it is like for a modern human to see the color red does not match external reality. And the same applies to the factors of heat from fire. Ripeness may still play some role because when we go to the supermarket we judge ripeness of certain fruits by their redness. But it does not play the important role that it did in prehistoric times.

So, if we considered these factors of blood, ripe fruit, and fire as being required to be out in the world in order for our experience to be real, then our sensation would simply not be real.

Yet, the situation is more complicated ("scientific" investigations are the same as are our daily lives – they're always more complicated than at first we think). Let us go back say 20,000 years, to our ancestors with spears, stealthily tracking a boar. To the misfortune of the animal, a spear flies toward it and embeds deep in its side. Blood flows. Immediately four more spears impale the rapidly extinguishing physical life form. A common day's fare in those times. Gathering around their food, the hunters see the red of the blood splashed over the dead animal. Does the red the hunters experience at that moment exist outside their minds?

Certainly the part that concerns naked blood exists. But the other parts, blood of oneself, of one's fellows, the heat of fire, lava, the ripeness of fruit, all these, that we are presuming are also part of that logic that is the experience of red in the consciousness, none of these are present. Hence, the red that the hunters experience, at that moment, has connections to elements that are not in external reality at that moment. *Indeed, this example shows that never is the red that one of these prehistoric humans sees all out in reality <u>at that moment</u>: never is the light that causes the experience of red coming from something that is simultaneously fire, and blood, and ripe fruit.*

So what is the red that these prehistoric people see? It has some kind of connection to the average of the experiences from seeing blood, in various kinds of situations, and of seeing ripe fruit in various kinds of situations, and of seeing and experiencing fire, also in various kinds of situations, and so on. Through evolution, it has a connection to the summing of millennia of experience. That is what they see.

At the instant they look at red, they see this average over millennia.

But is that *average over time* something that exists in external reality at the particular moment they are looking at some example of something red.

At earlier points of the journey we have talked about a logic as if it is "at one point in time." At later points we have talked about a logic as something that can "exist across a period of time", especially evolutionary logics. We now revisit this theme. The issue is simpler if we think of a logic as the *meaning* of a sentence of words (though we must be careful – as pointed out in Part 2 on animals – logic existed before any words). If the sentence is about a single point in time, then the logic is at one point in time. If the sentence is about the world over a period of time, then the logic is over a period of time.

You are seeing a logic that goes across a period of time, but your experience is of the whole logic experienced at a single instant in time. That color that we experience, what we are actually seeing with our experience of that color is logic over an enormous evolutionary time span of our ancestors. That is what we see when we experience red. We are seeing logic that goes on over an enormous periods of time, but we are seeing it all at this single instant in time.

Is that logic across that vast period of time something that exists in the external world at this instant in time? If it is, then what we see exists out in the world. If it isn't, then what we see is not real but just a kind of common group genetic memory.

Thus the question comes down to ontology: what exists? I think that such things do exist out in the world, because they can have a life of their own, even if they are rather complex in terms of issues of time. More than that, generally, I think that what exists are the meanings of sentences about the mathematical space that consists of physical time and physical space and physical matter, and possibly even of more.

A mathematical space is simply a set of points, with structure (and the points can be absolutely anything, from numbers, to functions, to whole spaces themselves – mathematics never mentions the awe of how large these spaces are – early on, math jumped to infinity and has worked with it ever since). Thus there is not a ruling out of sentences just because the sentence is surprisingly sophisticated about some of the structure of the mathematical space, for instance about time. And no doubt, from a scientific perspective, all meanings can start to have a life of their own in the right circumstances. Hence, this seems to say that these more complex meanings exist. Further, from a perspective that goes beyond the current scientific one, from the perspective of sentience, logic is what exists, which statement was explored earlier in the book, in Part 3. These are the reasons why I think a logic that goes on across a period of time but where some kind of summation of the whole also occurs at the current point of time, exists out in external reality.

I know. To some degree this material has stepped outside precise compartments of reasoning in solidly constructed categories.

Nevertheless, if we say such logics exist, then what we experience when seeing red, exists out in the world and it exists at that moment.

General Statement

To determine whether the "red" that we experience exists is to investigate to what degree the "red experiential logic" is the <u>same</u> as some logic going on outside the head – to what degree is the logic out there at the thing we are looking at. This is what it means to ask if what we see is real.

This replaces a difficult, abstruse question, with a more concrete one. Let us make it more concrete yet.

What do we mean by "same"? The above statement talks about whether a logic in our head is the "same" as some logic outside our head. Interestingly, neurobiology approaches the perimeter of such an issue when it might talk about different kinds of electrical spike trains being merely different encodings of the "same" meaning or logic. If we move over to the area of mathematics, there is a concept of sameness that is well worked out. It is called *isomorphic*. A mathematician might say that two things are isomorphic if they have the same structure. *We* might say that two things are isomorphic when they have the same purely logical form; all the logical relations between parts or characteristics have the same form.

To ask if what we experience exists is to ask if significant aspects of the experiential logic corresponding to the experience itself are isomorphic to logics going on out in the external world. The more significant the aspects the more can one say that our experience exists and is not an illusion inside our head.

(See endnote 25 for a comment on isomorphism in this context.)

(The reason we say isomorphic is because terms like "summation" or "average" might turn out to be too glib. Maybe it is better to look at it a little differently. There is a sense, and quite possibly a mathematical sense, in which what we are experiencing is vast expanses of evolution, but of course the experience itself takes place at a single point in time – one could use the experience of seeing the color red as an example. Centuries from now we and our machines may fully analyze this logic that is the innermost experience itself of red, and it will be seen that this logic is, in significant ways, isomorphic to the logic that went on over evolutionary periods. It is an isomorphism that takes the logic that is spread out in time and converts it to an isomorphic one that can be displayed as a logic being activated at an instant in time, the activation being the experiencing of red. But the purely logical form of significant parts of evolutionary logic and of the experiential logic in the head are the same, are isomorphic.)

Other Issues

When does experience not exist

It is also instructive to ask in what ways "red experiential logic" does not exist – has no correspondent out in the world.

It would not exist out in the world if some foreign agent had physically intruded into the brain and altered one of the areas where the red experiential logic was going on. Two

such examples are these. A virus can change sections of
DNA. If that DNA is passed onto offspring and if that DNA
changes affected parts of the brain that had the going on of
the red experiential logic, then that would change the
experience the person had when they saw red, and the
changed part of their experience would not correspond to
what was out in the world. It would be due to a tinkering
with their brain. Another example might be that a strong
evolutionary force was causing one functionality in the
brain to take over others, one of those others being where
some of the red experiential logic went on. Then that would
distort what the person experienced as red, if other
connections remained the same.

It is not unlike if one had a formal language. The brain
is like a gigantic description or expression in a formal
language. The letters and so on of the language are physical,
moving parts. If you go in and mess up the letters in a
description in a language, then of course you are going to
damage the meaning of the description.

If one rephrased the last sentence in a more technical
way, it would go like this. If you go in and mess up letters
in an expression in a language, then of course you are going
to damage the meaning of the expression.

It would be like in formal philosophy, one has formal
languages, and there are descriptions (expressions) written
in those languages. It would be as if a child who had no idea
what they were doing came in and pushed around some of
the words and letters in an expression. It would damage the
meaning of the expression.

If such a physical change in the brain somehow
occurred quickly, how might a person describe it? Suppose
the person decided to visit a doctor.

"Well, no I still see red when I look at what I know is
red, but … Well, that's just it, I really don't see red. What I
mean is I do, but I don't. I really can't explain." The doctor
nods. The person continues, "You know doctor, it's really
upsetting, because I can't even explain it to myself. All I
know is somehow it isn't the same, when I see red, it's a
new red." The doctor asks, is it like some other color that

you see or used to see. The person responds, "Oh no, no. Nothing at all like that. No, it's not just another color, or at least no color I used to know of." The person grows quiet. Then a look of sadness crosses their face.

In such a case, we can assume that since the interference was purely a foreign agent making a physical change to the brain, that the new aspect added to the quality of red has no counterpart at all in the external world. The new experience itself would be real as an experience, but it would not be real as something that exists out in the external world.

This paragraph is for the reader who is interested in the evolutionary perspective, who is fascinated by the miraculous way in which the three thousand million plus letter-long blueprint that is DNA leads to the construction in our brain of the physical structures that will have the going on of the red experiential logic. This blueprint is always drifting, usually only a little, sometimes a lot − "drifting" meaning some of the letters are accidentally changing. After all, there are these three thousand million letters in the DNA of each cell − well, in almost every cell. It follows that our experience of what it is like to see red is also drifting and since the drift is in any direction, that means the children of different people might not see exactly the same red. Over generations these differences can build up. If some people experience red in a way that somehow turns out to give them an advantage in the surviving number of offspring they have, if that advantage maintains for enough generations, eventually all the people will perceive red in the new way. Interestingly, this last statement is the consequence of nothing more than mathematics. Yet this, like almost everything in evolution, is fraught with additional considerations and restrictions as one looks more carefully at the issues. A possible additional issue might be that it is not the kind of red that people experience but some other brain processing factor, which influences the kind of red a person sees, some factor that contributes in a positive way to the number of surviving offspring. For instance, a brain logic much more important in evolution, due to limited spatial and physical brain aspects deriving from the blueprint genetic code, would "push itself physically into the going on of" the red experiential logic in the brain and could thus have the same effect on that logic as a physical intrusion. So in some ways the experienced red would not exist out in the world.

What is the experience when a part of the brain is physically intruded upon

Our brain is a physical system. Therefore the overwhelming part of it in the overwhelming number of cases has come from the past environment through the process of evolution. Thus likewise the logics that come from the brain, such as our native perspectives and consciousness. This is an important point. These logics are so deep that we have a hard time even imagining their nature and consequences. In the emotional energy around these difficulties, it is easy to get distracted from the fact that these logics still come through evolution."

"Hmm," mumbled a hiker who conveniently appeared – even here – and then the hiker took a bite from a ham sandwich and asked, "Why? Why do they come from evolution?"

"The brain is a physical system. It derives from evolution just as would any physical system. So do the logics natively in it. This is true no matter how deep the mystery of those logics is to our limited capability of full understanding."

"Hmm," the hiker said, wiping some crumbs away.

"Given this, it is an interesting question."

"What is?"

"When a certain experiential logic goes on in a region of the brain, and when that region is physically intruded upon, either through some slow process, perhaps taking place even over generations, or through some sudden, fast physical process intruding on that region, what is the experience that the being associated with that brain has? Standardly, the experience a being has is a joining, or amalgam, or a one through the other, of something going on out in the environment right now, together with a logic going on in a long evolutionary period. But if the physical substrate of that experiential logic is shifted, what then can be the experience the being is having? The being is a logic derived from an evolutionary time span, which is reflected by the physical brain; yet here the physical mirror itself is being intruded upon, is being physically distorted. That is an odd issue. Perhaps the experience is nothing more than input to a logic from an even larger evolutionary scope."

38 Judgment: Logic

Introduction

There ensued several rounds of the pillars many planets tall, or gods, or whatever they were, telling me to "define logic," and I was repeatedly answering, "I cannot, I do not know how to."

"You haven't stated enough information on what you call logic."

I thought to myself that I have stated quite a bit, and I defined it by pointing, as something out in the external reality. I thought about how I had connected logic to meaning, that when you say sentences, those have meaning, and logic is the same sort of thing as is meaning. I thought these things as I also thought about how the pillars could quite easily read my thoughts. I had stated over and over that this was the limit of what could be said about logic. After all, it was not an easy concept. It was not directly graspable by us humans.

But now I saw an orange light spread over the landscape, and glancing upwards, I saw a great fireball.

Yet this time I thought, "I said what I had to say." I looked back down at the great glistening marble landscape, getting more brightly orange-colored by the second.

I could see the small mounds of charcoal scattered over the marble environment also getting more brilliantly orange. Did I care? Who are we? What is our being? To be happy means to be free, and to be free means, finally, to be brave (endnote 21). I could now feel the fire heat burning down everywhere. From the charcoal lumps, smoke started to rise. We are human beings. I am a human being. And now I spoke out loud.

"We are beings of the universe. We are sentience, and we will not be intimidated," I said looking out on the rapidly deteriorating landscape.

I could not help feeling surprised at my words, yet it was as if I were beyond fear; respect for one's being finally steps outside fear. The great fireball vanished. All was back to how it had been a little while before, and again I stood on a vast marble plane. Finally a pillar spoke.

"We demand more information about what you call logic," said the pillars. One pillar added, "But you can produce this report in phases. You can produce it in pieces. You can produce this report however you have to. Why don't you start by telling us what is new in the overall modeling of this logic and the universe?" I looked upward toward them. They were so huge.

Suddenly I was walking back on Earth, in a pleasing part of the desert, with some wild dark-green grass, and a few desert willow trees, with mountains rising in front of me. It was rather beautiful, which is odd because I still could see in my minds eye the little mounds of gray, burned-out charcoal scattered here and there over the marble "chess board." But it did not seem to matter. I found myself just enjoying the easy place, and I calmly started to think.

Always I had been pointing and saying, "There is logic going on." A few times I said, "Logic is the same sort of thing that meaning is." But I left it at that. I said nothing else. It seemed impossible to say anything more at humanity's current state of understanding.

But now it came to me. Mathematics and formal philosophy are not without tools that can be applied to the situation. And such tools, being within mathematics, are certainly within the perspective of science, and hence within our journey.

So I wrote the following report.

Back at the columns rising into space, the report moved from my hand upward toward the central pillar. A voice intoned the report's words.

A Mathematical Model of the Universe

Throughout this journey I have mainly pointed out into external reality, pointed and eventually spoke the word "logic", as in "there, there is logic going on." You seem to not like my pointing. You seem to want more. Consider then, that the philosopher Martin Heidegger states that pointing is the first origin of things.

In Heidegger's book *What is a Thing*, if we look toward the end, at Eugene T. Gendlin's analysis of Heidegger's material, we find this on page 258.

"Heidegger argues that words such as 'this' and 'that,' the demonstrative pronouns, should not be called 'pro' nouns, that is, substitutes for nouns. The use of the words 'this' and 'that' is the most original and earliest mode of saying anything and thereby selecting and determining a thing (25,19). Only after our interplay with things do they come to have a resulting nature of their own. The noun becomes possible only on the basis of our pointing. Our demonstrative definitions precede more developed definitions, i.e., 'things' are only in the context of their relation to us and our pointing them out."

In other words, our knowledge of the most fundamental aspects of the universe starts from pointing out into that external reality mentioned throughout our journey. Did I do any different?

I am not saying that I agree with, or disagree with, or even know about other elements of Heidegger's philosophy, unless they appeared during our journey. I simply indicate the above significant statement from his ideas. What I have done throughout much of the book is point into external reality, and after a time, start to use the noun "logic" in connection with the pointing.

Nevertheless, you, the pillars, want more than pointing. After some effort, and pressure, here it is. In a far less developed form, some of this has appeared earlier in the journey of the book.

Let us consider an alternate perspective of the universe, some would use the words "alternate model".

The usual physical universe of science might be shown as

$$P$$

Figure 22 The Physical Universe (Usual Perspective or Model)

P is the standard, whole physical universe of science: space, time, matter, and we include electromagnetic waves (which is to say, light waves and radio and TV waves and all the electromagnetic waves).

We propose extending, or modeling, the universe, so as to be slightly different.

$$P \leftarrow L$$

Figure 23 The Universe (Alternate Perspective or Model)

We will write this as $P \leftarrow L$. Perhaps we might call this the "L to P" perspective, or the "L to P" model, or the "L to P" view, or the "L to P" framework.

Here, L is *all* logics, and they are being "mapped" to parts of the physical universe, P. They are being mapped to the places where they are going on. In this chapter we say so many things about L that it might be possible to lose track of the fact that L is the logics of the universe. That reality.

Let us briefly look at the perspective from mathematics. In mathematics applied to the real world, one applies some mathematical structure to some aspect of the real world. In mathematics applied to the real world in this context, L is the mathematical space of logics, and L is being mapped to the physical or material world. L is a very big space, but so are many mathematical spaces, even with those that are mapped to the physical or material world. Because of the nature of logics, perhaps there is an element here of turning on itself what mathematically and scientifically is.

The P ← L is a model of the universe whether or not one looks at it from the perspective of mathematics.

Let us look at some examples from P ← L.

The logic that is an automobile would be mapped to all the collections of matter where that logic is going on – where, when this logic is going on in a collection of matter, we say in our speech, "there is an automobile". The logic that is human sentience would be mapped to all those regions in human brains where that logic was going on. The logic that is increasing phase differential and falling amplitude, in electric fish, would be mapped to all those brain regions in electric fish where that logic was going on. The logic that is a certain dynamic and number of electrons, protons, and neutrons, arranged into three atoms, in the right way, all forming what we call a water molecule, would be mapped to all those places in the universe where the logic of a water molecule was going on. (The number of electrons is slightly varying in a water molecule). The logic that is a bank account would be mapped to all those regions where a bank account was going on (each such region would consist of widely separated and rapidly varying areas in space). A logic that was a logic of evolution would be mapped to where it was going on.

Each logic in L is mapped to (the region) where (in time and space and matter) the logic is going on in P. Actually, as will be elaborated below in the section on the "Match Function", each logic is mapped to where, *and* to how (the manner in which), that logic is going on. (The "how" is handled in a little different form than stated above.)

In the standard perspective or model, P, of the universe, the logics are hardly there. They certainly are not explicit. They are rarely pointed out in a clear way as logics. However, in the P ← L perspective, they are out in the open. They are explicit.

Does this new model attempt to now scientize and mathematicize the whole universe? Not to get frightened. This alternate model loosens up the usual one. To take just one example, some people believe in angels. In the U.S., for some years before and after 2000, the belief in an entity called "angels" increased. In the usual model, there would be a scientific predisposition against such an entity. In the alternate model, one just adds the entity to L. There are plenty of issues about doing such a thing, but the point is, a first step in this direction is easy. (One would then have to explore and state ways that the angel logic mapped to different regions in P. One could become involved with issues of the laws of physics that constrain P). Or to take another example, consider the sentience in the "All Souls" thought experiments earlier on this journey. In the new model, we can have a sentience logic in L, but it is mapped to a region that consists of two separated time segments, with an intervening time period where it simply is not mapped (the sentience would be mapped to the time up to when Aunt G was cryogenicized, and to the time period after which Aunt G's physical body was successfully brought back from the ultra-deep freeze; the time that she was in a ultra-deep freeze would be a region simply not in the map of the sentience). These are examples of how the new model frees up our thought processes regarding logic and especially regarding sentience.

This journey is from the perspective of science: can sentience be understood from a scientific perspective; or if not totally understood, at least more understood than now. All people use all kinds of models to give structure to their thoughts. Scientists use mathematical physical models: a meteorologist will have a model for certain characteristics of air and moisture movement in some part of a country; a physicist will have a model of objects and forces, interacting with each other in certain general ways.

All that the P ← L model does is this. Compared to our usual perspective of the physical universe, the alternative P ← L separates out and emphasizes logic. The essentials of the usual scientific model of the universe are still there, but in that model, logic was an obscured, hidden-away, nebulous derivative of the physical world. However, from this alternate perspective, logic stands out in the open, with the option of more freedom in how we can look at logic's inherence in the fabric of the universe. It can help us to better see how logics have a life of their own, and this contributes to our awareness of the substantiality of logic. Such issues are important to us. After all, our sentience is solely in logic.

One of the ways we seek to prove the strength of a formulation, even a formulation of the universe, is by comparing it to variations on that formulation. The above alternate perspective or model, along with its match function below, allow a wide number of variations, within a mathematical framework, because L and the map from L to P and the match function can be chosen within a vast range, as will be seen below. Thus, even though we are dealing with the universe, which is the totality of all, the $P \leftarrow L$ still gives us the needed capability of exploring alternatives.

(See endnote 22 for whether a cross product would be more appropriate than $P \leftarrow L$.)

The Layout of this Report on Logic

The layout has two main points. The first was to recast the perspective of the physical universe and logic. Instead of the P perspective, we focus on the $P \leftarrow L$ perspective. In the P perspective, logic was a nebulous sort-of-something which, when we thought about it at all, derived directly from the physical universe. In the $P \leftarrow L$ perspective, logics have an up-front place and are mapped to the physical universe. This is one of the main points and was covered in the previous section.

The second main point, which we will shortly cover, has to do with what is logic itself. We have touched on this several times throughout the journey. But it always remained nebulous. And I thought that in our time in history it would not be possible to do better. However, with the tense pressure I was under from the pillars, it dawned on me that mathematics already has tools of the right sort. (This is not to say that these tools tell us what meaning is, but they give us a way to mathematically work with it, and they make it far less nebulous, even though meaning itself still remains just out of reach; and mathematics is readily accepted in the scientific perspective.) The tools will need to be adjusted, but their form is basically already there.

In the following, central ideas will be the same as they are in mathematics and formal logic. However, for the sake of explanation, some perspectives, issues, and parts of definitions have been shifted a bit. When we move substantially away from the standard mathematical terminology, that is pointed out. As the chapter proceeds, some issues are made more precise.

Logics. Formal and Natural Languages. Meaning

In the area of formal philosophy, in the area of mathematical logic, indeed, in the area of programming languages, in all these areas there is something called a *formal language*. We will say that a formal language consists of two parts, the language itself, and the *meanings* of the things you could write in the language. As for the things that you could write, they are sequences of letters, words, punctuation, and so on, and generally we will call all these *symbols*, sequences of symbols, that is what you could write a piece of paper. When a person reads that sequence of symbols that person gets a meaning – provided they are consciously or unconsciously able to connect meanings to sequences of symbols in the language. Meanings themselves are strange things and their capability of going on in the thousands of billions of electric spike trains and nerve signals and the equivalent, darting around on the nerve fibers and the equivalent of our brain and such is already present before the language.

Technically, these sequences of symbols themselves are meaningless, for it is only when for instance someone reads such a sequences that a meaning is got.

Let us introduce some terminology. It is laborious to keep saying "sequence of symbols' or "sequences of letters, words, punctuation, and so on". Instead, we will often speak of a *description* or *expression*. A *description* or *expression* is just another word for one of these sequences of symbols, or sequence of words, letters, and so on. A description or expression is the stuff you could write on a peace of paper. It is technically without meaning, for it is only the person who reads it or writes it – provided they "understand" the language – who *gives* the description or expression a meaning, or who had a meaning in their mind before they wrote the description or expression on paper; for the reader the meaning comes from their brain but is triggered by the sequence of symbols on paper.

A language and meaning. A language and meaning. On one hand you have the language, with its descriptions and expression, which you could write down on paper. On the other hand you have the *meaning* of descriptions and expressions. What is on these two hands is two different things. Language and meaning. Language and meaning. Two different things. Very interconnected, but they are no way one and the same. The descriptions and expressions, those sequences of words, letters, and so on are one thing. But something of a wholly different kind in the fabric of the universe is the *meaning* that we get when we read those sequences. Language versus meaning. Language versus meaning.

When we were always *pointing* and saying, "logic is what is going on there, for instance in the brain of the electrical fish," and then we were mentally pointing at something going on in the fish, and we started innocuously calling it logic that was going on in the fish, but we kept using the word logic more and more; now we define logic.

Logic is meaning.

There is a point that could be of minor confusion: the rest of this chapter talks about meaning and a description in a formal language as if the two were interchangeable. Yet we just spent most of this section emphasizing how meaning and language were completely different things. The explanation for this apparent contradiction is this. Usually the only way for scientists to talk about a logic or a meaning, is by using a description in some formal language. When someone reads the description a meaning comes into their mind and that is the meaning the scientist is talking about.

(This section closes with technical comments on terminology of formal languages and on the possible role of mathematics in the fabric of the universe.

For those who want to compare the presentation in this section with an official terminology, usually, the term *formal language* is used only for all the issues that have to do with these sequences, whereas all the issues having to do with meaning are moved off to a separate concern labeled *semantics*.

All the stuff about the language itself, with its descriptions and expressions – sequences of symbols – this area is referred to as *syntax*. As for meaning, in the standard mathematical logic model, it is not defined. Instead, there is for each language, a way of going from sequences of letters, spaces, and punctuation, to the object or thing thus referred to by the sequence (or to 'true' or 'false' if the sequence should be a sentence). The object or the 'true' or 'false' is relative to a specific situation, the situation often being called a *structure* or *model*. Our interest will be to emphasize the term 'specific situation' rather than the technical term 'structure'. Everything about how one goes from the sequence of symbols to the object referred to, or to the 'true' or 'false', is referred to as *semantics*. In the expert's way of talking, a formal language is just the syntax, not the semantics; whereas we have made the formal language have both syntax and semantics. For our purposes, this seemed an easier and more natural way of explaining these concepts.

Some readers may be aware of the famous theorems by Goedel and Chaitin, deep mathematical theorems about the relation between formal language and meaning. Goedel's theorem is about how formal languages, once they are beyond a certain minimal strength, will be unable to have a proof within the language, to the effect that the language is consistent. These theorems show the offerings, sometimes, of the mathematical approach to some of the deep issues of the universe. Could the surprising successes of mathematics not only in these theorems but in analyzing and predicting the physical universe be due to mathematics' reaching toward pure logic and to logic being a real part of the fabric of the universe?)

Idealization versus reality: "come at" real logics via languages

Throughout the book it has been indicated that a logic is something that exists out in the external real world, independent of the human mind.

It is as if we could *come at* the real logic from *different directions*, each direction being (the meaning of) a different formal language. For instance one kind of (formal) language might be a simulation language - it might specify how to simulate what one hypothesizes is going on in a brain. Another kind of formal language might be solely involved with for instance the mathematics of interference waves, just the pure mathematical properties of interference waves, combined with some physics, and so from these mathematical properties one could deduce the kind of "circle" detection explained earlier for electrical fish; nothing about simulation at all. Some languages might be programming languages that specify how processes are executed (and indeed some simulation languages are programming languages since they specify how to simulate something on a computer). And there are quantificational predicate languages, which one finds in the area of mathematical logic, and which are also in an area called formal logic, but is also an area of philosophy. Furthermore, one can come at the real logic from many variations of any one of the kinds of formal languages.

Examples of formal languages

There are a wide range of kinds of formal languages.

Example - simulation language for JAR

One kind of formal language might be simulation languages for parts of the JAR logic in the brains of electric fish. One description of an assemblage to be used for simulation might be

S = ((100 Spherical)   (150 basilar-pyramidal)   (200 nonbasilar-pyramidal)
        (100 amplitude-coder) (100 phase-coder)   Z)

Maybe in this language this description means an assemblage consisting of 100 spherical neurons, 150 basilar pyramidal neurons, 200 non-basilar pyramidal neurons, 100 amplitude code neurons, and 100 phase coder neurons. Z is some description of the connections between these cells. Z might also contain statistical characteristics of the electrical spike trains between the cells, firing conditions, and the like.

The "S=" is just meant to indicate that the whole conceptual connected assembly would be called "S". Of course this does not have to be the only syntax for this piece of meaning. It is given only as an illustrative example.

(The above cells types were arbitrarily taken from Figure 4.9 in Heiligenberg *Neural Nets in Electric Fish*.)

Example - current most common formal languages

In our period of history the most commonly occurring formal languages are computer programming languages. The meaning attached to these formal languages consists of several levels of conceptual execution (sequencing of steps) and several levels of objects of data structure that are being transformed during execution.

Example - predicate language

Another important example of formal languages is that of predicate languages. These are from mathematical logic as well as from the areas of formal philosophy and formal logic. Though their meaning can be anything for the predicates in the language, most examples of such language have no "process" or "execution" aspect, and in this important aspect are unlike programming languages. For instance, the description

For each x there is a y such that y = x+1

The meaning of this statement is that whatever number a person considers, there exists another number which is equal to the sum of one plus the first number. Does this seem excessive verbiage for so simple an idea? Sometimes that is the nature of statements in precise languages. Yet sometime such statements become elements of a solid foundation for deep theory. And profound results.

Here is another example of a description within a predicate language

    for all f
        f(0) and
                for all x (f(x) => f(x+1))
        =>
        for all x f(x)

This statement is typically interpreted to have the meaning of the principle of mathematical induction: if f is true of 0, and, whenever f is true of an integer it is true of the next integer, then, f is true of all (non-negative) integers.

Again, the meaning here has nothing to do with anything like execution of steps in time, or of anything even to do with time.

(In these examples we are not making the distinction between first and second order predicate languages, or indeed any order of them.)

Natural language in place of formal language

L is all the logics. In the methodology of science, we settle on a single word, whether or not what that word is intended to refer to in the fabric of the universe is fully understood, and then we stick to that word as the official term, and we treat other words that might be synonyms in the common vernacular for the word as less formal, more free floating, common but imprecise ways, of referring to what the official term is referring to. These common vernacular synonyms are also used as ways to help us get at what the official term is referring to. At this point in our journey in this book, we have come so far that now the word "logic" is an official term. Other common vernacular synonyms are "meaning" and "idea", which help us get at what the official term "logic" is referring to.

All this is by way of clarifying the use of words in the following main topic of this section: natural language is also allowed as a way of indicating a logic.

A natural language is the language people speak in their everyday lives; for instance, the languages spoken by the people of the earth.

In the model $P \leftarrow L$, we allow not only formal languages as a way of indicating meaning (logic), but also natural languages.

Certainly, the more one moves to a full scientific perspective, the more a formal language will be used. But all ideas start out expressed in everyday language (in fact, they start even earlier than that as some kind of primitive forms going on in the ocean of electrical signals in the brain). So if one is using $P \leftarrow L$ as a scientific model to work with, even then, it seems that allowance should be made for natural languages sometimes as a way of talking about L. Even if the idea or meaning is real precise, natural language is often used in science as the way of indicating the idea or meaning. For instance, in the most precise scientific papers and textbooks, most of the ideas and meanings are indicated using natural language.

Here we are only saying that one of the languages for getting at logics (the elements of L) is natural language. In this book, our journey is from the perspective of science, and the only way that science can get at L is through languages, and while ideally those languages should be formal, in practice that may not always be 100% possible.

Finally, it should be noted that natural language itself may be nothing more than an extremely advanced formal language.

For all these reasons, whenever we talk about formal languages, we include the possibility of using in some cases natural languages too.

Scientists' activities

Here is one of the ways that the activities of scientists fit in with the P ← L framework. Scientists will say or think something like, "let such and such . . ." and they have a conception of the such and such, and then they go ahead and work with that such and such as it occurs or goes on in the physical world, and by "work with" we mean that they use mathematical equations and deductions from the equations, and with deductions from their conception of the such and such, but all these deductions are in a relatively "tight" form, as in deductions done within axiom systems (axiomatizations). The such and such is a point in L. The results of the working-with are the ways the point is mapped to P.

The Match Function

We return to the main idea of this chapter: the alternate perspective or model of the universe, P ← L. We have been discussing logic or meaning, that is, L, and we spoke about formal languages.

Now we look at how those logics or meanings of L are mapped to the physical universe P. In effect we look at the "←" part of "P ← L". We introduce what we will call the *match function*.

The match function tells us to what degree a specific logic is going on at a certain place or region in the physical universe. In the real world (external reality), it is often not the case that a certain logic is completely going at some place, or is completely not going on at that place. Rather, it is degrees to which something is going on. It is degrees to which a logic is going on in a certain region.

The degree could be measured with a number or with some more complex mathematical entity. The degree could even be restricted to 0 and 1 for certain logics, 0 meaning that the logic occurs not at all at a certain place or region in the physical universe, and 1 that the logic occurs there completely. This is allowed. However it should be noted that even something like the logic whose going on humans refer to as a car, even this logic can have degrees *between* 0 and 1. For instance, if the car is in an extreme accident, one might say that there is not totally a car there, but to some degree, yes, a car is there. That degree would presumably be shown by a number *between* 0 and 1. Or if the car goes off to the junk yard, over time the degree to which a car is there goes in steps from 1 down to 0, some steps being larger as when someone takes away the engine, some steps being pretty small, as when someone takes away a piece of chrome, and other steps being miniscule but capable over time by themselves of "continuously" going from 1 down to 0, as when it rains, during which a number of atoms are swept off the surface of any car, and in enough hundreds of years the collection of matter, where the logic went on, would be completely gone, so that toward the end, in the little bit of matter that was left, the degree to which the logic was going on would be 0 or very close to 0.

If you consider the JAR (Jam Avoidance Response) logic in the electric fish, you will find the logic is taking place – going on – in a region consisting of several areas of the fish's brain, and, it is going on with a certain degree of strength, for presumably in some fish, the logic is going on quite fully and strongly, while in other fish, the logic might not be completely, fully as all there. The match function captures – expresses – handles – all these aspects.

Consider an area of any brain, fish, human, or otherwise. What logic is going on there? That certainly depends on how you interpret the nerve signal encodings and how you interpret many, many other things going on there. These are part of the *way* in which the logic is going on. There are many logics going in any specific area, in the brain, or outside the brain, or in the environment, or across evolutionary time spans. All kinds of logics are constantly burbling everywhere in the physical universe, and they are all potentially relevant because their going-on's can potentially interact with the going-on's of other logics. In the Vinegar Bog we saw how quite different logics from the ones we humans pick up can be going on in the same region (the creatures perceived one logic, humans perceived other logics, but all in the same place). In addition to different logics going on, similar logics can be going on but with different degrees of strength. The match function incorporates all these issues.

Mathematical formulation of match function

Thus we have

match_fn (wsp, p) = degree

where p is a specific logic, and wsp is a region *and* a way, and the match function tells us the degree to which that specific logic is going on in that region in that way.

As stated earlier, the degree could be anything from a value in a discrete set of two values of 0 and 1, or could be from the set of continuous values from 0 to 1, or could be anything up to some complex mathematical structure indicating the degree to which the match was successful. (If one really wanted to get complex, the degree could even contain information about how p occurs at wsp.)

The details of all this will develop over a long time and will come out of the way researchers think and talk about how a certain class of logics occur in some place.

Calling the parameter "wsp" above is due to thinking of it as the "way in space" that the logic is to occur. Here "space" means *not* the space of physics but instead a mathematical space that incorporates physical space, time, maybe matter, as well as further information as suggested above (electromagnetic radiation and so on). One should not unthinkingly restrict the nature of the wsp region. In its most general form, wsp could be any mathematical subset of the points in the mathematical space. For instance, in the thought experiments on Aunt G's cryonicization in the chapter on "All Souls," the region where Aunt G's sentience logic goes on could contain *two separated* periods of time, one consisting of the time up to cryogenicization, the other consisting of the time after she was successfully resuscitated; the period when she was frozen would not be included because the sentience is simply not going on at that time. Using the mathematical conceptualization of match_fn, Aunt G's sentience or being might match with degree 1 in this region and 0 outside it. (This does not take into account such issues as Aunt G's birth and her final death and so on.) Generally, in all matters about wsp, p, and the match function, we want to be motivated by how people speak about these things, when they speak in their usual, everyday way. People have a great deal of helpful thought processes that go on automatically, and we do not want to detrimentally interfere with, or distract from, these valuable oceans of activity in their brain, by introducing artificial restrictions. If a neurobiologist says such and such a logic is going on across such and such areas of the brain of a certain creature, then generally the match function and wsp should express that. Nevertheless, sometimes the speaking and thinking that people do might be positively influenced afterwards by the addition of the perspective of "mapping logic to regions of time and space and the like" and of the "match function."

(From the perspective of the strictly analytics of physics and mathematics, one reason that logics are so important is that they so strongly affect what might have gone on in the past and what might go on in the future. For instance, if cat logic is going on in a region, that has totally different implications about what was going on, and what will be going on around that region, compared to if tree logic is going on in the region instead.)

The mathematical formulation of the match function is
match_fn (wsp, p) = degree

where p is a specific meaning or logic. The only way science can get at meaning or logic is with a formal language. So if one were looking at equations involving the match function, p would likely appear as some writing in a formal language.

Which match functions and formal languages

Which match functions and formal languages can be used?

Any kind of match function and formal language is fair game as long as it works. The point is not to place any restrictions ahead of time on what kind of match functions and formal languages are acceptable, so that scientists and others are free to naturally develop different versions.

Another point is that it is completely alright that the construction of a formal language and match function comes out of our everyday ways of talking and thinking. Indeed, that would be a common source of such construction.

Could one person, talking to another about electric fish, say that certain aspects of the pure mathematics of interference waves are going on in the brain of the electric fish? Sure they could. You could picture two people carrying on such a conversation, and one says that the mathematics of interference waves is going on in the fish's brain. The other person says, "how is that?" The first explains how different groups of nerve cells are actually sending out signals that are encodings of the usual mathematics of combining interference waves. The second person says, "That is interesting. Yes, now I see that the logic of the mathematics of interference waves is going on in the fish's brain and I see in what way it is going on. Cool."

And if people agreed that the logic was going on in that way, then that indeed would constitute a legitimate mapping and match.

In short, any kind of formal language (or natural language) with any kind of conceivable idea of match function is fair game. Of course it must be reasonable, in the same way that one person speaking in regular language to another about the logic (for example, what logic is going on in the electric fish) must be reasonable in the sense that people who heard the explanation would say, "yes, I can see how that logic is going on in the fish."

It is very much just as we today handle languages, as well as truth and statements in those languages. Different groups of people in society have different degrees of how solidly their words (statements) should match some part of the external world. Scientists have a pretty high degree of solidity. They require that a fair number of other scientists accept a match and accept that some statements are true in that match, before the statements are labeled as true. Compared to other groups of people, scientists will frequently also develop formal and semi-formal languages, which they use in addition to their natural language. None of this should be construed to detract from the reality of the logics going on out in the external world independent of the human mind.

As for sentience logic in the brain, as thinkers and researchers gain more knowledge over the next years, decades, and centuries, they will see what kinds of formal languages are good for describing this kind of logic. The artificial languages that humans, or their machines, will develop to describe sentience logic will develop incrementally over time. The whole picture will not come all at once. At first humans will start out with limited and relatively simple languages to describe basic logics in a few simple creatures. As time goes on and humans create more advanced such artificial languages, they will gain more insight into a larger picture of meaning and external reality. In time, machines will join in the work.

Our understanding of what can be understood advances with our experience. This is shown by quotes earlier on our journey, from Nicholls et al, *From Neuron to Brain*, third (not fourth) edition, page 2.

"...the modeling of neural circuits becomes a sterile exercise unless it incorporates the known properties of nerve cells."

For a time, some wondered whether properties of individual neurons would yield an explanation of such things as visual depth perception and pattern recognition. But as scientists gained more understanding through experience, they unexpectedly came to see that these properties of individual neurons could be used to pretty much explain such things (see page 2 of the above book, also see the earlier chapter "Astounding Sentience Logic," the section on "Neurobiology").

Many meanings, matches, possible sentiences

Perhaps a virtually infinite number of quite different logics can be going on in the same matter. For instance, the Vinegar Bog creatures existed in a physical substrate of mathematical energy characters. The logics going on out in the world and causing them to have experiences were fully different from the logics going on in those same regions that cause us to have experiences. All these logics are going on in the same region of matter. All these logics match this region.

We might consider another example. If we focus on a certain area in the brain, certain logics may be going on there. But we notice that if we changed what meanings were encoded in the signals, that another logic would be going on, a very different one. In this example, these two logics are going on in the brain in that region. These two logics could be mapped to that same region, only depending on the encodings.

Let us consider another example. We notice a certain logic going in some region of the brain. One day, to our surprise, we notice that even with the same encodings and same signals, it is possible to see a radically different logic going on in exactly the same limited region. This surprises us, or at the least, fascinates us. Both these logics are going on in the same limited region of the brain in the same signals. (Presumably, some other part of the brain that made use of these signals would only be using – treating – the signals as if they were from one of the logics.) Both logics match the region. Both logics map to the region.

Here is a further example. Let us consider a region of the brain. It is not only that there are signals and there are meanings in those signals. Even in an individual nerve cell, if we could see into it, or even in a group of cells, if we could look generically into them, we would see a lot of meaning in terms of the dynamic structures of protein chains, proteins, and even molecules in a protein, and so on. Any statement that you could make about what you saw and that was basically true – that statement has a meaning, and that meaning is going on in that region. That we said the statement was basically true, is that its meaning is basically going on. The degree the meaning is going on is given by the match function. Note that this example applies to regions anywhere in the animal's body, not just in the brain. In fact, the same idea applies to any region in which there is matter. Even if a region had just two atoms moving about, there are many things specialists could say about those motions, and all of those said statements are a meaning going on in that region. Nevertheless, in any region, each kind of specialist will concentrate only on certain kinds of meanings (going on in certain kinds of ways).

Now let us move to a more important example. Is a sentience logic going in a region of matter? This is important because if there is, then there is a being going on there. *Any* match of a sentience logic, *any* match in *any* way that expresses the sentience logic, would mean that a being is going on there. The match might not even match our time with the being's time. Maybe its time is one of our spatial dimensions and our time is one of its spatial dimensions. *Any* mathematical match in terms of set theory and structure is a legitimate candidate, and if you know how sophisticated mathematical functions and transformations can get, that is saying a lot. The being's time could be in a mathematically profound way in what is time and space for us.

Given that the first part of this section stated that virtually an infinite number of logics (meanings) are going on almost everywhere, maybe one might conclude that the same applies to sentience logics, and that beings are going on in regions all over the place. But a sentience logic of the degree of a cat or dog or human has a tidal depth and complexity of structure of logics, including all the internal relations of experiential logics and support and all the relations to logics outside the body. The extensive requirements for a match, no matter what kind of match you were trying, would *severely* restrict where such logic could be going on.

As for the issue of time, note that with the Aunt G thought experiments in the chapter on "All Souls," that the sentience that was Aunt G could be thought of as mapped (matched) to time that consisted of two separated time segments, one going up to the point where Aunt G died, and the other starting at the point where Aunt G was successfully resuscitated from cyrogenicization. As for the period during which Aunt G is "deep frozen," the match function is simply not mapping her sentience to that region of time. (Mathematically speaking, the time to which the sentience is matched – mapped – could be any mathematical set and is not required to be time-contiguous.)

The reason we needed the match function is that there are an infinite number of logics going on everywhere. Even if we restrict our attention to logics that are significantly different from each other, there are still likely an infinite number of logics going on in every region everywhere, perhaps at almost every point. In practice it is not possible to have a mathematical function that gives all the logics going on in a specified region, because in practice that is too open-ended. All that we can do is say to what degree a certain logic is going on (in a certain way) in a given region. That is what the match function does.

Universals

How do universals fit into this conceptualization?

Someone asks you to describe the logic that is going on over "there," in that collection of matter, as they point to Jack Smith, standing next to a tree. You might answer with words such as the following. "That is a human being going on in that matter, the logic going on interacts in an extended social logic in many ways one of which is that it is called

'Jack Smith', and its physical substrate is the matter over there" and you point to that spatial location over there.

That is an example of one particular logic.

Now someone talks about a human, or about the human body (the logic going on in a collection of matter for it to be a human body). That is *one* logic. Today, on the planet Earth, that logic is going on (matches) about six thousand million different places on the Earth at the same time. We express this in our speech by indicating any one of these places and saying, "there is *a* human." (Technically these places are not completely separate, because as when people physically touch, if one looked closely enough at the atoms, one would see a good deal of atoms interchanging between the two physical loci; in fact any time people are in contact with any object, animate or inanimate, a good deal of atoms are leaving that collection of matter that is defined as where the human body logic is going on.)

Describe an oxygen atom (describe the logic that is gong on in matter for something to be an oxygen atom). And someone gives a description in a formal language or in natural language. That one particular logic is going on at places in matter throughout the whole physical universe. Whenever this logic is going on at a place in the universe, we say "there is an oxygen atom." Note also, that the collection of matter that constitutes the physical substrate of even this logic can be changing, as an electron may easily leave or be added to the atom, or even be shared with a nearby atom; so that even at this level of incredibly small spatiality, the material substrate of the logic can be changing with time.

This logic whose going on at a place can cause us to say, "there is an oxygen atom," is *one* logic. But this one logic has come to be going on at a virtually inexpressibly large number of places in the universe.

What happens if two descriptions felt to be contradictory

What if researchers see that two logics, A and B, are going on in the same region of the brain but that the researchers feel these two logics are contradictory? The signals from one small area of the brain realize two contradictory logics. How would researchers deal with this situation?

Well, they would do what all humans do. On one hand they would think more deeply about what they meant by "contradictory." On the other, they would look more carefully at what was going on in the external world – what is the situation in this region of the brain? Perhaps they would find that only one other part of the brain made use of the signals from the questionable area, and that part used the signals as if they were from logic A. So now the researchers are happy. Or maybe they find that several other parts of the brain use the signals from the questionable area, and that some use them as if they were from logic A, and some as if from logic B. Again the researchers are happy. In both cases they learned how complexly surprising reality can be, how strange is the inner nature of these logics.

The main point in this section is that many logics can be going on in the same place, even seemingly contradictory ones. People do what they always do when contradiction seemingly appears; they look harder and they think harder.

Generally these issues enhance the feeling of the separateness of logic and the dynamics of matter, which makes it all the more interesting that we are purely in logic.

Further Discussions

Meaning. Frege

This section is not related so much to the mathematics model but to meaning and language in general.

There is a difference between the "real" meaning, the meaning that is out in the world, and the formal meaning of any formal language. The formal meaning is nothing more than the meaning that is formally constructed by us humans to go with some formal language, which we have also constructed. The "real" meaning is, well, the real one out in the external reality. One is a construction of our thinking, the other is what actually exists out in external reality. The formal construction in our thinking will be an approximation to, and hopefully a good representative of, the actual meaning that is out in reality. And in different circumstances, and as time goes on, there may be several constructions in our thinking (that is, several varying formal meanings), that are different kinds of approximation, and to different degrees in various ways good representatives, to that real meaning out in the world.

Researchers in formal logic have at times looked directly at meaning, but these results are not as widely studied. Possibly the first person to do so, and at the same time to look at formal languages as themselves mathematical objects, was Gottlob Frege in his book *The Basic Laws of Arithmetic*. He made the greatest advance in two thousand years of formal logic, with the concept of quantifiers ('for all', and 'there is some'). Not as well known, his system also looked directly at meaning (one aspect of which he referred to as *sense* – as in the *sense* of an expression – as opposed to its reference or denotation – which is what typically is concentrated on in the results of mathematical logic today). It was not only as opposed to reference or denotation, for his work looked at meaning as it occurred out in external mathematical reality, whereas in today's formulation, mathematical logic looks at issues of meaning as they are related to something called a structure or a model, which is a controlled, formalized idealization of a part of external reality.

Over time, as humans start to have different approximations, or representative formal meanings, we will also start to develop more insight into what are the relevant aspects of the real meaning, and after a while, those relevant aspects will be addressed in the formal meanings.

Many Issues Lurk in Match Function. Metaphysics

As we get more knowledge of what logics are going on in the brain and which of them are related to sentience and which are not, we will develop a deeper understanding of the match function. The real understanding of the metaphysics of the universe may come from such a strengthening comprehension. And this will be true even though the function is an artifact of the particular way of our trying to get at logic.

When people first start to think about sentience and what is the miracle going on in the brain, they focus on the signals. But the signals are nothing. Or people look at the physical foundation of the signal, the connections and functioning of the neurons and the changes in the functioning of the neurons. All of that is nothing too. It is the logic that is going on via the signals and their foundation. That is where it is at.

Many issues lurk in this match function: the issue of whether an inactive part of the brain can still be involved with sentience; the issue of (sentience) logic going on not only in the brain, but also more diffusely in the body; the issue of (sentience) logic not only going on in the brain, but to degrees outside the body too in the great regions of evolutionary time and space; the issue of perception logic going on both in the brain but also, in a time transformation, outside the brain; the issue of the relation of logic and space; whatever issues are related to the period of time during which Aunt G is cryogenicized; the statistical and continuity aspects of how the goings-on of logics develop and move about in space, and in the evolutionary mathematical space consisting of time and regular space. Furthermore, being cognizant of this function, leads one to look at these issues, especially for scientists, who, from their perspective, tend to skip past them.

(See endnote 23 for the match function and evolutionary psychology; also see the chapters on "All Souls," "Teletransportation," and "Standard I, Non-standard I.")

Formal functional aspects of logics in the brain

Viewed functionally and abstractly, logics going on in the brain may have time-related states. For instance, whenever the 'red experiential logic' in the head is in a state that we might decide to label 'activated', the being we associate with the body, of which the head is a part, has the experience or awareness that we call "seeing red." Issues such as these can be set up in a formal logic meant to capture the logic that is in the brain. Perhaps one could even cover a lot of ground by assuming the logics were mathematically not messy but very clean.

Whether a logic has states is merely the same matter as whether you choose to have a display device portray a logic as having states (See chapter "Devices to See Logic"). It is one of the ways that many logics can be thought of – can be displayed. That is, one way to think of certain logics is with the help of states. But that is certainly not the only way to think of them. And if a device is sophisticated enough, it is not the only way it could display the logic.

Similarly, logics going on in the brain may be thought of as having different kinds of componental or relational relations to other logics (structure of logics). Eventually science will determine the various kinds of such componental relations. As one example, consider 'red experiential logic' – that logic in the human brain that is the very experience itself of seeing the color red. That logic may have certain kinds of componental relations to logics in our brain related to the perception of blood, logics in our brain related to the perception of ripe fruit, logics in our brain related to the perception of fire and so on.

For all this to be so, it is only necessary that the dynamics of that incredibly huge, three-dimensional forest, crammed and squashed against itself, the brain, function as if it were so. As far as the physical brain, nothing deeper need be going on. There need be no "understanding" in the brain as to what it is carrying out. Indeed, all that is needed is that the brain function as if the various logics where going on, each logic shining through other logics, in spectrums. If the brain functions as if the logics were going on, then they will be going on. (This is provided the match function is not too restrictive – for the match function could be made so restrictive as to force detailed behavior down to the neuron level.) Furthermore, this business that the brain need only function as if the logic were there is similar to Turing's assertion that if a computer could *function* as if it were a human then it would be a human. Turing's assertion deals with a big logic, that of the whole human, and causes questions. But there should be no question for much smaller logics. As long as the brain *functions* as if the logic were there, then it is there. This is one of the characteristics of logics.

(Actually, Turing's assertion was that if a computer could respond in a manner indistinguishable from a human, in terms of text going into and coming out of the computer, then the computer would have (human) intelligence. But the idea is the same as above.)

(I wonder if for a number of significant, deep logics going on in the brain, whether the brain is like a mirror that has no idea of what it is reflecting. It blindly picks up much logic, width and depth, but has only repetitive mechanical processes applied to them, repeated over and over on top of each other. To be sure, in ways, some of the processes are sophisticated, but more importantly, they occur in astounding numbers of repetitions. This blind repetitive character likely comes about from some repetitive character mark in the building procedures of the brain, encoded in the DNA blueprints, with the character mark occurring in places in the DNA code, each single mark causing many repetitions of neural structure to be built.)

Substantiality of logic

One of the special points on our journey was when we discovered that logic, or meaning, existed independent of the human mind. And that was from the perspective of science.

(One cannot help have the feeling, when digging into these issues, that meaning must truly be disconnected and separate from matter, in some sense not currently definable. In human history, feelings have been right, and feelings have been wrong. If we want more certainty on this issue, we must wait till we achieve it.)

When we consider that our sensation of seeing red is a logic that comes from our evolutionary history; when we consider that the same is true of what it is like for us to experience hearing, or experience seeing the shapes of objects (our experience is not that of mathematics and geometry); when we consider that our childhood experiences of royal kittens among the bright dandelions and green grass are real – and are logics; when we consider that our experiences of other beings are real – and other beings are logics; when we consider that our experience of our own being is real – and is a logic; when we consider all this, we conclude that logic is what is real.

All this supports the substantiality of logic.

The math model presented in this chapter moves logic from being buried in un-explicit derivatives of the physical universe. It moves logic to a conceptually independent place where it is mapped to various regions of the physical universe in various ways. The math model with match function fully incorporates this.

We sometimes speak of this as contributing to the *substantiality* of logic because logic is moving away from an absolutely dependent and confused derivative conceptual relation with the physical universe. It is becoming more substantial. And the term "substantial" is appropriate because in the math model, logic is coming to a level of importance on a par with the physical universe, and the physical universe is certainly substantial, so therefore it is natural to speak of this as the increasing of the substantiality of logic.

Moving to even greater substantiality, though in the speculative chapter, "Brain Mirror Theory," we jumped to having logic exist independently of the physical universe P. (Also see endnote 24 on brain mirror and P ← L models.)

The ultimate proof of the substantiality of logic may be us. We, the sentience logic, whose going on is in the brain, exist. In fact many people believe their existence to be surer than anything else, including that of the physical world and science.

The brain is a language. Is every physical thing?

The brain itself is literally a language. While the languages that one typically studies are composed of letters in an alphabet, with expressions composed from those letters, the brain is composed of moving parts and the expressions are composed of those moving parts, but meaning is assigned to those expressions just as meaning is assigned to the expressions of a formal language. It is rather strange to think of a language as having components that are moving parts, but there it is.

Could one speak similarly of all physical things?

Ways we come at logic

In explicit scientific analysis of logics, about the only way we come at them is via languages. There are other ways though. Outside of scientific methodology, we come at these logics by having an idea – a meaning – a logic – in our thoughts. Finally there is the most important way that we come at meaning. That is by being it, in the case of sentience logic. Our very awareness is sentience logic. We are it and it is us. While the physical body is biologically alive, the sentience logic goes on there. Thus there are three ways that we come at logic.

Example math model – volcano

We have an understanding of what is a volcano. This is a logic, this is a meaning, the logic or meaning of what is volcano, this is the logic that needs to go on in matter for us to refer to that area as a volcano.

No doubt, if researchers today did a survey, they might present different pictures to person A, who had to say to what degree they thought a volcano was going on in the picture. Without going into the detailed issues (issues of quotes and names and what is being named), we might say that p1 is "Person A's idea of a volcano – in other words, person A's ideas about what constitutes a volcano." Let us suppose that wsp1, wsp2, and so on, are various areas of land. Then match_fn(wsp1, p1), match_fn(wsp2, p1), and so on, would be the degree to which person A thought a volcano was going on at wsp1, wsp2, and so on.

In this example we have taken wsp to be "an area of land," and there are likely alternate ways to take wsp. But in a general way this shows how logics, in this case the volcano logic, are going on at different places wsp1, wsp2 ....

Example math model – paper clip

There are as many logics as you can think of, and many more. What are some of the relations between those logics? Generally, they are the relations that you think of in your natural, regular thinking.

Petroski's book, *The Evolution of Things*, describes the evolution of for instance the tin can, or of the paper clip. There have been different versions of the paper clip, and they didn't look like the ones we have today. Thus, among various logics, we have one for each version of the paper clip. There is relation between these different logics: one version of logic of a paper clip being much "better" than another one. This may be hard to define in words, but it definitely exists out in the complex real world, and it is the existence of this relation that clearly drives the evolution of the paper clip, and has driven it to its current form. In the versions of anything, in those cases where there is a characteristic out in the real world of versions of a certain kind of logic clearly being better than others, then a process of evolution of that logic is not only possible, but absolutely a given, just as it is for the paper clip, from its primitive first versions to now.

One might ask, what logic of a paper clip is one talking about: the first version, some other version, the current version, all versions together (the paper clip viewed as a totality of all it versions) – and there are many other logics too, that hover around our phrase "paper clip." The answer is, all these logics are ... well ... logics. Each is a single point in the mathematical space of logics (L) – it is a very big space. The first version of the paper clip is a logic and it is a single point in L. Any intermediate version is a different logic, and is a different point in the mathematical space of logics. The same is true with the logic that is the current version of a paper clip. Also, the logic that is all versions of paper clips – the logic we think of as a paper clip without thinking about versions – that also is a logic, that also is a *single* point in the mathematical space of logics.

One might ask, can there be so many logics out in the external world? The answer is that all these logics are already in our mind, and each logic in our mind is a point in the mathematical space of logics (L). And there are many more logics in L than those in our mind.

As for the relation between these different points, you ask, "what is the relation between the logic that is the first version of the paper clip and the logic that is our idea of a paper clip generally?" The answer is that these are two points in the mathematical space of logics, and the relation between them is the relation that you struggle to define when you work to define the relation between the idea that characterized the first version of the paper clip and the idea the characterizes paper clips in general.

(As stated earlier, the mathematical space of logics is continuous, similar to the way in which there are an infinite number of numbers between say 43 and 44, varying continuously from 43 to 44; although these numbers are one-dimensional, whereas the mathematical space of logics is infinite-dimensional.

There are lots of logics. The mathematical space of logics is a standard kind of mathematical conceptualization of the logics that are going on out in the external world; ideas in our mind are logics out in the external world that are logics in the brain, attempted to be brought close to those logics outside the brain; while from a perspective quite different from our native one, perhaps the brain is a bottleneck gateway from regions of evolutionary time and space.

Logics would also include such as social forces and economic forces, for they are logics going on in the society overall.)

Evolution of logics

All kinds of logic, no matter how few or how many levels of abstraction appear to us to be between our directly sensible world and them, are going on throughout the whole physical universe. In other words, these can occur in ways that seem extremely abstract to us, but still fully exist in physical reality, as much as anything in physics.

Such a logic, if it is going on at a number of places on the Earth, or throughout the universe, or in some environment, no matter how concrete or abstract that environment appears to us, and if there are a couple of such logics, and if their relation with the environment and each other can be defined in terms of an axiom system that has logical primitives similar to those in a theory of evolution, then these logics can be defined in terms of the axiom system as "competing," and it could be proved, using this axiom system, that over time some of the logics grow stronger than others, given the random background of jiggling or random perturbations that are always taking place everywhere in the universe, and it might be possible to prove, depending on the axiom system, that other logics would cease after a time to be going on in the environment.

Typically researchers think of evolution as in a certain kind of space-time-logic environment wherein DNA is a highly particularized kind of physical carrier for the logic that will be going on in various places. But this is just one example of one particular kind of axiomatic perspective.

Mathematics first?

Already in 1569 during the intellectual fracas of whether to place the sun in the center of the planets, the French philosopher and mathematician, Petrus Ramus, also known as Pierre de la Ramée, (born 1515, killed 1572 in the St. Bartholomew's Day Massacre), said that we should let our explanations of things play a lesser role. Instead we should take observations and the mathematics of those observations as primary, and we should go with those, and establish those clearly. Then at some future date we could come up with the explanations that had lead to these observations and mathematics. That fracas started with Copernicus around 1540 offering a system of the planets where the sun was the center.

"The first warning that astronomy would have to be cultivated in a totally different manner came from the French mathematician, Pierre de la Ramée, or Petrus Ramus, Professor of Philosophy and Rhetoric at the College Royal at Paris, who had from his youth been a determined opponent of the Aristotelean natural philosophy. He published at Basle, in 1569, *Scholarum mathematicarum libri xxxi*, … [in which he said that] Astronomy is involved and impeded by the many hypotheses from which it can be liberated by mathematics." The idea was first to get the mathematics right, and then maybe at some future date, someone would propose explanations (hypotheses) to explain that mathematics. "If only Copernicus had proceeded without hypotheses, … ; and it was to be hoped that some distinguished German philosopher would arise and found a new astronomy on careful observations by means of logic and mathematics, discarding all the notions of the ancients."

(Quotes are from pages 358-359 of J. L. E. Dreyer, *A History of Astronomy from Thales to Kepler*. The second quote is Dryer's description of what Petrus Ramus wrote. These quotes were written by Dryer before 1906.)

I believe that the same ideas were espoused several centuries later by the French mathematician Henri Poincaré (1854-1912), concerning the issues around the Michelson-Morley experiments, which were in turn the driving experiments of Einstein's theory of relativity. Let us go ahead and develop and fully accept the mathematics of the situation. Later, someone can come up with an explanation.

This suggests that since we are so far away from explaining sentience from a scientific perspective, maybe we should first establish possible frameworks, using various axiomatizations, based on what we may more directly intuit, without worrying about how these things could be in nature. Then at some future time, we can explore hypotheses as to why things are this way.

Martin Heidegger, in *What is a Thing*, discusses a central role of mathematical thought in our developing understanding of the universe. In the section "The Metaphysical Meaning of the Mathematical," in the subsection "Descartes: *Cogito Sum* ; 'I' as a special subject", Heidegger brings up the pivotal Descartes, who lived 1596-1650, which it should be noted is the same general time period as the above Petrus Ramus. Recall that Descartes is not only famous for the "Cogito ergo sum" ("I think, therefore I am") in philosophy, but for the Cartesian coordinates in mathematics, coordinates that mathematize location in space and time and so on. Heidegger writes, "It is no accident that the philosophical formation of the mathematical foundation of modern *Dasien* is primarily achieved in France, England, and Holland ... [in that time frame]" Heidegger says that the usual idea that Descartes' greatest contribution was solely the "cogito ergo sum" and the "I" is not so. This idea is at best like "a bad novel" . In fact, in that period of history there was the development of the idea "that the mathematical wills to ground itself in the sense of its own inner [nobody else's] requirements. It expressly intends to explicate itself as the standard of *all* thought and to establish the rules which thereby arise." Heidegger then characterized Descartes as follows. "This simultaneous advance in the direction of a foundation of mathematics and of a reflection on the metaphysics [deepest originating principles] above all characterizes his [Descartes'] fundamental philosophical position." (This material is from pages 98 to 100 of Heidegger's book.)

In the preceding paragraph I have added the square brackets, and some contain my own definitions of terms. But returning to the theme of this section, maybe we should involve a mathematical approach to the study of sentience. And what better way is there of involving the inner requirements of mathematics than through the exploration of different systems of axiomatizations? Descartes himself wrote "*Omnia apud me mathematica fiunt*" ("With me everything turns into mathematics") (http://www-groups.dcs.st-andrews.ac.uk/~history/Quotations/Descartes.html).

We need not go overboard. There is a spirit in this method. That is what we must lay our hands on and use as one of the long-term approaches to understanding sentience.

Conundrum like quantum mechanics

As for the equations of quantum mechanics, physicists can mathematically explore *aspects* of them in order to find further properties of the universe (as for instance I believe Stephen Hawking does). But if they try to understand the inherent nature of the whole equation itself (as for instance I believe Roger Penrose does), then there is a great deal of struggle and controversy.

This chapter has offered a definition of logic, and its relation to the physical world via the match function. But this formulation will have a similar problem to quantum mechanics. On one hand there is a scientific-mathematical formulation that is acceptable in mathematics and formal philosophy. On the other hand is the problem of trying to understand what that formulation in context could possibly indicate, and there one enters into issues that elude a solid grasp, issues that cause one to have feelings that oscillate from, this is scientifically and logically trivial, to, this is a profound secular miracle.

Fundamental equation of quantum mechanics

There is an equation of physics from which all laws of physics can be derived, at least so far. It is called the fundamental equation of quantum mechanics.

The trouble with this equation is that it is often very difficult to derive results from it in a general form. Might it be possible that if our underlying space and time were conceptually modified to include something of "$\leftarrow$ L", the equation would become friendlier?

Might there be logic or information that could be incorporated into, and produce, a new geometry of space and time so that when the fundamental equation was recast in terms of this new geometry, the equation would become more amenable to being worked with.

Ultimately our ideas about the world derive not surprisingly from our native perspective, if indeed one may not view some of the ideas as synonymous with our native perspectives. Those ideas include our "containering or structuring" conceptualizations that we refer to as space and time.

A variety of issues, some having parallels to here, are developed in Poincaré's writings, especially the essay "The Relativity of Space." Among other topics, the essay talks about different geometries of space, and how they ultimately derive from our sensations and mind.

Two hikers appeared. One seemed to have ideas about what kinds of logics to build a new time and space on.

"As a rustic in these matters, I would guess to look at quantum entanglement, whereby particles at arbitrarily far distances are instantaneously connected but have no capability of passing information between them. I wonder if that might suggest something to incorporate into a new geometry of space and time, one that would somehow incorporate logic too. Well, when generalized."

"What do you mean by 'generalized'?"

"Of course, I don't know. I claim not otherwise."

"It is true. These days knowledge is so specialized and demanding."

"And from the safe place of being a rustic, I would suggest something else too. Single electrons being fired through a double slit act in some ways as if they were not individual particles but rather a small part of a wave, even though there is only one individual particle at a time. Clearly this has a kind of informational transfer across time and space, where our native perspective does not see it in this way."

"Dear rustic, it may be that these things are obvious, or then again, maybe irrelevant, to the issues. I certainly don't know. We are looking for changes in geometry that make it easier to work mathematically with the fundamental equation of quantum mechanics."

"When the equation is recast in the new geometry. These examples – the electrons and the entanglement – present themselves as odd in terms of our native understanding of distance and time, and logic too. So that might suggest a 'space and time' that incorporates but extends the one we have used so far, one where distance and logic are tangled together. When the fundamental equation is recast in terms of this new geometry, it might become easier to work with. Just as the theory of relativity extended our previous geometry of space to include time, so too this would extend space-time to include some kind of aspects of logic. The theory of relativity introduced an amalgam of time and space: time and space are not as separate from each other as appears in our native perspective. In the perspective of this amalgam, time and space are not as separate as appears in our native perspective."

"Yes."

"I may be a rustic, ..."

"So am I."

"I may be a rustic, but it would be interesting to see some special new geometry, maybe a 'reduced' geometry, with all kinds of special new laws of physics applying in this geometry, the results being even more mind-boggling from our native perspective of space and time than the theory of relativity."

"Well, we don't know if it would be that wild."

"To what extent are, perhaps, what we take as geometric conceptions existing out in the external world, in reality, from our native perspective. By the way, the terminology would probably have to be changed."

"What do you mean?"

"We couldn't just say 'space-time' anymore. We would have to say something like 'space-time-logic', indicating that this new containering is an amalgam of what appears to us as space, time, and logic. In this new space-time-logic, the fundamental equation of quantum mechanics would become much more amenable to solutions. Our native perspective of time and space and so on may be only that, our native perspective."

(For one reference on quantum entanglement, see for example, Martin Gardner's article on "The Guided Wave Theory of Louis de Broglie and David Bohm." For a description of the strange results when an electron gun fires individual electrons through two slits, see for instance, page 11, of Hey and Walters *The Quantum Universe*.)

(Is it possible that looking at a minimalist axiom system for plain old Euclidean space might lead to a better formalizable understanding of how much more there is in "reality," than in our native perspective. For some examples of minimalist axiom systems for Euclidean space, see for instance Schwabhäuser and Szczerba, "Relations on lines as primitive notions for Euclidean geometry.").

A different kind of mathematics

(1) If one approaches sentience from the direction of the scientific perspective, I believe that one will have the sort of mathematical model of the universe and the sort of match function as described in this chapter. One may not call it a

mathematical model, one may not conceptualize it in exactly the same way as the match function. But it seems hard to avoid some kind of conceptualizations equivalent to those presented here.

(2) The kind of mathematics used in an area of scientific investigation goes hand and hand with what becomes the area's fundamental character.

The mathematics of the match function and of formal languages and of their meaning is different from the mathematics of traditional physics, which was primarily mathematical Analysis applied to a situation where everything deeply derived from points in space-time. In the model and match function in this chapter, it would seem that Analysis will not play too much a role. Instead, the mathematics of logic and formal language and meaning comes strongly to the fore. This is a very different kind of math than Analysis.

Another difference is that the role of points in space-time as originating foundational concepts is replaced with the much different concept of regions in space-time, along with the "indirect" nature of the match function.

Quantum Mechanics has moved away from so strong a notion of points in space-time as in traditional physics. Instead one works with functions over space related to probabilities that such and such will be at any point. But this function is couched in terms of mathematical Analysis. As far as I can tell from my understanding, there is another difference with "old-fashioned" non-Quantum physics. There is a strange conceptualization called "wave collapse" (Although in as least some reformulations of Quantum Mechanics all these differences have been removed – see for instance Martin Gardner's article on the Guided Wave Theory (GWT) of de Broglie and Bohm.) In addition to the mathematical Analysis used in traditional physics, Quantum Mechanics mathematics reaches into another area, Group Theory (For instance see chapter 10 of Byron and Fuller, *Mathematics of Classical and Quantum Physics*). Group Theory can arise from problems of analyzing structure of logical form. Perhaps that might suggest that someday the model in this chapter will give rise to uses of Group Theory.

The mathematics of logic and of formal languages occurs not at all in traditional physics or in Quantum Physics. Thus we may expect the character of the scientific study of sentience to by fundamentally different than that of physics, either traditional or quantum.

(3) There should be appropriate axiomatizations of the P ← L model that lead to theorems to the effect that logics, as they occur in matter, have a life of their own. In particular, that logics come together with other logics to create even larger logics. For instance, in the electric fish, there are sublogics of the JAR logic that come together to form larger sublogics of the JAR which finally all come together to form the whole JAR logic. All of this is in the context of the P ← L model and the ubiquitous but strange match function. The theorems would state mathematical, statistical tendencies for logics, presumably with the same match function, to start going on in matter in a coordinated way with each other so as to result in larger logics going on in the area. The reason that there should be such theorems is that there are sublogics coordinating with other logics to form larger logics everywhere, all the time, in the physical world, inside and outside the brain. Why is so much meaning going on in matter almost everywhere? Are there theorems answering this question? Are there other theorems addressing the depth of that meaning?

This may be the main place where the traditional model of the universe that is physics interfaces with the alternate model of the universe presented in this chapter.

Weight of the soul

In the library the checkout person noticed some of my notes.

"I heard that in a number of studies there is always found to be exactly a 24 gram decrease in the weight of the body after a person dies. In a number of different situations where they went through the trouble to check these things, that is what they found."

I don't remember the person's exact words, and I don't remember if it was 24 grams, but it was something like that.

It is interesting to think that the soul or our being or a logic would weigh something. After all, you could change the electrical signaling in the brain so that there was no being at all, and it would not change the weight of the brain, or the body. So sentience logic weighs nothing. And the same applies to any logic. If you had a toy house built of blocks (that is, the logic of a toy house was going on in those blocks) how much would the logic weigh? What is the difference between the weight of the blocks scattered about, and the weight of the same blocks arranged into a toy house. That is how much the logic of a toy house weighs. Nothing. Zero. The going on of that logic added no weight at all.

And yet our language, the way we talk when logics go on somewhere, is neither that way nor that simple. For instance, when we ask, how much does that toy house weigh, and we point to the toy house of blocks, the answer, by the usual way that we approach questions for this kind of logic going on, is the weight of all the blocks that make up the house. It is the weight of the physical substrate of the matter in which the logic of the toy block house is going on. Likewise when we say how much does a certain human body weigh, we mean the weight of the physical substrate of the matter in which the logic of the human body is going on.

But many logics, and the human body is no exception, go through time. There is possibly a different weight of the physical substrate at different time slices. So the being associated with a certain body might say, "now I weigh 180 pounds, but two years ago to the day, I weighed 160." When we speak of the weight of a certain logic going on, we have certain shared ideas about what physical substrate or substrates might be under consideration.

But what about the weight of a soul or of a sentience? The sentience logic goes on in the brain. So by analogy with the block house, or with the human body, we might ask what is the weight of the physical substrate of this logic, and that would pretty much be the part of the brain where the sentience logic is going on. Buy it may turn out that sentience logic is a structure of logics, with different logics belonging in different ways to the sentience logic. So there may simply be no longer a clear-cut collection of matter that is the physical substrate for the sentience logic. There is more than that. Since sentience logic can be seen as a transformation of, and hence the same as, certain logics going on outside the human body right now, combined also with certain logics that have gone on outside the body at some time since birth, and combined with certain logics that went on at times in the environment of the human bodies of all the ancestors of the individual, perhaps the physical substrate of all these should be used. If a device that showed logics could show all this, we would see the logic in the brain as being parts of the same logics going on in all these areas of space and periods of time. The device would show something like, this logic is composed of logics x, y, z, and so on, and x is a transformation of the following logic going on in the environment of its ancestors 15,000 years ago for a period of about 1000 years. And similarly for y, and z, and much other logics. If the device showed time as going from the left to the right across the screen, it would be like a kind of branchy tree top at the right (that would be the logics going on in the brain at present), but the same logics would string out into the environment, and also back in time.

If now we asked what is the weight of sentience, and for an answer we looked at the matter in which the logics go on, then it could be trillions of tons. As to what the weight would be, the answer is difficult, because the logics in brains, much more than other logics, in time and space fan out to other logics in the brain and to other logics outside the skull. There would be no exact border where you could draw a line. (Interestingly, if we did weigh sentiences in that manner, most people would share almost all their trillions of tons of weight with other people, except for the pound or so in their brain or few hundred pounds in their body right now.)

The central point is this. When certain logics go on in matter, we say and think a sentence like, "there is a such and such," to indicate that that kind of logic is going on in the matter. The issue of the force of the pull of gravity on the matter in a certain physical substrate can be central in many ways to the logic going on in that substrate of matter and it can be important to other logics at times interacting with the logic. So when we say that the weight of a such and such is such and such, this corresponds to gravity pulling with that force on the physical substrate in which the logic is going on. When gravity is pulling on a set of blocks in which a toy house is going on with a force of two pounds, we say sentences like "there is a toy block house," and "that house weighs two pounds." Or if the logic is that of a human body going on in a part of a cloud of about 7,000 trillion, trillion atoms, we say and think a sentence like, "there is a human body," and if those atoms weigh 180 pounds, we say, and think, a sentence like "that human body weighs 180 pounds." This is for certain logics, typically for logics whose going on we refer to as "physical objects". For other logics, and these have as much full-fledged reality and existence, there is no longer a clear-cut, obvious set of atoms to take as the physical substrate that the logic is going on in. For a bank account logic, the way the logic goes on in matter, there simply is no clear-cut single candidate physical substrate, even at a single point in time. Sentience logic, on closer examination, strongly fans far outside the skull, not only in space, but in important ways it also strongly fans out backwards in time. So here too, from the way the logic goes on in matter, there simply is no single candidate physical substrate.

(A technical note. Appearing out of nowhere, two hikers roamed over. One said that when the blocks are arranged into a house, they actually weigh an incredibly small amount more. Immediately the other hiker and myself decried such an utterance, pointing out how sometimes the obscure ideas of modern physics, when not carefully used, can mislead people into falsehood.)

The notion of proof. Seeing logic.

The pillars insisted I address the following topic. If I had been less drained, I would have explored my feeling with them that this was an odd question. What would it be like to see directly all logic?

I wrote the following answer.

I think if you can see all logic directly it means you are instantly, unconsciously aware of the *results* of all proofs, all reasoning. That means of all proof and all reasoning *that exists*. Aware in the same way as that when you see a tree, or when you see the color red, you don't have to stop and analyze and say, let's see, this is true, and this is true, and oh, that is true too, so therefore such and such, and also such and such, and ah, so therefore this is a tree, or ah, so therefore this is red. We don't do that. We know just like that (at this point, I clicked my fingers, and I instantly knew that the pillars accessed that action for humans and knew what it meant). We just perceive that something is a tree, or is red, like that. (I clicked my fingers again). We don't have to think about it. We don't have to prove it. And it is the same if you could see all logic. I think. Just like knowing right away that something is a tree or is red, so too you would know the results and consequences of all proofs without having to come up with a proof or having to think about it or having to work to see it. You would just naturally see it. That's what seeing logic would imply. I think.

If two meanings (logics) were provably the same, then of course they would be the same, and when you "looked" at them you would see one and the same thing. In other words, for instance, if you looked at two descriptions written out in some language, and if it turned out that those descriptions had the same meaning, then you would just inherently know the one single meaning that they both had, no matter how complicated the descriptions were, and no matter how different the descriptions seemed to be.

(The previous paragraph is informed by experiences in the field of mathematics. Two mathematical statements may appear completely different. But sometimes it turns out they have the same meaning. For a mathematician, the explanation that they are the same consists of theorems and proofs. Similarly, two things can appear as if they would have no connection or relation whatsoever. But sometimes there are astounding relations and connections between them – well, at least they are astounding to us humans who have limited vision and don't natively see the connections. Is it a fact that the essential societal function of mathematics is to reveal the purely logical connections natively unseen by humans?)

Anyway, what is a proof? For us humans a proof is just the drudging, laboring analytical part of the brain trying to come up with steps that show some inherent part of logic already in the fabric of the universe, but of which, natively, we have no grasp. We are blind to it, in our native perspective, so we have to struggle, feeling around for steps of reasoning, to make the connections that we do not natively see. If you could fully see all logic, you would see all of that right away. You wouldn't have to grapple around trying to feel connecting steps.

(There is a theorem in mathematical logic that states something like the following about any two descriptions in a language.

(In any situation, two descriptions refer to the same object)
if and only if
(there exists a formal proof that the two descriptions have the same meaning).

The detailed version of this would have further complexities. However, sometimes it is beneficial to go back to this kind of non-formal language, for it gives the underlying idea. At any rate, the theorem indicates the deep connection between meaning and proof.)

Closing Statement

Adding to our conceptualization of the universe something like $P \leftarrow L$ may seem too simple – it may seem like a nothing. But there is within it the seed of much. It is amenable to the perspective of science and mathematics. Some day our ideas will be highly developable in that perspective. In the past, scientists have got goose bumps when they saw what immense aspects of the universe were predicted by their playing with little squiggles and formulas on paper, squiggles of the mathematics of traditional physics. Some day, the P ← L model will be a rich field of scientific and mathematical development leading to scientific philosophical advances in understanding of what is logic (meaning), sentience, and the nature of the universe.

If from one direction, P ← L seems like a nothing, from a different direction, it may seem like too much, and one may criticize that it adds too much. But science often adds considerable *mathematical abstraction and structure* for the purpose of analyzing and understanding and predicting the physical world. The notion of force and how it is used in analyzing problems in physics, is quite an abstraction and is a requirement for all the rest of physics. Other mathematical sophistications and abstractions of colossal analytical and predictive power are potential energy, momentum, conservation of momentum, angular momentum, Cartesian coordinates. I believe there are also techniques whereby physicists add to the dimensions for space and time extra dimensions of energy characteristics, and these additions allow questions to be mathematically answered that would be extremely difficult if not impossible to answer otherwise. P ← L is just another such addition in the analytical perspective of science. There is though a little trick under the surface. Though the addition seems like a nothing, it sometimes appears to be finding a substantial something. It is finding the substantiality of L.

As humans, we individually have always been able to come at logic by being it (from most scientific perspectives, we are synonymous with the sentience logic going on in our brain: we are that logic). We can also come at logic by experiencing (when we experience, the experience itself is some logic going on in our brain: there is a kind of overlap with sentience logic.). But science cannot come at logic by either of these approaches. Science can *only* come at logic by using formal languages – symbols, expressions made out of symbols, and so on – even though the meaning of any such formal language went on a long time before the language – science can only come at logic by using formal language to name logic or meaning, and from there, to investigate logic, sentience, and the nature of the universe.

If one looks at our being in terms of that slice of reality shown us by science, it is hard to see how one can avoid the issues presented in this chapter as well as in this book. While the notions and terms of being and sentience and soul may not be so clear, the notions of sentience logic in the brain, and the structure of sentience logic in the brain, will, one way or another, be foremost. For the only way to analyze what goes on in the brain is to discover the logics going on, and to discover the relations between those logics, that is, to discover the structure of those logics. The foremost nature of something called sentience logic, or the structure of sentience logic, is traceable to the fact that we are capable of talking about the deep-most part of our being. Talking comes from muscle movements of the vocal cord and mouth area, and those muscle movements are fully traceable *solely* to the logics going on in the brain. This is as according to physics, and this book is from the perspective of science. Talking comes *solely* from the logics going on in the brain, and this is true even far into the future when science can explore the massive detailed control of the ocean of signaling going on in the brain.

As the book pointed out earlier, almost everything is a logic going on, whether or not it is sentient. The match function discussed in this chapter is an inherent part of logics going on in the physical universe. Whatever terms and conceptualizations we will use one day to characterize these areas, they will not be able to avoid the issues associated with the match function. Nor can they avoid the metaphysical issues that follow from such a function. It reduces the central role played in our thoughts by the concept of single points in time and space. It contributes to the substantiality of logic. Mathematics can be used to express the relevant structure of the universe whether or not we understand why the universe is that way.

Our perceptions, the experiences themselves, our awarenesses, come from the logical form of our evolution. From the perspective of science, this means that the perception and the experience-themselves logics in our brain are partly isomorphic to the logics that went on over the large time and space region of evolution – the evolution-logics. The match function would be key to mathematical formulas expressing this.

Finally, from the perspective of science, native sensations and awarenesses are limited compared to what is actually there. This is due to the physical limitations of our sensory neurons. But an even more profound blindness comes from the way we are laid out in space and time and in the fabric of the universe.

Whatever spiritual view one has, if one takes the external reality as presented by science as a real part of the fabric of the universe, then it seems that basically many of the same issues as presented in this book must appear.

All the details in this chapter may be as so many trees that preclude us from seeing the forest. It is a beautiful forest. There is the space of logics that map to the physical universe. The logics do not map to single points in space and time, but to regions in space and time, as well as to degrees in these regions. All this suggests the substantiality of logic. That we are sentient, further suggests it.

Thus went the last part of my last report to the pillars. As I stood in the witness stand, I started to say something to qualify the above. Though I was not sure what I would say.

Immediately my words were stopped from coming out. I looked upward toward the pillars. They were clearly accessing the reports. The great columns rose in majesty above the black, gleaming marble plain. Being unable to make any words, I started to just relax and study the physical forms of these beings. They were so big compared to me. Then I looked beyond them at the stars and nebulas and galaxies of the universe. This time went on for a while, during which I was free just to be and take it in.

Suddenly I could tell the judgment was done.

In that same instant I was streaking, my voice howling, back to Earth, faster than a bolt of lightning.

39 *Relation to Other Work*

And so I was bolted toward Earth, like lightening. Yet in this midst, everything stopped. Outside of my body had been plenty wild, but inside I knew that I was alright. The reason I stopped? By now I could read well enough some of the thoughts put in my mind. The idea was suddenly there that I had to relate this journey to other books and works. After that, the bolt to Earth would continue. Unlike hikers, I had more awareness about what was going on. Here is the report.

In this chapter, the title of this book, *The Soul and the Fabric of the Universe*, is referred to simply as *The Soul*.

Plato and Aristotle

In relating this work to others, it is most honest to go back first and foremost to Plato and Aristotle, and then the atomists Leucippus, Democritus, and Lucretius. Plato gave forms the first order of being, with all else a reflection thereof. Aristotle reacted, letting forms exist only derivative from matter. Clearly forms, whether of Aristotle or Plato, have similarity to the often mentioned logics on our journey. Our forms – these logics – straddle both sides of the fence, with Plato on one side, Aristotle the other. In places they are like Aristotle's forms derivative from the dynamics of matter. In other places, metaphysically, the logics are fully Plato's forms, especially when considering sentience and inner-most experience, sensation, and awareness.

Already at the beginning of our journey these logics started to assert their substantiality by affirming their independent existence outside of human thought. Thereafter they repeatedly expressed statements consistent with substantiality, till finally at the end, an option was established for their complete independence from matter, in the P ← L framework. But this independence required no

439

struggle against Aristotle's position. Rather, such controversy was moved to the side, and the domain of mathematics was given the task to gestate the future. But this mathematics is not the kind used in physics. It is the kind developed, and so far used, by formal mathematical logic and formal philosophy.

As for Aristotle, I always had a hard time understanding his repeated, alternating rungs of matter and form, that great chain of being, ascending alternately through one and then the other. In our journey, there are no "alternating" rungs, but instead logic everywhere going on, indeed a virtual infinite number of logics, going on in every region. Many issues are still to be worked out here. But these are to be worked out partly by looking at how people normally talk about a logic going on in the various areas of human investigation, with gradual formalization of the languages and with gradual creation of languages to encompass various particular languages.

(Plato, about 427 to 347 B.C. Aristotle, 384-322 B.C.)

Atomists

On reading the atomists, especially the developed expression of Lucretius, about 55 B.C., one cannot help but be impressed by description of the first rank of power for that period in history; it explained our world in terms of a multitude of these tiny pieces of matter – what they called *atoms*. In one way our book's journey has been a modern continuation of those ancient atomists. We know so much more than those brave pioneers. They were as babes in the woods. Nothing did they know back then of molecules, atoms (as we know them), forces, algebra, (Descartes') Cartesian coordinates, calculus, formal theories of language and meaning, and the mathematics of logic and formal languages. They had no awareness neither of physics' laws of motions nor of statistical characterizations of the motions and interactions of great numbers of atoms. In fact they had not the imagination of a mathematics of statistics; they did not even have the decimal point, zero, or Arabic numerals. They knew not of the developed concepts of evolution.

They had nothing of our induced mentality through decades of exposure to computers. They did not even know the special role the brain played in behavior. Even as late as 1748, La Mettrie, expositing how we are just machines, (*Man a Machine*) could only invoke various philosophical attributes of matter, without having any way whatsoever of tying those in a concrete manner to the physical world, because he knew nothing at all of the function of the brain. From one view, our journey is simply using all this new modern knowledge, and awareness, to further that journey of the atomists.

Though the atomists were on the right path, they placed forms in a worse location than did Aristotle. Soul or mind was literally composed out of the minutest of the atoms.

(Leucippus, somewhere around 450 B.C. Democritus, about 470 to 380 B.C. Lucretius, about a little after 100 B.C. to 55 B.C.)

Contemporary Ideas about Mind

For the relation to contemporary ideas about the philosophy of mind, we somewhat arbitrarily select two works, based to a degree on the amount of their usage: E. J. Lowe's *An Introduction to the Philosophy of Mind* and Chalmers' *The Conscious Mind*.

Lowe

Lowe's book explores a wide range of the ideas in the philosophical area of the mind (what *The Soul* calls sentience). I will only make the following comparison.

Philosophy wants, almost needs, to look at meaning from every angle, and some of those angles are intricately compounded meanings; so it is a challenging activity. I believe that the motivation is that the real nature of the universe is at this level of exploration. Science has a different approach, and that is all it is; it is not a denial of the philosophical approach, it is just a different way. As to what the difference is, it is not so easy to say. But generally it has to do with postulating or divining ... "a material world," "a physical external reality," "a common external reality." Then from there, science uses "tight reasoning" to build up on what it already has earlier built. This book moves in that approach.

This is the general structure of the approach of science. There are many details left out. For instance, in our particular historical period, some people stress that science is always ready, if the approach needs it, at any point in time, to demolish earlier of its ideas. However one can also point out that the point in time is in fact always a period in time, and that the term "demolish" is a relative concept as to what is being destroyed and more importantly what is not being destroyed.

Another detail in the approach of science is what it is that is focused on when people look out into that external reality. Writing on William Harvey's discovery, in the early 1600's, of how the blood circulated through the body, Herbert Butterfield notes that in that period of human history, "Concerning some of the writers of the sixteenth century, it has been discovered that, though they talked of the importance of seeing things with one's own eyes, they still could not observe a tree or a scene in nature without noticing just those things which the classical writers had taught them to look for." (Butterfield, *The Origins of Modern Science, Revised Edition*, page 51.)

As for modern philosophy, is to be so kind to it truthful? Is there something in the Presocratics that is being lost already in Plato and Aristotle, with that something more than just brevity? Such topics however are afield of our journey.

In terms of the framework in *The Soul*, scientists can posit almost any abstraction (which will then be a logic in L), and then deduce characteristics and results of that, and that constitutes the mapping part of the P ← L framework.

Chalmers

Chalmers book roams into many issues and I will only comment on two. Further, the comments below are based much on my reading of Chalmer's paper, "Facing Up to the Problem of Consciousness," which content appears in the book, and seems to address our key issues. (As of 2004, the paper is online.) My comments are on functionalism and dualism.

I had read along with the material, and then I read "There is an *explanatory gap* (a term due to Levine 1983) between the functions and the experience, and we need an explanatory bridge to cross it." (See the paper "Facing Up to the Problem", about four eighteenths of the way through; or see the book *The Conscious Mind*, the first two entries in the index, for "Explanatory gap.") To me, the gist of this section, as I had read along, is that function cannot offer an explanation of experience itself. I thought, yes, indeed I agree, I can see that.

And yet, earlier, when I was jumping around in the paper, I came across that same point, and I could not see it. It was the lead-up of material that allowed me to see it later. And that is natural; the nature of talking and reading is that communication leads up to a point. However, the trouble is with the word "functionalism", or as we shall call it, "functionality".

The word "functionality" is no kind of ordinary word, for what does it refer to out in the external world? The lead-up in the book suggested by general points and examples that "functionality" refers generally to a certain wide range of things. Yet that wide range of things could be the smallest pinpoint of what the word "functionality" actually can refer to.

Indeed, no matter how you formalize the concept of functionality, the collection of all functionalities is basically the same the collection of something called *computable functions*, and the collection of computable functions is immense in breadth, and most importantly, *in depth*.

Unless the lead-up examples are without bound in terms of depth, in every sense, then the examples will encompass only a pinpoint compared to what the word "functionality" encompasses, and this would be according to any kind of measure, including any measure of depth of understanding. (Here we talk about when examples are not able to reach out to all functionalities; the understanding that goes on in the thought processes taking place in the brain may be far more reigned in than that, as possibly suggested by the proof sketch in endnote 26 "Fallacy That We Know That We Know.")

Now one needs to flesh out these ideas, flesh them out at the interface between the regular concepts we use in our natural language (the way we naturally speak and write) and the definitional formalized concepts in areas such as computational complexity, which are involved with mathematics, mathematical logic, and the like. But the gist is valid. Whatever relatively more concrete examples we look at of functionalities, there are functionalities that go way beyond those in depth and in *every other way* too. And that includes those that go way beyond what we presume today.

The second issue is that of dualism. One example of dualism is that which occurs in at least one part of *The Conscious Mind*, but the modern version of the concept goes back at least as far as Descartes.

Dualism takes the position that sentience cannot be (fully) explained in the world of physics or in the material world. There must be another quality or aspect to the universe. Hence the term, dualism.

It seems that a theory of dualism, either from the start correlates sentience with certain things in the physical world, or if not from the start, then after time, it seems anyway to come full circle back to some such correlation. For example, at one place (page 302 of the book, the section "It from Bit"), *The Conscious Mind* wonders whether sentience might be correlated with information as it is conceptualized in "it from bit" (Wheeler, "Information, physics, quantum: The search for links.") Whether one uses the word locate or correlate, the journey in *The Soul* locates (or correlates) sentience inside an aspect of the universe I call logic. *The Soul* digs into these logics, drawing them out, both inside and outside the head, finding them in evolution, and looking at issues of time transformation and isomorphism. (In the dualism of Descartes there is also a kind of correlation when he states that this other aspect of the universe has contact with all parts of the physical human body – see the section below, on Descartes).

Yet as soon as dualism posits such correlations, it risks flirting with non-dualism. As part of this positing, one starts to ask why are these particular aspects of the physical world correlated with sentience while other aspects are not, and further, why are some of these things correlated with certain of our experiences while others are associated with different experiences?

There is an exquisite, profound intricacy and quantity of various experiences, and also of their corresponding logics in the brain. People who embrace dualism will have the same problem that non-dualists have: to understand this relation and thus to increasingly stand in awe at the extent of its nature. The dualists feel the relation is merely a correlation. Dualists do not deny that a physical understanding of the correlation will offer insight into consciousness, but they feel that the insight will be of a limited sort not able to deeply explain consciousness. Non-dualists, on the other hand, feel the correlation to be the strongest one possible, identity.

It is not clear that the approach of *the Soul* can at present be classified as dualist or non-dualist. *The Soul* looks out at the external material world as having full reality and as having the only scientific gateway to an explanation of sentience. This approach, in turn, forces our beliefs about the metaphysics and ontology of the universe to be different than what our native perspective presents.

Many ideas in *The Conscious Mind* and in *The Soul* are not at opposite poles. For instance, the concept of information, around the idea of "the bit is it", is not at opposite poles from logic in *The Soul*. (Nevertheless, logic is likely different than information: see the chapter "Initial Notes on Logic." Also see note 4, for chapter 8, of *The Conscious Mind*, that states that the strong binary yes/no of "the bit is it" perhaps does not apply to reality – this is my reading of the note.)

General

The approach of *The Soul* takes the view that the scientific mystery of sentience will be solved with scientific advances within ten centuries. The way of science is to look out into the world as the ultimate confirmer, to build on what is already built, to build using certain fairly tight reasonings. *The Soul* treats sentience as one type of logic. In what way that logic is not just in the brain (or the body) is left as an open issue. In line with the particular approach, *The Soul* merely points out that for science, the brain is the gateway to understanding sentience. Even as for treating sentience as a logic, *The Soul* is at places cautious, simply because of the largeness of logic and because of what must be many issues that we cannot be able at present to see.

In terms of comparison with modern approaches to sentience, whether approaches from philosophy, science, or otherwise, the approach of *The Soul* is that of when, throughout history, science has moved into new areas, always picking out certain competitively stronger observations, with this time, the approach especially forcing us outside our native perspective, even outside the possibly usual "scientific" native perspective, and seeming to involve, from mathematics, *tools* from formalized languages and meanings, with these tools sterile only as are a surgeon's instruments, with these tools to step over parts of our deepest native perspective, in order to see better; all as a way to begin to approach this new idea.

Descartes, Heidegger, Ramus

There are many philosophers over the last several hundred years. With them, some points of comparison can be made here and there. Considering the range of the math model and match function in *the Soul*, that should not be surprising.

It is interesting that in talking about mind, Descartes writes that it "has no relation to extension, nor dimensions," and that we cannot "conceive of the space it occupies." The term "extension" here refers to what today we might describe as the three dimensional space along with the time and matter that is the human body. But then he went on to say that it "is really joined to the whole body and we cannot say that it exists in any one of its parts to the exclusion of the others." This is from page 122 of Burtt *The Metaphysical Foundations of Modern Science*.

Logic, in *The Soul*, is similar to Descartes' conception of mind in that it is easily thought of as existing apart from time and space. Descartes has it only as mind having this property; we have all of logic having this property, with mind (sentience) being one kind of logic. When Descartes says that mind is really joined to the whole physical body and that one cannot say it exists in any one part to the exclusion of the other, this is functionally equivalent to the match function mapping logics to *regions* of space, in this case the region occupied by the physical body. The same element of wonder applies to the joining, in Descartes' description, and to the mapping, in *The Soul*.

(Could one set up two languages and axiom systems, one for Descartes' system and one involving the math model with match function, with some kind of intertranslatability between the two: whatever could be proved in one would have some version provable in the other, and vice versa. How would the "cogito ergo sum" – "I think therefore I am" – be handled? Would it imply some statement about the substantiality of logic?)

The Soul contains three pretty much different items related to Heidegger's ideas of around 1936.

As stated in the text (see index), Heidegger indicates a pointing, via the words "this" and "that", as the first form of indication, before one defines a thing in words. This is what *The Soul* has as the initial way of indicating logic.

Heidegger points out that, in the historical period and geographical area of Descartes, mathematics sought to become the central foundation stone of all thought. *The Soul* has a much less strong, but parallel statement that mathematics should be used as a starting point to structure our knowledge of sentience, given that our understanding of sentience is so uncertain. Petrus Ramus (Pierre de la Ramée), in the general period of Descartes, expresses a similar sentiment about the usage of mathematics in trying to understand the motions of the planets through the sky (see name index under Ramus).

Heidegger analyzes what humans find to be a thing in different historical periods. *The Soul* answers that a thing is a logic.

(Petrus Ramus, 1515-1572. Descartes, 1596-1650. Heidegger, 1889-1976.)

Hobbes

The philosopher Hobbes, 1588-1679, tried to reduce everything, including thoughts and feelings, to motion of matter (see the internet but also for instance the latter part of page 204 of Burtt's *The Metaphysical Foundations of Modern Science*). In looking at the relations of the match function to these conceptions from earlier philosophers, we might wonder just how much is in this match function. From the modern perspective of science and mathematics, typically we have no concern when we set up a mathematical abstraction that has at least a certain validity in covering the nature of things, as in P ← L.

Alfarabi, Avicenna

One of the most famous of medieval philosophers was Avicenna (Ibn Sina, 980-1037, Persia). Though a person with his own ideas, there were many ways in which he followed a slightly earlier philosopher, Alfarabi (around the year 950, Baghdad).

As for Alfarabi, "He is notable as having made the epoch-making metaphysical distinction between essence and existence, ..." "... he drew the metaphysical conclusion that the concept of essence did not include that of existence." (quotes from Knowles, *The Evolution of Medieval Thought*, first half of page 196).

"Avicenna followed Alfrabi in many ways, ..." though certainly was independent in others (Knowles, page 198). Avicennna wrote that "*Being* is applied also to objects of consciousness even though they have no counterpart in the world outside of consciousness e.g., fictions of imagination." (Weinberg, *A Short History of Medieval* Philosophy, page 113). "*Being* cannot be defined because there is no concept more general in terms of which *Being* could be defined;" (Weinberg, *A Short History of Medieval Philosophy*, page 112).

I read these all and put them together, in terms of the P ← L model, as follows. They would state logic is separate from the going on of logic in the physical world. That essence or being does not necessarily imply existence in the physical world would correspond to the P ← L framework in which there would be points of L not mapped to P. Further, being (i.e., part of logic) cannot be defined in its most general form. Interestingly, note that in the P ← L model, L is not defined. It is only postulated as like a huge mathematical set. It is stated that science can only get at meaning (logic) through the mechanism of language. Even the way mathematical logic gets at meaning is a little stretched, postulating structures and functions, and reducing meaning to "primitive" meanings, undefined except for a basically relational axiom system. (Nevertheless, these are valuable tools from mathematics; and from the perspective of science, we accept mathematics.)

Areas of Possible Interest to Mathematics and Religion

Possibly of interest for mathematicians are several sections. One is related to physics, mathematical analysis, and maybe more: see "Chapter 18 Closing Points," the section on a math theorem .

The other issue is more general. Maybe it is trivial and already known. Maybe not. How can there be so much meaning going on in matter all about us. This is in the "Chapter 38 Judgment: Logic," in the section toward the end, "A Different Kind of Mathematics," the part marked "(3)."

The axiomatizations of chapter 32 on "All Souls" and chapter 33 on "Teletransportation". These chapters might be of interest as far as constructing initial examples of axiomatizations, either inside or outside, the math model and match function of "Chapter 38 Judgment: Logic."

Also perhaps of interest is the pure theory of logic entering matter. See chapter 29 on the definition of evolution, the section entitled "Abstraction of environment. Logic passing transformed into creature." It would seem that this would be proved outside the math model and match function of chapter 38.

In the area of religion, spirituality, or theology, it seems the following sections might be of interest. The definition of, and comments on, evolution, in the opening section of chapter 29 on the definition of evolution.

Possibly of background interest are some of the other comments in the current chapter on Plato and Aristotle, and on the medieval philosophers, Alfarabi and Avicenna, since they were influential in the formative years of parts of theology.

Certainly of interest is chapter 32 on "All Souls". Maybe of interest is the theoretical background on the math model and match function and substantiality of logic, in the large chapter 38, "Judgment: Logic."

Though technical, the evolutionary logics of I, in chapter 33 on teletransportation, strike me as leading to thought-provoking issues about the nature of being, especially the time-slice analysis not of Y, not of the unusual I, but of the usual I. This topic occurs more in chapter 34, "Standard I, Non-standard I."

Of interest is "(1)" and "(2)" of the subsection on "A Different Kind of Mathematics," toward the end of "Chapter 38 Judgment: Logic."

Toward the end of the chapter on "Deep Physics," there are some small isolated points about Newton's relation to theology.

Software Concepts

There is a certain approach to the mystery of sentience that looks at techniques used in current computer software. Computer interrupts, stacks, hill climbing, process, subroutine, and the like. Years ago there was hope that somehow feedback loops in engineering would lead to a kind of being. There was a palpable sense of excitement that here at last was sentience or something like it, that all we needed was to build some little machine, just a small little contraption utilizing these mechanisms, then let the machines loose, and the awesome, reverential mystery of sentience would start to appear before us. No such thing happened. As for the general issue of software, while what goes on in the brain is logics and what goes on in computers is logics, this by itself is not much argument for similarity, since just about everything is logic going on.

Sometimes quoted along with a discussion of such issues is what I will call principle K.

K: If a person knows x then they know that they know x.

Sometimes it is then intimated that K could be part of the explanation of sentience, or part of what distinguishes sentience from non-sentience or humans from other animals or from simpler animals. As a matter of fact, it can be shown that K is false (see endnote 26 "Fallacy That We Know That We Know").

When looking at concepts from current software, and when we focus too narrowly just on the concept, we find all of them fundamentally unsatisfying as an answer for what I-ness or sentience is. Perhaps such lookings are premature attempts to "locate" sentience in certain kinds of informational processes. Instead of looking too soon for such correlates or explanation of sentience, *The Soul* looks directly at sentience, then directly at the brain and then tries to look in the most general way possible at what are the relevant dynamics or meaning or forms going on in the brain that are relevant to the issues. (This leads to looking farther afield than the brain, though the brain is still a fundamental gateway to scientifically understanding these issues. *The Soul* eventually introduces a mathematical framework.)

Abstraction

After one of the last times that I spoke with the god, I did an internet search on the terms "reify" and "reification", only to discover that their meaning has come to be primarily one in political philosophy. Exploring sites further, link to link, through a Wikipedia web page, I came across exactly what I was looking for: an article in philosophy on the notion of abstraction. After all, one of the issues running through our journey has been the existence of abstract "things": for instance, consider the chapters on "All is Logic," especially the chapter "All is Logic: Abstract Objects"; or consider logics, in all their aspects, or sentience, which from the scientific perspective is a logic, or the notion of "a logic going on", and the big chapter "Judgment: Logic." All these are intimately tied up with existence, abstraction, and reality. Here is the article.

Stanford Encyclopedia of Philosophy, "Abstract Objects," by Gideon Rosen, and referring several times to David Lewis. As of February, 2004, it is online at http://plato.stanford.edu/entries/abstract-objects/ (See bibliography for this URL dated).

This article lists a wide range of points, points that contact with issues of our journey in *The Soul*: non-spatiality, non-spatiotemporality, abstract/concrete distinction, way of example (related to "pointing" used in *The Soul*), existence of mathematical functions, even Frege's ideas of meaning (which *The Soul* mentions in a high-level way), quantum mechanics, existence "outside of time" (my phrasing), existence in spacetime.

The perspective as developed in *The Soul*, for instance in the chapter on logic, "Judgment: Logic," or in the chapters on "All is Logic," seems to me, after not too detailed an analysis of the article, to offer a single framework for most of the issues, except possibly the following: the relation of *The Soul* to causal inefficacy is not obvious; possibly *The Soul* has some connections to the way of abstraction, but that also is not obvious. Another apparent difference is that although the article deals peripherally with the physical world of science, the framework of our journey has a strong tie-in to that world.

These are some of the relations of the ideas in *The Soul* with other works.

Suddenly I was continuing the journey back to Earth.

40 Final Flight

I saw the planet of my species racing toward my feet, then I saw my country, then the desert and mountains, all in less than half a second, then instantly I sitting on a log.

My friends the willow trees, some Joshua cactus, and barrel cactus were about. Purple and white flowers bloomed from one cactus. A mild breeze was sliding down the side of the northwest mountains. And the white and rust-colored rocks lay peacefully.

And there was the god.

It was sitting on another gray log. The god was all in one piece!

"Great!" I beamed, "you look much better."

The god shined back. There followed our last conversation.

"Next, you're going on a trip ... but not unexciting," said the god, in good spirits.

"Now what! After the pillars, and judgment! Wasn't that enough!" and I looked down at the charred part of my hand. "That's not what is so bad. It's that the whole thing took a lot out of me." But in truth, I couldn't feel too bad.

"Don't worry. That's over. You're home free."

"This next part," the god continued, "will be very different. The test is over. Well done. You can now go back to a more leisurely experiencing of existence." The god put its hand toward itself and said, "And I passed too."

"You?" I was sitting a little too far back, and even though I threw my arms up as if trying grab some imaginary branch, I slipped over backwards onto the ground next to a group of little barrel cactuses.

"Lost my balance."

"To be serious now. This is special. And not bad." The god looked away in the direction of the horizon. "You are going to experience parts of the universe that are outside the perceptions of your human body. The senses that are of your physical body are going to be greatly extended to your sentience. Remember the zoomoscope?"

"Yes. That was a lot of fun," I said as I pulled myself up from the ground.

"The zoomoscope. What you saw was just as real as your native perspective. With a long enough zoomoscope even the stars would be near. Or remember how your native eyes are blind to almost all light (your eye picks up wave lengths of electromagnetic radiation *only* in the range of about 100 to 200 billionths of an inch – 400 to 800 billionths of a meter – 400 to 800 nanometers – 4000 to 8000 angstroms). On this trip, eventually your perception will pick up *all* light."

"Me?"

"These will not be the big changes. After all, these changes leave your physical body laid out in time and space pretty much the same."

The god explained how eventually my being, along with a kind of suitable physical support, would be literally laid outside of time.

Whenever the god spoke the word "logic", it sounded as an echo from increasingly far.

"Logic has many times and spaces. Indeed it has anything that is logically consistent. And now you will be laid out differently in the structure and form of the universe. To describe it is difficult because your human words and concepts have grown up for you as you exist on Earth. Even more limiting, your ideas have grown up for you as you are natively laid out in your own particular way in this structure of the universe. Profoundly blind are you to anything outside.

"The hardest part," the god continued, but by now almost everything the god uttered was a haunting echo from far out somewhere. The hardest part, the god went on, was extending my sentience, my being itself, so that I could process – understand – receive these larger sensations. That! was a challenge.

The god started to disappear, not in the usual way.

We exchanged a few more words.

"But I don't like discomfort or pain," was the last thing I said to the god.

I could have sworn the god got a smiling look. Then it turned directly toward me and good naturedly chuckled.

"Keep up the journey. It will be grand."

This time after the god faded out completely, I knew I would not see it again. Not on this trip.

I cast about in the sunny afternoon, sad, yet excited at all that had happened. How much distance had been covered. Completion and happiness and sadness always seem to go together. But it was a good journey. Just as a monstrous rain drop splattered on the stone in front of my foot, I was filled with excitement, but fear too, of what the next part would be.

But now the sky grew cloudy quickly – not common for the desert. No sooner than it clouded, it got dark, then darker yet. In alarm, as I prepared to head back, I saw above two immense volumes of air across the dome of the sky, colliding into each other, forming an untamed upside-down anvil, with the peak starting to plunge toward the ground, toward me. It covered about a quarter of the heavens. "This could be some storm, I better run!"

No sooner however, the cloud was overhead and I was yanked towards the heavens.

The planet Earth fell thousands of feet below as I let out a long groan and was dashed about in the twisting maelstrom of wind with pelting rain and ice. Since I knew I would be alright, it felt exhilarating. Then all went black.

After a while a clump of dirt appeared out of the blackness only to return.

A point of light appeared. Then a few more from the opposite side, then all around. They were stars. Then whole fields of stars.

A rock appeared, 8000 miles in diameter, with nothing holding it up.

Even ancients had a hard time with that. One of these tiny little ancients, Anaximander (580 B.C., very approximate, see endnote 27) said that since everything was the same in all directions, that the earth did not need to sit on anything. Some earlier thinkers had said the earth sat on a turtle, or on water. "Nothing to sit on? – that's crazy."

Even today the tiny little humans on that rock cannot emotionally grasp their situation, even though two thousand four hundred years after Anaximander they have teeny little books that say the rock is just there. It doesn't sit, it's just there

Now I saw some astronauts – those tiny little humans that have got off the rock for a wee little while – and because of the way their infinitesimally small sense organs work, they finally become more aware that this rock is just out there, with nothing "under" it or anywhere. Well, their teeny little books had said it wasn't sitting, but their sense organs ... that was something else. For a few, their minds spun. Nothing was on any side of this smooth round rock covered in swirling patterns of blues, cloud whites and yellow browns.

They were transfixed, directing their little sense organs out the portholes. I saw these tiny little astronauts shake their little heads back and forth.

Blackness was all about the rock. No! It was their teeny sense organs that made it black. But I saw the reality. It was the blindingly, golden light of billions of diamond like stars, dense across the firmament.

I would have blinked but those were not the eyes I was using.

And I looked at the 8000 mile wide marble, and down there, in 580 BC, Anaximander said that the earth didn't need to sit on anything because once you talked about the whole earth, everything was the same in all directions. That implied there was no up or down. Axanixmader struggled for words and concepts that didn't exist then, would not exist for millennia. Yet he spoke what reason presented.

The other little humans could only wonder at such a thing to say.

But it came about when his words were more than words. Later, after 2400 revolutions of the rock around the fire ball sun burning with a blinding yellow, these miniscule astronauts had got a little off the rock, and now it was more than in the mind's eye. They saw it with all their senses. There was no up and down.

But 2400 years earlier they had to grasp it on faith that there was no up and down. Now, on television, the tiny humans could see their compatriots, out in space above the rock. Their compatriots just floated, and although there were elaborate detailed theories of why there was no up and down off the rock, the people were always in an awe to see the astronauts just floating. The astronauts experienced it first hand. How worthy of sacredness of the mind's eye that one of these little creatures on the rock saw the same thing 2400 years earlier, if only as if through a glass darkly.

What does our mind glance through the dark glass today?

Such patterns of logic wafted through my sentience as I looked at the tiny creatures and astronauts, their senses looking out at the universe, boggling their brains.

Now the galaxy in the background was receding though at the same time getting closer. There was talk about going to a much higher level. But at the same time there were feelings of difficulty, with those of ease.

Next I was both inside the galaxy and far from it, and all distances in between. But most important, and at the same time least important, in some sense I saw, well I didn't see, somehow I was reaching to the edge of the whole universe, and even the far-into-the-past and far-into-the-future started to merge. Words. It can be done. Struggle, back toward the original galaxy. That was my galaxy. Totally imperfectly though.

Here goes. Suddenly "next", I was falling through a rainbow tunnel.

What I saw was intensity deep, clarity exquisite, and colors three-dimensional and radiating. I heard a roaring cacophony of the planets, then of the solar system, then of galaxies, then of I do not know what. All was a roar, all was the most profound and beautiful quiet, as if I had truly found peace and had escaped from all the seeming eternal cares of the world. The cacophonous universe and the peace and rest that surpasses all. They were both there at the same time, at no time, at one time, and neither too. And "eventually" even more and more beauty came into my hearing. Golden voices singing in increasing choruses of whole segments of galaxies, "then" of all being. Human choirs, sentience choirs, then different directions of being. As for almost all of what I heard and saw, it is outside our words. It is outside our being. It is outside myself. I suppose if I were somehow seized in total overwhelming by being, and I fell finally in massive swirling yellow whirlwinds, at least I could give peaceful form to myself. But even that was far behind.

Eucalyptus and Cinnamon and lotus blossoms and Jasmine breezes wafted, but of the whole universe itself.

And I experienced the full three-dimensional shape, whereas in my old human form I could catch only the two-dimensional *surfaces* of shapes and forms. From boxes and rocks to suns and galaxies, I didn't even see "through" something, I simply experienced the whole three-dimensional object at once, inside and outside. I started to experience as a "single point in time" shapes that were going on across time and space jointly. It's not that any mind's eye was needed, the perception was just there. And other logics I could see more clearly too. I did not simply see what I used to for red. Now I saw all the aspects of the logic for red, and I saw in what ways those logics were inside the matter in a brain, and what ways the same was outside through the long time and space of evolution. The perception was just there. Time and times were there, space and spaces too, and they were part of much more.

I could see logic itself, and experienced without thought its sameness inside with outside the brain, spreading over the whole earth, and over time that eventually spread hundreds of thousands of years back, but that in turn spread throughout the universe in one direction, and back to the beginning of the universe in the other. And "then" even that was only a start.

Shape and form and color and feeling and thoughts, I experienced, far. They became so intense and wondrous, separated from the logics that had got tangled up in them. But rather than being diminished from having the logics that appeared through them, they appeared in their own enchanting joy.

And the cats in the diamond green grass of our front yard were there. With all else, not separated.

Diamonds jumped out of intense backgrounds. Not stars. The diamonds of light were from brains. No! Not even that. The brains were only catching and reflecting a light.

And more, and more. But I started to see so many other points of diamonds.

And then I noticed just a few which seemed to be colors more like me – for I was one of the points of light too. Especially the closer relatives suddenly gave a sharp twinkle in my direction. Deep twinkling. We entered a honored resonance of beautiful color. By itself even, this was worth everything.

– children mirrors being born and older ones dying – but the light was there all along before and after - it only not being reflected now. But not one after another, nor as such major parts, but all as many parts of the fabric of all being in its perfecting summation.

Eventually even remembering and sureness became irrelevant.

I started to shake. But after that it was as if a hand reached out, a hand of the universe.

The hand stayed.

... various parts of my visual field. then I saw only the infinite blackness and profound piercing light of the stars.

Then, in a perfect instant, all became absolute blackness and peace.

After a while I was just sitting with a few friends, true friends. The three main cares I had in the world – and we each know what those are if we dig down inside honestly – were simply gone. And the myriad of toiling wariness didn't really make any difference either. It was such a happy – I'll just call it "evening". I never knew what it was like to be so at ease.

Then there were "times" where that wasn't so.

Now again there was another perfect instant wherein all became absolute peace and absence. This was something I didn't want to leave. But then everything started to go in reverse, the lights, then the galaxies, and then regular time reappeared, and then the 8000 mile rock, and the tiny little astronauts, and then they weren't tiny at all, we had all become regular-sized, and I streaked back toward Earth, through the atmosphere, through the clouds that had been the terrible storm, streaked like a meteor to the surface of my planet.

41 End

I lay on the desert ground, alone, heaped on painful stones and dirt, and I looked up to a green brown willow waving in that familiar warm breeze. I glanced back down at the stones next to my eyes. The rocks shined in the wetness after the rain. Into my nostrils drifted the sharp desert moisture.

"Oh my gosh!" I thought, "did I lose any more of my body?" I got myself stood up. "Nope, everything OK." It was still only the missing thumb and forefinger, with a little charred edge. "Thank goodness."

So that was it. Yes, I roamed around the desert a bit, but this time I felt relaxed. The cactuses, willows, mesquites, even the red and yellow and white pebbles and red cliffs, and the blue sky and a white and gray cloud, and the sun, and the mountains – I just soaked it in.

Oh, there was one throwback. It felt good seeing the plain old light and sky. But then intermixed with that a picture of galaxies and strange colors flitted through my brain, and I got a horrible pain that tore my stomach and chest. Yet after vomiting, I was back to my current self.

Our planet and the totality of logic happening on it is some event!

In truth, it is at least a three-way jig-saw puzzle, pulling in every creature that is or ever was, covering the whole evolutionary history of being on Earth, with all the changing *functionalities* of all the creatures, all their changing *genetic codes*, and all the changing *environments*. The functionalities come from the genetic codes, and the genetic codes come through evolutionary distillation of the meaning going on in the environments of the whole history of the Earth, and that meaning is impacted by the meaning going on in the universe, and that meaning at all times includes all results true in pure logic.

The functionalities, the genetic codes, the environments, each is intimately correlated with the other. Using future data bases appropriate to the task, experts will constantly go back and forth between these three areas. Did there seem to be a major shift in the environment in some geological time period? If so, that would likely have implied a slowly changing functionality of a certain creature as it evolutionarily reacted to the new environment. If that was so, then that in turn would only have come about through certain changes in brain functioning and structure (future experts will know many such details). Maybe those brain changes could only have come about through a small set of possible changes in the genetic code. The experts look at the molecular structure of the genetic code of the creature. Why yes, such molecular changes would have been easy, almost nothing. Conversely, this change in the genetic code would likely have implied several other changes in the brain and consequent further changes in the creatures functioning. And isn't that interesting, because also for several other creatures on this continent, there is evidence of similar changes in functioning in this time period. And all of that would seem to imply other changes in the environment that we currently cannot detect.

Thus the pieces of an immense three-way jigsaw puzzle fit more and more together, a puzzle that will be the story of the life of logic on Earth.

On the fourth day I started hiking up that final mountain trail again.

This time I got so far so fast – I had already gone on the first parts so often before – in no time I was far up the mountain side, marveling at the view. And then in no time, for the first time, I was at the top, the pass that saddled over to the other side.

On the way up I thought about infinity. And about us humans.

Tossed into the world without understanding, having no choice but to try to struggle through, we tiny creatures on a planet, which is a spec in a solar system, which in turn is a spec in a galaxy, which is a spec in the universe, we tiny creatures, with an understanding that matches our size, are required to make it through what is called life. That is the way.

For every human on the Earth, there are ten thousand, thousand, million suns – together with all that swirls about the sun, from dust to planets. And the physical universe is as nothing in extent compared to the universe of logic.

Let no one unfairly put bounds on the scope of all.

The little humans fall down over their own words and understanding, their brain the size of a large grapefruit. Still they strut about, these glorious, tiny forms, stumbling through centuries and millennia. For that is the way.

Always before I had looked up at the trail winding ever farther, ever higher. Now I was finally at the top. Pink and gray clouds crisscrossed the far horizon. For all their accomplishments, humans can hold their head high. From here the sky stretched far indeed.

I looked down at the sandy, purple valley of our journey. Far across further, more hills rose, turning into mountain ranges, and beyond that, ranges continued one after another toward that past horizon. I pictured my father working across the valley previous to mine. And I thought of so many over history, working their way through all the valleys before that one.

How long that was.

From there my gaze settled on where my thumb and forefinger used to be, and then I looked at the ground, then at the shrubs about, and then onward to the future, in the other direction, with its own valleys. Maybe I might go a little into the next valley. But other generations would go farther, us humans with our not too large brains, generation after generation, valley after opening valley, mountain after mountain, range after range. For that is the way.

New generations burgeon forth, so one generation is always dying while another is rising up in its place. And they each make forward with their ideas.

Appendix: Numbers and Computations

This section describes the sources for various numbers used throughout the book, or it describes how those numbers were computed. It lists internet sites.

Neurons and Glial Cells in Human Brain

There are between 10**10 and 10**12 neurons in the human brain, as stated for instance, in the 4th paragraph from the bottom of page 10, of Nicholls et al, *From Neuron to Brain, Fourth Edition*, 2001.

I chose 10**11 as an approximation. 10**11 is 100*1000*1,000,000 which is a hundred thousand million, a convenient number to remember.

There is another kind of cell in the brain and nervous system, possibly 10 to 50 times more numerous than the neurons, called *glial* cells. They serve several important functions but are not implicated in the processing of information. See internet sites for "glial", for instance
http://www.sfu.ca/~ablaber/lec06.htm
http://www.hubin.org/facts/brain/texts/the_neuron_en.html
http://training.seer.cancer.gov/module_anatomy/unit5_2_ne
rve_tissue.html
Also see Nicholls et al *From Neuron to Brain, Fourth Edition*.

Neurons Related to Human Eye

There are 100 million primary (rod and cone) electro-receptors in each human eye. Yet there are only about 1 million nerve fibers from each that go out to the brain. This is from paragraph 1 of page 318 of Nicholls et al, *From Neuron to Brain, Fourth Edition*, 2001.

The bottom of page 381 has a good picture of the nerve layers at the back of the eye, along with the nerve fibers.

The bottom of page 383 has a good drawing of the rod and cone. And yes, they actually look like a rod and a cone.

Neurons Related to the Human Ear

Our hearing takes place in a maybe half-inch long thing that looks like a wound-up cornucopia, called the Cochlea. Behind the eardrum are a few of the smallest bones in the body, and these transfer the vibrations of the air waves into the fluid in the Cochlea in which tiny hairs pick up the waves or vibrations, which in turn cause electrical signals to be sent into the brain.

One description of this working can be found on page 366 and onward, of Nicholls et al, *From Neuron to Brain, Fourth Edition*, 2001.

The number of hairs in each cochlea is about 15,000. This information I found using an internet search for "cochlea" and "hairs". One such site was www.sciencenet.org.uk.

A site with excellent drawings, some in motion – copyrighted material – is associated with the name "Tony Jefferies."

Neurons Related to the Human Nose

There are about 100,000 sensory neurons in the human nose or olfactory system. Such information can be found for instance on page 347, paragraph 5, of Nicholls et al, *From Neuron to Brain, Fourth Edition*, 2001.

Number of Atoms in a Cell, Cells in Body, Space Taken by Atoms and Cells, 36000 Mile Tall Behemoth, Red Blood Cells, and the Like

A wide variety of these and related calculations start from the following observations.

In terms of weight calculations, the proton and neutron are virtually identical. Furthermore, in terms of weight, the electron can be taken as 0, when compared to the proton and neutron.

Hence, let us use the term *eutron* to indicate a proton or a neutron. Virtually the total weight of any physical object about us comes from the number of eutrons in it. This simplifies a variety of calculations.

How many eutrons are in a 150 lb person?

A Eutron weighs 1.674*10**-27 kg. This only need be divided into the weight of the person, to get the number of eutrons (taking into account the number of kilograms in a pound or vice versa). Any spread sheet program will do this.

In the table on page 21 of Starr and Taggart, *Cell Biology and Genetics, Ninth Edition*, it gives the percentages of the different atoms in us. Oxygen 65%, carbon 18%, hydrogen 10%, nitrogen 3% calcium 2% phosphorous 1.1% potassium .35% sulfur .25% and sodium .15% and so on. I make the assumption that these percentages are by weight.

Thus, that oxygen is 65% of our weight means that 65% of the eutrons in us are in oxygen atoms. Using the earlier calculation of how many eutrons are in a person, you can figure out how many eutrons are in oxygen atoms in that person. Then, to get the number of oxygen atoms in the person, divide by 16 (since there are 16 eutrons in each oxygen atom, the 8 neutrons and the 8 protons, ignoring isotopes).

You can do this with each oxygen, carbon, hydrogen, and so on, getting the number of each of these atoms in the person. Add them up and you have the number of atoms in the person. All this can be done fairly easily in a spread sheet.

You can use various methods to estimate the volume of a human body. I used my own body, dividing it into hands, feet, legs, arms, hips, torso, and head. Divide the number of atoms into the total body volume, and you get how much volume, on average, an atom in the human body has allocated to it. (One can do this computation even though atoms are of different sizes, and even though there is a lot of empty space between them. Ah, the oddness of mathematics.)

Starr and Taggart, *Cell Biology and Genetics*, page 148, paragraph 4 (that is paragraph 4 of chapter 9) states that there are roughly 65 trillion cells in the human body. I made the assumption that this was for a 150 pound person. Dividing this 65 trillion into the *number of atoms* in a 150 pound person gives the number of atoms, on average, in each cell.

Suppose that you want to know how big a human will be if each cell is increased x times in height, width, and depth. Well, then the whole human is also increased x times, in height, width, and depth. And vice versa from human to cell. And the same of course applies to atoms.

Starr and Taggart, *Cell Biology and Genetics*, page 55, paragraph 1, states that red blood cells are about 8 millionths of a meter across. A finger or thumbnail is in the range of ½ inch across. Using a magnifying glass, a ruler, and clippings, I estimated that the thickness of my middle finger nail, and of my thumb nail were 1/50 of an inch and 1/40 of an inch, roughly. From these numbers it follows that about 2000 red blood cells could be laid width-wise across a fingernail or thumbnail, roughly, and that 65 laid next to each other would be the thickness of a fingernail other than thumb, and 80 the thumbnail thickness.

A 1000 Mile High Behemoth

This is computed with fairly straight forward estimates except for one. That one is computed as follows. A neuron is a two-directional tree, about 150 feet in one direction and 150 feet in the other, in the behemoth perspective. We take the total volume of the brain and divide it up among the neurons. Then we make the estimate that each of the two trees of the neuron (neuron in the behemoth perspective) gets this many cubic feet: (150 ft) * (30 ft * 30 ft). There are certainly different assumptions wrapped up in this estimate. But onward to the next part of the calculation.

Since there are a hundred thousand million neurons in the brain, that gives the brain (in behemoth perspective) as having so many cubic feet. (The current estimate of the number of neurons is between one tenth to ten times this number: see Appendix.)

We assume the brain is a sphere (no it is not, but we make that approximating assumption). From the above numbers, such a sphere in the behemoth perspective would be required to be about 70 miles in diameter.

In our native, non-behemoth perspective, we assume a brain volume size approximated in volume roughly by a 4 or 5 inch diameter sphere, and a person about approximately 5 to 6 feet tall.

This gives a person, in behemoth perspective, in the range of 1000 miles tall.

A spread sheet can be set up to carry out the above calculations.

Number of Galaxies and Stars

As of March, 2005, the following site
http://imagine.gsfc.nasa.gov/docs/ask_astro/answers/021127a.html
gives a range of estimates for number of galaxies. We will take a conservative number of 100 billion galaxies.

For the number of stars, the following site as of March, 2005,
http://rst.gsfc.nasa.gov/Sect20/A5.html
presents an article about the work of the astronomer Simon Driver and team estimating the number within the range of our telescopes on Earth to be 70 sextillion. That is 70 thousand, million, million, million stars.

Speed of Air molecules, Pressure Waves (Sound). Humans, Bats, Pianos

Waves of air pressure (i.e., sound) travel at 758 miles per hour in air that is 68 degrees Fahrenheit. The speed goes up a little as the air gets warmer. Such figures can be found in many places. I found these in Asimov's *Understanding Physics*, vol 1, page 164. Pressure waves in water (not electrical voltage waves in water) travel at 3240 miles an hour (an underwater eardrum perceives these the same as air pressure waves, i.e., sound). In steel, 11,200 miles per hour. (These are on page 166 of the same book.)

That individual air molecules travel over 1000 meters per second (2200 miles per hour) is found on page 8 of Averous' *The Atom*.

Pages 179 and 180 of Asimov give the frequency range that most humans hear (20 cycles per second to 15,000 to 20,000 cycles per second, depending on age of human). Also there are frequencies of selected piano notes, and bat shrieks can be 40,000 to 80,000 cycles per second.

Car Length

We arbitrarily take 15 feet to be the approximate length of an automobile.

Weight of All Bacteria

As of March, 2005, the site
http://www.wsu.edu/NIS/Universe/microbes.html
has an estimate that if you took all bacteria their weight would be 10 times that of all the humans, mammals, and birds.

CAT Scans, PET Scans, MR

Internet searches for these produce many sites. Here are some of the ones looked at for this book. These sites were in effect as of March, 2005.

For PET scans:
http://www.nationalpetscan.com/petovew.htm
For CAT scans:
http://www.colorado.edu/physics/2000/tomography/
http://imaginis.com/ct-scan/
For MR
http://www.cis.rit.edu/htbooks/mri/inside.htm

Endnotes

Endnote 1 References for Fizeau

Sources for the information about Hippolyte Fizeau are: Olson, *Biographical Encyclopedia of* Scientists vol 2, Elliott, *Electromagnetics*, Asimov, *Asimov's Biographical Encyclopedia of Science and Technology, 1964*, and Porter and Ogilvie, *The Biographical Dictionary of Scientists, vol 1*

Endnote 2 Religion and Knowledge

Newton was a firm believer in his religion, and in fact that may have been a factor in allowing him to make required breaks with some of the scientific thinking of his day. (Newton, Cohen, Whitman, 1999).

The great scientist, Michael Faraday, 1791 – 1867, deeply committed to a humble religion, twice turned down the presidency of the Royal Society, and turned down an offer of knighthood (Guillen, 1995).

In the U.S., the Puritans, profoundly committed to knowledge and higher education, were among the strictest of religions. They founded Harvard University after only six years of the founding of the Massachusetts Bay Colony (Vaughn, 1997, page 248; page 245 in the 1972 edition).

Endnote 3 References for Ancient Greeks

There are many excellent books on ancient Greece and ancient Greek philosophy.

As for the Presocratics, eminently understandable, somewhat detailed for the non-specialist, and certainly consulted by specialists, are the two volumes by K. K. C. Guthrie. Also for the specialist is the book by Kirk and Raven. If you want remarkably full overall analysis and detail, these are the sources.

I think of a specialist as someone who wants to get into the material in a fair amount of full-fledged detail.

Also, any of the books by Wheelright are beautiful and authoritative.

Lloyd's book, *Early Greek science: Thales to Aristotle*, covers ancient Greek science.

Also good is Lindberg's book, *Beginnings of Western Science*.

Information on ancient Greek astronomy can be additionally found in:

Berry, Arthur *A Short History of Astronomy from the Earliest Times through the Nineteenth Century*; and

Dreyer, J. L. E. *A History of Astronomy from Thales to Kepler*.

Endnote 4 Structure of Logics. Further Comments. A Stronger Statement

The book has just talked about how there is something called a structure of logics in brains. One could make a stronger statement than this. Whether the stronger statement is true is not certain. This stronger statement is not used in the book but it is one of the ways people, consciously or unconsciously, think about what goes on in the brain, and so I mention it for the sake of completeness.

When we look for instance at the JAR logic in the electric fish brain, we see that the *meaning* going on at each neuron is the same as parts of the logic, or of a suitable logic, going in the environment outside the fish's body. This suggests that this is the case with the vast majority of neurons in all brains. The meaning going on at that neuron is part of some logic (i.e. meaning) going on in the environment of the fish. The meaning encoded in the spike trains of the neuron is a meaning going on in the environment of the fish. For example, the meaning encoded in the output spike train signals of a specific neuron might be this: the difference in strength (amplitude) of the electric water waves as it hits a certain point A on the skin versus some other point B on the skin of the fish. This is a meaning. The output of the neuron is just spikes of electrical energy. But this meaning is encoded in the spike trains. The spike train is not a meaning, it is a bunch of electrical activity. But meaning is encoded in those spike trains. And one could say that that meaning is going on at the neuron. One could more generally say that in this way meaning is going on at every neuron. One could also say that this meaning is a meaning going on in the environment of the creature. This is the stronger statement that we do not use in this book. But it might be true. It depends on the meanings and usage of words like "meaning". It also depends on knowing enough about brains and environments to know if it is true.

(As a separate issue one might wonder about the word "encode", when one speaks of spike trains encoding a meaning. From one mathematical perspective, an encoding can be viewed as nothing more than a mathematical association or correlation between spike trains and meanings. From another perspective, spike trains seem to have some general structure of pure form independent of encodings, and the specific structure influences the number and kinds of meanings that can be associated or correlated with the spike trains of one neuron.)

(A further issue in the text is that we say logic goes on in a region, not at a point. This way of conceptualizing where logic is going on comes about from looking for instance at the JAR logic of the electric fish. The JAR logic is clearly going on in a whole region of the fish's brain. Some people would say it is being "computed" in a whole region of the brain. There are specific neurons whose output spike train encodes large aspects of the JAR, or indeed, toward the end of the whole process there is a neuron whose output encodes the whole JAR. But even if you decide to look at the brain from the perspective of computation, those neurons are at a computational apex, so to speak, with the outputs of many neurons being fed into them, directly or indirectly. Whether one speaks of the JAR logic, or of parts of it, it is completely natural to speak of it as going on in a region.)

Endnote 5 Meanings Encoded in Spikes, Some Sophisticated

The book *Spikes,* by Rieke et al, addresses a number of issues about the meaning that is encoded in spike trains. The book may not use the word "meaning" but rather speaks of encodings. The following are my interpretations of some parts of the book. To me, the book looks at the meanings going on in the brain from a strongly statistical perspective. It looks at meanings or logics that are going on at a neuron and that are statistical logics. Page 19 and 21 are on the language of meaning. Pages 261 to 263: how the meaning or logic of the environment may be more complex than we thought. Page 257, cricket insect system picking up information about air velocity and so on. Page 235 and 238 and 63 to 64, how bad the fly's visual system is and how quickly it must still make important decisions. Page 181, correlation in time and space. Page 231, bats and electric fish. Pages 3, 4, 6, 9, 10, 12, 19-22, foundational material and some basic statements and also cat vision. Though the concept of *homunculus* is used in the brain of the fly, it seems to be to achieve a certain perspective of analysis.

Endnote 6 Books on Electric Fish

Heiligenberg's *Neural Nets in Electric Fish* is a classic on the information of the kind presented in this chapter, what I call the JAR logic. In 1994, he was unfortunately killed in an airplane crash. His book is a classic for a relatively complete connection between a more advanced creature behavior and the underlying neuronal activity. Peter Moller's *Electric Fishes/ History and behavior* presents information on electric fish and specific information on its electrosensory receptors (what I have called the "electric eyes").

Biographical information may be found by searching the internet for Walter Heiligenberg.

Endnote 7 Books and Information on Neurobiology and History

Nicholls et al *From Neuron to Brain, Fourth Edition,* 2001, is a solid textbook presenting a wide range of topics in medium depth.

The book by Simmons and Young, *Nerve Cells and Animal Behaviour, Second Edition*, presents much current understanding of the logic of animal behavior.

An excellent, expert discussion of the history of the discoveries, and theories, of neurons and of the nervous system, is in Gordon M. Shepherd's book *The Foundations of the Neuron Doctrine.* Here you will find in detail the history, eminently readable with a high level of scholarship, of Golgi and the stain he discovered (pages 79-93) and of Ramon y Cajal using the stain method to see into the nervous system (page 136 and on). You will also find how these two people continued to differ over what was seen in the microscope (page 260). The book also has valuable perspective on the history of science in this area, tracing the development of ideas of the neuron (see for instance pages 271 to 283). In fact, one can compare the ideas at the time Shepherd's book itself was written with those of today, where the neuron is seen in a little looser way and where certain general principles are no longer required to apply to every single neuron.

An internet search of approximately the conjunction of nerve cell, neuron, processes, fibers, glial, also leads to much helpful information.

Endnote 8 Logics of Animal Behaviors

Peter Simmons' and David Young's *Nerve Cells and Animal Behaviour, Second Edition*, is a delightful book on animal behaviors in terms of presenting the logic of the behavior. This is an excellent book, pleasantly readable by someone not acquainted with the terminology of neurobiology. Discusses in understandable terms the techniques used in the brains of bats and owls to determine where something is. Behaviors of many other creatures are also discussed.

Pages 20-41 has a good description of neurons and signals. Pages 129-142 has a good descriptions of owl auditory locationing, and page 139 has some description of phase locking.

Endnote 9 Device Displays of Logics

There are several issues about the similarities and differences between the two ways of showing the logic. The presentation as little green squiggly lines moving all over the place versus the presentation as lights moving along straight line segments neatly laid out as in an engineering diagram – both presentations show the same logic, provided one understands enough about the squiggly lines moving around. One would understand enough if one had studied and analyzed the squiggles over a long enough time. Yet the same can be said, though to a lesser degree, of the green lights moving around the engineering diagram. What then is the difference?

One difference between the two presentations is that it is much easier for us humans to see where moving lights are going when the paths are laid out in a "neat, orderly" engineering diagram. It is easier for us because of the particular dynamics, in the human brain, for quickly grasping issues of straight lines, even when there are a lot of them going in different directions, as long as the different directions of different lines are perpendicular or at least are fairly different (i.e. our vision-brain makeup unconsciously goes through the work, and "immediately" distinguishes the two directions). We need only notice the beginning of a straight line whereupon we can immediately jump to the other end of the line. We don't have any visual-mental apparatus to do such things with such ease for a wiggly mess of spaghetti-like lines (mental apparatus with such capability would be complex, but probably no more complex than much of the apparatus already in our brains). Thus, the reason it is easier for us to look at the engineering-type diagram, versus the squiggly moving green lines, is that the device has presented the logic as a picture that is conducive to our particular mental capabilities (our capabilities related to straight lines).

This leads into the second issue. Are the two presentations showing the same logic? Or, has the engineering-diagram presentation snuck in substantively more of the picture of the logic? As indicated above, if one analyzes for a long enough time the little green squiggly lines moving about, one will know the same thing as shown by the engineering diagram. For instance, with the JAR logic of the electric fish, one finds that certain squiggly little lines in certain places, only go to certain other places, and those in turn give rise to only certain other places, and so on There is perhaps the more solid approach, which on the basis of some line of reasoning, asserts that such and such mathematical transformation preserves the underlying logic, since all the relevant actions and interactions are preserved in the pure logical form. The text in the chapter simply posits that the device works, which is to say, that the engineering diagram is showing the same thing as the wiggling squiggly green lines.

Endnote 10 Bonaventura Francesco Cavalieri

Cavalieri, Bonaventura Francesco, Italian mathematician, 1598 – 1647.

An internet search will yield a number of sites about this person. Among several of which have extensive information and links is that of the University of St. Andrews, Scotland.

Endnote 11 Note on Galaxies and Sentience

The logic going on in a galaxy would seem to be pretty separable into spatial areas around each sun. Yet in the opposite scale, the logic going on in the brain is barely separable into the spatial region that contains a neuron; in fact, speaking extremely generally, perhaps one might argue that the whole thrust of the neuron and nerve signal is to overcome the distance that can be a hindering aspect of physical space, and in this way much more serious logic can go on.

Endnote 12 Copernicus, Kepler, Galileo, Especially Relation to Religion

There are many books that cover Copernicus, Kepler, and Galileo. Berry's book, *A Short History of Astronomy* is good.

For information about their interactions with religion see, Berry, Arthur *A Short History of Astronomy*, pages 125-6, 159, 169-71, 179-82, and 194-5.

For a great many facts and corrections to popular beliefs, from the Catholic side of the issues concerning Galileo's situation, see the entry for Galileo Galilei in Catholic Encyclopedia. As of June, 2003, the entry can be found online at
http://www.newadvent.org/cathen/06342b.htm
The following is a good website on the various characters in Galileo's situation, their background and motivation. Here we see categories of humanity in general.
http://www.law.umkc.edu/faculty/projects/ftrials/galileo/ke yfigures.html

Endnote 13 Catherine the Great and Musical Tone Awareness

I seem to remember reading in *Catherine the Great* by Oldenbourg that Catherine could not pick up on the tonality of music. But looking back at the book I could not find what I take to be the full reference, though on page 320 I found an indirect indication. There is mention of Potemkin's complex personality, including a turbulent mysticism. Referring to this mysticism and to Catherine, "This, like music, was a world into which, because of a natural insensitivity, she was incapable of following him."
I did locate a website that mentions that the article "The Empress of Opera." Civilization, 1 February 1997, 15, mentions her supposed tone-deafness.
http://www.kings.edu/womens_history/catherine.html

Endnote 14 Energy Character Fields

I made up the term "Energy Character Fields." The idea is that conceivably in mathematics there are rather sophisticated mathematical structures called "characters" which can be applied to conceptualizing energy fields in physics, not only on Earth but stretching through space. As part of the fictional aspect of this part of the book, it is assumed that these energy characters form a substrate that can and in fact do support sentient beings, the vinegar bog creatures.

Endnote 15 Goedel and Chaitin

There are many websites having information about Goedel, one of them is for the Kurt Goedel Society.
There are a number of websites having information about Chaitin. Some of them are

http://www.cs.auckland.ac.nz/CDMTCS/chaitin/
Also for his book "The Unknowable"
http://www.cs.auckland.ac.nz/CDMTCS/chaitin/unknowabl
e/

As I understand it, Chaitin's work is a set of certain sophisticated, powerful, mathematical tools, for digging, in certain directions beyond Goedel, into the nature of logic. Perhaps the tools could be used to prove various results to the effect that at any "level" there are issues that are not understandable by creatures – or sentiences – at that level. And such results would apply to any formal system that had certain kinds of axiomatizations of "understandable". Chaitin's tools might be useable on other issues of sentience as well. More generally, the work of people such as Goedel and Chaitin show that there is unforeseen depth in the nature of pure logic itself.

Endnote 16 Astounding Logic and Constructors

Viewing logics as something built out of constructors has the following aspect. In many situations we would like to look at the logic as if it were written down as some algebraic expression using all these constructors, and we would like the complexity of that expression to be some kind of indication of the depth of the logic. In some cases this does not work. Suppose that a particular constructor simply ignores its fourth component (i.e. fourth parameter). Then that fourth component could be a monstrously huge expression of constructors, making the overall expression also monstrously huge, but the overall logic would not be affected at all. So the complexity of the overall expression does not necessarily, in all cases, correspond to what we would think of as the depth of the overall logic as it would be out in the world.

Yet we have a strong feeling that if one limited oneself to the "right" kind of expressions for a logic, then the complexity of the expression would basically match the depth of the logic as it would be in the world.

Perhaps one tries to solve this issue by considering all the ways that one might write an expression for the logic. Maybe one then uses the simplest of those as a kind of name for the logic. (Mathematically speaking, one takes the expression of minimal complexity among all those expressions that express or describe the logic.)

Perhaps another approach is to take the minimal expression subject to some condition C, but leave it open as to what C is, but adding bits and pieces to C as one goes along in one's analysis.

Another approach might be, going back to the example of a constructor which ignores its fourth argument, to consider that fourth argument of such a constructor as a full part of the meaning, in some sense, only not expressed.

Maybe we should look directly at the brain and consider the levels and amounts of neural connections as something like the constructors. If we note that even within the retina of the eye, in about only 4 levels of nerve connections, various features such as the beginning of the detection of line segment directions are taking place at all points across the whole visual image, we can imagine how much deeper the logic must be after possibly tens or hundreds or thousands or hundreds of thousands of levels. Yet even if we consider the constructors to be something like levels of nerve connections as the signals move into the brain, even here there is a sloppiness of match between the logic described as an expression of constructors of this sort versus the logic itself. For the brain is a great big squished jungle, with no doubt much sloppiness and randomness in its construction. The logics in there have been created by a statistical pressure of the logics of the environment. So here too, the description of these logics as an expression of constructors is in places likely excessive and distorted due to the messiness and imperfections of the jungle itself.

In spite of all this, still one feels that it can be of value sometimes to think of logics as being described by, or correlated with, something like an algebraic expression made out of constructors.

Endnote 17 Panpsychism and Problem with Non-sentience Logics

This chapter states

"In panpsychism there is no special, astounding aspect that characterizes sentience logic, only ever more complex, extensive, machine-like logic, of a deeper and deeper nature, having more and more *degrees* of what sentience can be, with ever more complex "perceptions" corresponding to the depth of logic going on in the brain. *In this approach, all logic is sentient.*"

Yet if this is true, what about the logics in our brain that do not produce any experiences for us – the ones that if you tweak with a present-day electrode or a far-future massive adjustment of signals, result in no experiencing for the person. Perhaps in some the tweaking will speed up the heart or cause an eye twitch or some more generalized adjustment of leg balancing or such. If these logics are sentient, then that implies there are lots of sentiences/beings "in our head", but all except the main one are of a level of being as low as for instance an automobile. In these matters, it seems there are issues of connection between logics and possibly absorption of one logic by another, although these are some of the many terms that are not clear, at least at our point in history.

Endnote 18 Evolutionary Psychology Also Uses in a Central Way Ideas of Evolution

As I understand them, the terms *sociobiology* and *evolutionary psychology* pretty much mean the same thing. Further ... whoa ... what's this?

"Hello! I'm a hiker."

"Here in the Endnotes? You are not supposed to be here."

"I'm not sure what you are talking about. But I have some things I'm dying to tell you.

"Sobiobiology is involved with social issues, human actions in society, and though its goal is scientific, it seems to be profoundly involved and *shaped* by clashes with other academic areas that study society, and this leads to intense arguments over evaluative issues. To me, this seemingly eternal clash appears to profoundly affect that area in every way.

"Two links for a summary of evolutionary psychology are

http://www.anth.ucsb.edu/projects/human/evpsychfaq.html
http://www.psych.ucsb.edu/research/cep/primer.html

"But the perspective in this book is so different concerning the use of evolution. I read some of the book's reports while I was coming here." For a second the hiker looked as if some thought from far away peeked and then went back into hiding. "You come at this material from a wholly different direction than does evolutionary psychology. Indeed, off hand, I think philosophy would sink in a mire if it got too involved in trying to duplicate evolutionary psychology's current foundations of theory, though interestingly, when you read articles in the area, the reasoning itself has convincing elements of validity. Maybe the approach of this book has a chance to produce some deep, powerful, mathematically solid results, based in language and meaning (logics), in a very abstracted, pristine form, away from this eternal, distracting clash."

"Who said we're going to duplicate their system? Who said we are even talking about their system?"

"You could call your area 'Evolutionary Philosophy', or maybe, 'Evolutionary Sentience'."

"Or maybe no 'evolutionary' at all! Evolution logics are just one kind of logic, going on over vast time spans in the creature's (ancestors') environment. It's only one part of several equations. Our journey is indeed different."

Thereupon the hiker got thoughtful and spoke quietly.

"I read the reports of this journey. Everything comes not *from* evolution, but *through* evolution."

"Yes. Evolution is the *means* by which logic comes to be going on in matter, enters into matter. Over long enough times, more and more."

"It enters out of nowhere."

"That's from our perspective."

Endnote 19 Humans do not Require Deep Understanding to Create Neural Net People

In Kendrick Frazier's article, "A Mind at Play, An Interview with Martin Gardner," page 37, top of second column, *Skeptical Inquirer*, March/April, 1998. Martin Gardner is quoted as follows.

"I can say this. I believe that the human mind, or even the mind of a cat, is more interesting in its complexity than an entire galaxy if it is devoid of life. I belong to a group of thinkers known as the 'mysterians.' It includes Roger Penrose, Thomas Nagel, John Searle, Noam Chomsky, Colin McGinn, and many others who believe that no computer, of the kind we know how to build, will ever become self-aware and acquire the creative powers of the human mind. I believe there is a deep mystery about how consciousness emerged as brains became more complex, and that neuroscientists are a long long way from understanding how they work."

It is quite possible that we will be able to understand enough of the going-on's of the signals in each very small area of the brain so that we can duplicate its logic on neural chips, and thereupon duplicate to a sufficient degree all the logics going on in the human brain so that sentience appears in the neural chips. This would not require on our part understanding of the larger picture. It would only require that our machines can sufficiently deduce the logic at spatially small areas of the brain, and that we proceed from one small area to the next, eventually covering the whole brain, duplicating it all in chips.

Endnote 20 Seeing Red

For myself, if I look at a big fat red-lettered word, and above or below, is a big fat blue-lettered word, and it is all on a black background, I have a perception of heat from one and cool from the other. And if the red letters are a darker hue, I feel a heavier mood. Of course, this last could be imagination. And even if all these associations generally do occur in humans, this is far from scientific evidence that these characteristics are somehow inherently connected to the experience logic, going on in the brain, of what it is like for a human to see red. It is a problem as to what kind of results from ethology would show that the experience logic that is the experience itself of seeing red would demonstrate that such aspects as blood or fire or ripe fruit are somehow in the logic. We just have to wait for the distant future when we can see well enough into the brain, and understand what we see.

Endnote 21 To be Free Means to be Brave

This is almost a direct quote from Edinger's translation of Thucydides' report of Pericles' funeral oration, winter, 431 BC, for the first wave of dead Athenian soldiers from the Peloponnesian Wars. It is toward the end of the 43rd section of Thucydides' writing.

"It is for you to make these men your models. Be convinced that to be happy means to be free and that to be free means to be brave. Therefore do not take lightly the perils of war. For it is not the wretched and unfortunate who have the most reason to fear death, for they have little hope for better days, but rather those who fear a complete reversal in their lives during times of hardship and crisis. For a man of self-esteem, humiliation because of cowardice is more painful than death, when it comes unperceived while you maintain your self-confidence and shared hope." Page 39 of Edinger's translation.

An online translation, by Richard Crawley, of Thucydides' work is at

http://classics.mit.edu/Thucydides/pelopwar.1.first.html

Endnote 22 Model of Universe Set Up As Cross Product Rather Than Map

Instead of P ← L, would it be better to have the model be the cross product P x L. Mathematically, the set P ← L is a subset of P x L. In the subject matter however there is so much freedom as to the details of how P should be defined, that it is unclear if there is a difference between P ← L and P x L.

One difference that the two "suggest" is this. The same logic can be going on in many different collections of matter throughout the physical universe. Consider the logic that constitutes the going on in matter of a water molecule. In the last several decades science has discovered that the going on of this logic is very common in the current time region of 3-dimensional space of the physical universe. P ← L implies that there is only one element of P that any given L can be mapped to. But here, the single element "the logic of a water molecule" is going on at not one but many places of the physical universe. If P is taken in such way that this means many different elements of P, then P x L would be needed. However P, the physical, can be mathematically defined in many ways.

I chose P ← L over P x L because I thought of logic as being able to be mapped into the physical universe. Moreover, it should not be a total map but a partial, since the framework should allow the possibility that there could be logics that never map to anything in P. We want a model or framework that has a lot of flexibility because we want to be able to explore a wide range of possibilities as to the nature of the fabric of the universe.

The main advantage of the map P ← L over the cross-product P x L is that the map is the going-on of the logic. The map tells where and how that logic is going on in the material world. Setting up the formulation in this way is intuitively satisfying and natural. Further, it may encourage us to see how these logics have a "life of their own" as they exist and move through the universe, possibly relating to and interacting with other logics.

In summary, one could say that mathematical structures such as P ← L and P x L are fairly good at suggesting the range of mathematical structures for model of the universe, but the real emphasis is on conceptually removing logic from being absolutely subordinate as it is in our current scientific conception of the material universe. At the same time, the emphasis is to maintain some standard mathematical and logical structure of P and L and whatever relation there is between them.

Endnote 23 Evolutionary Psychology and the Match Function

A concept that might be called "is consistent with" occurs to some degree in most experimental and observational sciences. I wonder if it occurs especially strongly in evolution and for instance in evolutionary psychology. Perhaps in some cases this concept is the same as the match function.

To see how much this concept occurs in evolutionary psychology, search the internet for instance with "sociobiology" and "consistent with" both having to be true in the search. An April, 2005 search produced a listing of over 23,000 entries.

Endnote 24 Brain Mirror Model and P ← L Model

As for the brain mirror model, although it appears to be only a specific version of P ← L, might it also be logically derivable or constructible from P ← L ?

Endnote 25 Isomorphism, Time, Evolutionary Logics, Experiential Logics

The concept of isomorphism in the book is left vague. But it definitely implies the preservation of "logical form."

One aspect is how it pertains to the experience itself that we have. Our experiences are basically instantaneous: they are basically at a point in time. But the experiences come much from evolutionary logics – logics not instantaneous but rather going on over vast periods of time. Somehow there is an isomorphic transformation of some substructure of the evolutionary logic, and the result of that transformation is some activational experiential logic in our head that takes place at an instant in time. Perhaps the following is kind of illustration.

Think of a computer display that shows a moving picture. For the sake of this example, we will assume that the person at the computer cannot interact with, or affect, the picture display once it has started and that the display continues for a fixed amount of time and then stops. Mathematically, that display through time can be thought of as a single, static 3-dimensional object, two of the dimensions being the length and width of the display, and the third being time. A 3-dimensional object can be rotated, in many ways, and the result is the "same" object – the result is isomorphic to the object before the rotation. The 3-dimensional object, having been rotated, would correspond to its own moving picture on the display. This is one example of two computer displays over time being isomorphic to each other.

Consider an evolutionary logic. It is taking place in 3-dimensional space, across a (large) span of time. This results in a 4-dimensional object. A 4-dimensional object can be rotated in plenty of ways, each rotation producing an isomorphic version of the original. Many such rotations will take the time dimension of the original and transform it into something that is a combination of time and space, or transform it even into something that has no time aspect at all. At any rate, each of those results of rotations can be projected, in various ways, to the 3-dimensional object that is then shown by the computer display. One thing this illustrates is that the time dimension does not need to be the same between the evolutionary logic and the computer display, and therefore this might be insightful as to how the instantaneous experiential logic in our head might be related to the evolutionary logics.

Obviously these ideas are of a sketchy, incomplete nature. The hope is that they might speed up someone else's thoughts.

Endnote 26 Fallacy That We Know That We Know

Some areas of study assert the statement as an axiom that if a person knows something, then they know that they know that something. This sounds obvious, which is its trouble, because it turns out to be false. To see this, consider a something X that a person might know. Then supposedly

They know X
They know that they know X
They know that they know that they know X

Ad infinitem, with these being the first three elements of a sequence that goes on for infinity.

The summary of the proof that this cannot be goes like this. All these elements in this infinite sequence would have to correspond to different configurations in the brain. But the brain, having only a finite amount of matter, can only have a *finite* number of configurations of that matter. And that is the proof.

We can always analyze some aspects of a proof in more detail.

The matter in the brain is in a continuum of location and motion, and a continuum is infinite in number, so maybe the argument breaks down. The answer is that it does not break down.

Each of the above elements in the infinite sequence must be sufficiently distinguishable in the way that they occur as logics in the brain, for if a person cannot distinguish knowing something from knowing that they know the something, then we are not talking about any meaning of the word "know" that I am aware of. Even though many aspects of the brain are a continuum, the number of *distinguishable* configurations, according to any notion of *distinguishable* applicable to this situation, is *finite*.

Going back to the summary statement of the proof, one can make a stronger statement. Since the brain can have only a finite number of configurations, only a finite number of the elements in the above infinite sequence can appear as configurations. Therefore, after some finite number of initial elements in the above sequence, *none of the elements correspond to configurations in the brain*, and therefore the person *knows* none *of them after that point*.

Let us leave aside the above mathematical proofs. If someone knows something, it gets harder and harder for them to know that they know that, and to know that they know that they know that, and so on. Knowing implies having a relatively full and natural understanding of the consequences of the thing you know and of its relations to other statements, and natural probably implies some quickness. This probably gets very hard for a person to do after just the first or second elements of the above infinite series. It would seem this could be shown with some clever thought puzzles demonstrating just how strained it is to reason about such knowledge.

The mathematical proof showed that after some incredibly huge number of elements in the sequence, we know nothing. The observation above, which included clever thought puzzles, suggests that as a matter of fact we know hardly anything after just the first two or three elements.

(It would certainly be desirable if our mathematical understanding of the situation advanced to the point where we could prove that the above infinite sequence of knowing significantly degraded after only two or three levels instead of being limited to a mathematical proof that it ends only after some hundred zillionth level. Maybe a general method would eventually come out of the above clever thought puzzles. For presumably such puzzles would eventually lead to an understanding of just how many statements and sophisticated deductions a person must have readily in mind in order for us to claim that the person knows the first two or three elements of the series. But before this, someone, without any concern of these mathematical issues, must discover some puzzles that show just how surprisingly difficult it is to reason about statements above the first level.)

Incidentally, how did we come to believe this fallacy that we know that we know? In the "proofs" that we see, or in the "proofs" that we make up in our mind, we picture some example, for instance, "trees generally are green." We look at the principle in terms of this example, thinking something of the sort, "I know that trees are generally green," and, "I know that I know that trees are generally green." From thence, we imagine that the principle generalizes to anything. Turns out, this particular generalization process is false. Another kind of "proof" might be some kind of nebulous appeal to ego. In this case, the appeal leads to a false belief.

Endnote 27 Anaximander

Information about Anaximander will be found in any book on ancient Greek philosophy or science. Detailed expert source and analysis can be found for instance in Guthrie, *A History of Greek Philosophy*.

The 580 B.C. I got out of the Chronological table of Lloyd, *Early Greek Science: Thales to Aristotle*, combined with the time estimate in Dreyer, *A History of Astronomy from Thales to Kepler*, middle page 13. What Anaximander said is stated in somewhat different ways in different places. The idea is about the same, but one of the statements closest to my use is about the 5th line, page 14, of Dryer, "We know from Aristotle that Anaximander believed the earth to be in equilibrium in the centre of the world, because it was proper for it not to have a tendency to fall in any particular direction, since it was in the middle and had the same relations to every part of the circumference."

Bibliography

Abbreviations have been minimized; historical information sometimes added in parentheses.

As mentioned in the section on writing axiomatizations at the end of the chapter on teletransportation, learning the details of mathematical logic would probably not be helpful to the thinker investigating sentience. Yet the area is mentioned so often in this book, it seemed necessary to include references, and so, two popular textbooks are listed below.

Aristotle, *The Soul (De Anima)*, as for instance translated by J. A. Smith in *Introduction to Aristotle*, edited by Richard McKeon, Random House, 1947.

Averous, Pierre, *The Atom*, Barron's Educational Series, 1988 (translated from *L'Atome – De La Matière Inerte Aux Etres Vivants*, Paris: Hachette – Foundation Diderot, 1985).

Asimov, Isaac, *Asimov's Biographical Encyclopedia of Science and Technology*, Garden City, N.Y: Doubleday & Company, 1964 (*not* the 1972 edition).

Asimov, Isaac, *Understanding Physics*, 3 volumes, Barnes & Noble Books, 1993. (Presumably originally published 1966.)

Berry, Arthur, *A Short History of Astronomy from the Earliest Times through the Nineteenth Century*, New York: Dover Publications, 1961. (Originally published 1898).

Burtt, Edwin Arthur, *The Metaphysical Foundations of Modern Physical Science*, (revised edition), Doubleday Anchor books, 1954. (Originally published 1924, with the revised edition appearing in 1932.)

Butterfield, Herbert, *The Origins of Modern Science 1300 – 1800, Revised Edition*, New York: The Free Press, 1965. (Presumably originally published 1957.)

Byron, Frederick W. Jr., and Robert W. Fuller, *Mathematics of Classical and Quantum Physics*, New York: Dover Publications, 1992. (This 1992 copy is an unabridged, corrected republication of the work first published by Addison–Wesley Publishing Company, 1969 and 1970.)

Cavalieri, Bonaventura Francesco, *Geometria indivisibilis continuorum nova,* 1635.

Chaisson, Eric, *Cosmic Evolution: The Rise of Complexity in Nature*, Harvard University Press, 2001.

Chalmers, David J., *The Conscious Mind: In Search of a Fundamental Theory*, Oxford University Press, 1996.

Chalmers, David J., "Facing Up to the Problem of Consciousness," *Journal of Consciousness Studies*, volume 2, issue 3, 1995, pages 200 – 219. As of February, 2004, at for instance, http://www.u.arizona.edu/~chalmers/papers/facing.html

Cooper, Grosvenor, and Leonard B. Meyer, *The Rhythmic Structure of Music*, Phoenix Books The University of Chicago Press, 1966. (Presumably first published 1960).

Dreyer, J. L. E., *A History of Astronomy from Thales to Kepler, Second Edition*, revised with a forward by W. H. Stahl, New York: Dover Publications, 1953. (Originally published 1906, under the title *History of the Planetary Systems from Thales to Kepler*.)

Elliott, Robert S., *Electromagnetics History, Theory, and Applications*, IEEE Press, 1993.

Enderton, Herbert B., *A Mathematical Introduction to Logic, second edition*, Harcourt/Academic Press, 2001.

Frazier, Kendrick, "A Mind at Play, An Interview with Martin Gardner," *Skeptical Inquirer*, March/April, 1998, page 34.

Frege, Gottlob, *The Basic Laws of Arithmetic, Exposition of the System*, translated and edited, with an Introduction, by Montgomery Furth, Berkeley and Los Angeles: University of California Press, 1967. (This 1967 printing is a correction of some minor errors in presumably the original publication of the translation in 1964. Originally published as *Grundgesetze der Arithmetik*, 1893.)

Gardner, Martin, "The Guided Wave Theory of Louis de Broglie and David Bohm," *Skeptical Inquirer*, May/June, 2000, page 9.

Guillen, Michael, *Five Equations That Changed the World: The Power and Poetry of Mathematics*, New York: MJF Books, 1995.

Guthrie, W. K. C., *A History of Greek philosophy, Volume 1, The Earlier Presocratics and the Pythagoreans*, Cambridge: Cambridge University Press, 1980. (First published 1965.)

Guthrie, W. K. C., *A History of Greek philosophy, Volume 2, The Presocratic tradition form Parmenides to Democritus*, Cambridge: Cambridge University Press, 1980. (First published 1965.)

Heidegger, Martin, *What is a Thing*, translated by W. B. Barton Jr. and Vera Deutsch, published by Henry Regnery Company, 1969. (The title in German was *Die Frage nach dem Ding*. Heidegger's preface, written in 1962, says the book is the text of a winter, 1935 – 1936 lecture entitled "Basic Questions of Metaphysics.")

Heiligenberg, Walter, *Neural Nets in Electric Fish*, The MIT Press, 1991.

Hey, Tony, and Patrick Walters, *The Quantum Universe*, Cambridge University Press, 1994. (First published 1987.)

Kirk, G. S., and J. E. Raven, *The Presocratic Philosophers*, Cambridge University Press, 1981. (First published 1957. One change and slight improvements made across the 1959, 1961, and 1962 impressions.)

Knowles, David, *The Evolution of Medieval Thought*, Vintage Books, A Division of Random House, 1962.

Lindberg, David C., *The Beginnings of Western Science: The European Scientific Tradition in Philosophical, Religious, and Institutional Context, 600 B.C. to A.D. 1450*, University of Chicago Press, 1992.

Lloyd, G. E. R., *Early Greek science: Thales to Aristotle*, New York: W. W. Norton & Company, 1970.

Lowe, E. J., *An introduction to the philosophy of mind*, Cambridge University Press, 2000.

Lucretius, *On the Nature of the Universe*, translated and introduced by R. E. Latham, Penguin Books, 1982. (This translation first published 1951. Originally presented to public about 55 B.C., by which time the author was probably dead. The author's name was Titus Lucretius Carus, the book was *De Rerum Natura*, and the title has been translated alternately into English as for instance, *The Nature of Things, On the Nature of Things, The Way Things Are, The Nature of the Universe*, and so on.)

Mendelson, Elliot, *Introduction to Mathematical Logic, 4th edition*, Chapman &Hall, 1997.

Nagel, Thomas, "What is it Like to be a Bat?", *Philosophical Review*, volume 83, 1974, pages 435 – 450. This article is reprinted in Nagel, Thomas, *Mortal Questions*, Cambridge University Press, 1979.

Newton, Isaac, *The Principia: mathematical principles of natural philosophy / Isaac Newton ; a new translation by I. Bernard Cohen and Anne Whitman, assisted by Julia Budenz ; preceded by a guide to Newton's Principia by I. Bernard Cohen*, Berkeley: University of California Press, 1999.

Nicholls, John. G., A. Robert Martin, and Bruce G. Wallace, *From Neuron to Brain, Third Edition*, Sunderland: Sinauer Associates, 1992. Note that on our journey in *the Soul*, almost none of the references to this book are to this 1992, Third Edition, but rather to the 2001, Fourth Edition, listed next.

Nicholls, John. G., A. Robert Martin, Bruce G. Wallace, and Paul A. Fuchs, *From Neuron to Brain, Fourth Edition*, Sunderland: Sinauer Associates, 2001.

Moller, Peter, *Electric Fishes / History and behavior*, London: Chapman & Hall, 1995.

La Mettrie, Julien Offray de, *Man a Machine*, LaSalle, Illinois: Open Court Publishing Company, 1912. Originally published as *L'Homme Machine*, 1748.

Oldenbourg, Zoe, *Catherine the Great*, Random House, 1965.

Olson, Richard, *Biographical Encyclopedia of Scientists*, New York: Marshall Cavendish Corporation, 1998.

Petroski, Henry, *The Evolution of Useful Things*, Vintage Books, 1994. (Originally published Alfred A. Knopf, 1992.)

Poincaré, Henri, "The Relativity of Space," in *Science and Method*, 1897. Also online, as of Jan. 2004, at http://www.marxists.org/reference/subject/philosophy/works/fr/poincare.htm

Poincaré, Henri, *Science and Method*, with a preface by Bertrand Russel, St. Augustine's Press, Oct 2001. This is an unabridged republication of the Dover reprint of 1952. The book was written in 1908.

Poincaré, Henri, *Science and Hypothesis*, with a preface by J. Larmor, Dover Publications, 1952. (First English translation was published by Walter Scott Publishing Company, 1905.)

Porter, Roy, and Marilyn Ogilvie, *The Biographical dictionary of Scientists, 3rd edition, vol 1*, New York: Oxford University Press, 2000.

Rieke, Fred; David Warland; Rob de Ruyter van Stevenink; and William Bialek, *Spikes: Exploring the Neural Code*, The MIT Press, 1997.

Rosen, Gideon, "Abstract Objects," *The Stanford Encyclopedia of Philosophy* (Fall 2001 Edition), Edward N. Zalta (ed.), URL = <http://plato.stanford.edu/archives/fall2001/entries/abstract –objects/>

Schwabhäuser, W., and L. W. Szczerba, "Relations on lines as primitive notions for Euclidean geometry," *Fundamenta Mathematicae, of the Polish Academy of Sciences, Institute of Mathematics*, volume 82, 1975, pages 347 – 355.

Scientific American, Special Issue on Nanotech, September, 2001.

Shepherd, Gordon M., *Foundations of the Neuron Doctrine*, Oxford University Press, 1991.

Simmons, Peter and David Young, *Nerve Cells and Animal Behaviour, Second Edition*, Cambridge University Press, 1999. (This second edition is an updated and expanded version of the first edition, published in 1989, and it does not seem, to the author of *The Soul*, to have the charm and directness of perspective found in the first edition.)

Starr, Cecie, and Ralph Taggart, *Cell Biology and Genetics, Ninth Edition*, Brooks/Cole, 2001.

Thucydides, *The Speeches of Pericles*, translation and commentary by Edinger, H. G., New York: Frederick Ungar, 1979.

Turing,   Alan, http://www.turing.org.uk/turing/scrapbook/test.html

Vaughn, Alden T., editor, *The Puritan Tradition in America, 1620 – 1730*, University of South Carolina Press, 1972. (If one wants to investigate these issues, probably the later editions, 1997 or on, along with later editions of another book by Vaughn, *New England Frontier: Puritans and Indians, 1620 – 1675*, have much relevant information.)

Wightman, William P. D., *The Growth of Scientific Ideas*, Hew Haven: Yale University Press, 1953.

Weinberg, Julius R., *A Short History of Medieval Philosophy*, Princeton, New Jersey: Princeton University Press, 1969. (Presumably first published 1964.)

Wheeler, J. E., "Information, physics, quantum: The search for links" in *Complexity, Entropy, and the Physics of Information*, Zurek, Wojciech H. editor, Addison–Wesley, 1990.

Wheelwright, Philip, editor *The Presocratics*, Indianapolis: Bobbs–Merrill, 1977. (Presumably first published in 1960.)

Name Index

Subject Index

Although some entries have many page numbers, one may compare those with the tables of contents for additional classification.